Progress in Pesticide Biochemistry

Volume 1

Progress in Pesticide Biochemistry

Volume 1

Edited by
D. H. Hutson
and
T. R. Roberts
Shell Research Limited, Sittingbourne

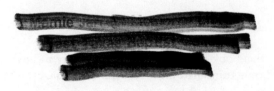

A Wiley–Interscience Publication

JOHN WILEY & SONS
CHICHESTER · NEW YORK · BRISBANE · TORONTO

Copyright © 1981 by John Wiley & Sons Ltd.

All rights reserved.

No part of this book may be reproduced by any means, nor transmitted, nor translated into a machine language without the written permission of the publisher.

British Library Cataloguing in Publication Data:
Progress in pesticide biochemistry.
   Vol. 1
   1. Pesticides—Physiological effect
   I. Hutson, D. H.
   II. Roberts, T. R.
   574.1'92   QP82.2.P4   80–41419

ISBN 0 471 27920 X

Type-set in Great Britain by John Wright and Sons Ltd.
at the Stonebridge Press, Bristol.
Printed by The Pitman Press, Bath, Avon.

Contributors to Volume 1

J. E. Casida	Laboratory of Pesticide Chemistry and Toxicology, Division of Entomology, University of California, Berkeley, California 94720, USA
J. K. Gaunt	Department of Biochemistry and Soil Science, University College of North Wales, Bangor, Gwynedd, LL57 2UW, North Wales, UK
J. A. Guth	Ciba-Geigy AG, Agrochemicals Division, CH-4002, Basle, Switzerland
B. D. Hammock	Department of Entomology, University of California, Davis, California 95616, USA
D. H. Hutson	Shell Research Limited, Sittingbourne, Kent, ME9 8AG, UK
J. B. Pillmoor	Department of Biochemistry and Soil Science, University College of North Wales, Bangor, Gwynedd, LL57 2UW, North Wales, UK
G. B. Quistad	Zoecon Corporation, 975 California Avenue, Palo Alto, California 94304, USA
T. R. Roberts	Shell Research Limited, Sittingbourne, Kent, ME9 8AG, UK
J. Seifert	Laboratory of Pesticide Chemistry and Toxicology, Division of Entomology, University of California, Berkeley, California 94720, USA
C. H. Walker	Department of Physiology and Biochemistry, The University, Whiteknights, Reading, Berkshire, RG6 2AJ, UK

Contents

Foreword ix

Preface xi

1 Metabolism and Mode of Action of Juvenile Hormone, Juvenoids, and other Insect Growth Regulators 1
  B. D. Hammock and G. B. Quistad

2 Experimental Approaches to Studying the Fate of Pesticides in Soil . 85
  J. A. Guth

3 The Metabolism of the Synthetic Pyrethroids in Plants and Soils . . 115
  T. R. Roberts

4 The Behaviour and Mode of Action of the Phenoxyacetic Acids in Plants 147
  J. B. Pillmoor and J. K. Gaunt

5 Mechanisms of Teratogenesis induced by Organophosphorus and Methylcarbamate Insecticides 219
  J. Seifert and J. E. Casida

6 The Correlation Between *in vivo* and *in vitro* Metabolism of Pesticides in Vertebrates 247
  C. H. Walker

7 The Metabolism of Insecticides in Man 287
  D. H. Hutson

Index 335

Foreword

Pesticides are designed and used to control unwanted organisms. This must be achieved with minimal disruption of the ecosystem and only transitory contamination of the environment. Pesticide chemicals should ideally affect life processes in only pest species, but in practice this selectivity is rarely if ever achieved. Most early pesticides required high dosages and acted essentially as non-selective poisons. Some recent insecticides, herbicides, fungicides, and growth regulators combine exceptional potency with remarkable specificity attributable to species differences in detoxification or the nature of the target site. These developments and continuing progress require thorough research on many topics including: structure optimization facilitated by quantitative considerations of the effect of changes in chemical structure on partitioning, reactivity, stability, and molecular conformation; biochemical or biophysical lesions associated with the primary mode of action, with secondary disruptions, and with any deleterious effects in non-target species; the use of pesticides as probes to explore new receptors or target sites and uncharted areas of biochemistry and physiology; metabolic activation of propesticides and detoxification of pesticides present at the wrong time and place.

Research in pesticide biochemistry has multiple sponsors and goals. It provides industry with an avenue to rational pesticide design. It lays the background for regulatory agencies to assess risks and define conditions for safe pesticide use. It places researchers in an interdisciplinary arena of almost infinite scope, at a molecular and organismal level. Ultimately, it enables the public to benefit from improved crop protection and health standards. I share the conviction of Editors Hutson and Roberts that this book series of critical reviews on *Progress in Pesticide Biochemistry* will assist in meeting these goals.

Berkeley, 1980 JOHN E. CASIDA

Preface

Primary publications on the biochemistry of pesticides are distributed widely throughout the scientific literature and the subject matter ranges from insect, plant, and soil biochemistry through to mammalian toxicology. This Series is one in which selected aspects will be reviewed and, where possible, be interrelated by experts in the various fields. We make no apology for introducing another review series into the scientific literature; a regular review series dedicated exclusively to the biochemistry of pesticides in the broadest sense has not been available. The reporting of pesticides research in the pharmaceutical/pharmacological literature, even though many of the techniques are common, is often inappropriate except in certain toxicological discussions. Moreover, we feel that the all too common juxtaposition of modern pesticides with 'environmental pollutants' (such as the PCBs) is unjustified and belittles the ingenuity and enormous research effort used in the discovery, optimization, evaluation, and safe use of an agricultural chemical. At a time when the traditional screening process of discovery is being increasingly complemented by the biochemical approach, an understanding of the biochemistry and mode of action of insecticides, fungicides, herbicides, and growth regulators is fundamental for both rational design of new compounds and their safety evaluation.

Within the scope of the series, the biochemistry of pesticides is deemed to cover the following areas: (1) mode of action (i.e. biocidal action in target species); (2) biotransformation in target species (which may, of course, be related to mode of action); (3) biotransformation in non-target species (which may include soils, bacteria, insects, plants, fish, birds, and mammals, including man); (4) environmental effects (e.g. effects on the ecology of treated areas); (5) environmental chemistry (distribution and fate in the environment); and (6) biochemical toxicology in mammals, including man.

In selecting areas for review, we shall take account of new or changing aspects of the science, including techniques, and of changes in the importance or use patterns of the chemical classes of pesticides. This is exemplified in the subjects covered in Volume 1 which range from a well-established class of herbicides, the phenoxyacetic acids (the mode of action and biochemistry of which is still by no means clear) to a relatively new class of insecticides, the juvenile hormone analogues which act as growth regulators rather than as acute poisons. A similarly broad range will be covered in each volume of the series.

Sittingbourne, 1980   D. H. HUTSON
                                                                                        T. R. ROBERTS

CHAPTER 1

Metabolism and mode of action of juvenile hormone, juvenoids, and other insect growth regulators

Bruce D. Hammock and Gary B. Quistad

INTRODUCTION	2
JUVENOIDS—MODE OF ACTION	4
Effects on JH Degradation	4
Effects on JH Synthesis	5
Effects on Site of Action	6
Effect on Macromolecular Biosynthesis	7
RADIOSYNTHESIS OF JH AND JUVENOIDS	8
JUVENILE HORMONE METABOLISM	11
Major Pathways of JH Metabolism	11
Insect Epoxide Hydrolases	14
JH Esterases	16
JHE assay methods	16
JHE changes during development	17
JHE inhibition	20
JHE regulation	23
Influence of JH Carriers on Metabolism	27
DODECADIENOATES	30
Methoprene	30
Insect metabolism	30
Mammalian metabolism	31
Environmental fate	33
Hydroprene	36
Insect metabolism	36
Environmental fate	36
TERPENOID-PHENOXY ETHERS AND RELATED COMPOUNDS	37
Insect Metabolism	37
Mammalian Metabolism	42
Toxicology	45
Environmental Fate	47
ANTI-JUVENILE HORMONES	50
Precocene—Mode of Action	50
Insect Metabolism of Precocene II	51
BENZOYLPHENYL UREAS—MODE OF ACTION	52
History	52
Inhibition of Chitin Biosynthesis	54
Secondary Effects	56

METABOLISM OF DIFLUBENZURON AND RELATED COMPOUNDS 57
   Insect Metabolism 57
   Mammalian Metabolism and Toxicology. 58
   Environmental Fate. 60
   Metabolism of Benzoylphenyl Ureas 62
FUTURE OF INSECT GROWTH REGULATORS 63
ACKNOWLEDGEMENTS 64
REFERENCES 64

INTRODUCTION

The insect growth regulator field emerged with the elucidation of the structure of juvenile hormone (JH). Based on the early work of Wigglesworth (1936, 1970) and other endocrinologists, the interesting report by Williams (1956) of high levels of JH activity in the abdomens of adult male *Hyalophora cecropia* moths, and the synthesis of methyl-10,11-epoxyfarnesoate by Bowers *et al.* (1965), the structure of JH I (Figure 1.1) was determined by Röller *et al.* (1967). Subsequently, two other JH homologues were identified (Figure 1.1) (Meyer *et al.*, 1970; Judy *et al.*, 1973) (for reviews see Wigglesworth, 1970; Menn and Beroza, 1972; Gilbert, 1976; Schooley, 1977). Williams (1967) predicted the advent of a new generation of pest control agents based on these insect hormones. As reviewed by Slamá *et al.* (1974), subsequent research produced numerous JH mimics or juvenoids, some more active and more stable than the natural hormones. The juvenoids with another group of compounds, the benzoylphenyl urea chitin synthesis inhibitors (Wellinga *et al.*, 1973), introduced a new group of insecticides termed insect growth regulators (IGRs). In this review, an overview of the mechanism of action of the insect growth regulators will be presented. The environmental fate and the metabolism of the IGRs by target and non-target organisms will be reviewed in greater detail and the promise of current and future IGRs discussed. A discussion of IGR metabolism can hardly be undertaken without a feeling for the metabolism of JH itself; thus, an earlier review of juvenoid and JH metabolism will be updated (Hammock and Quistad, 1976) and expanded to include other IGRs.

Such attention to the routes of IGR degradation is warranted because these compounds represent new structures for the metabolism chemist. For instance, several interesting routes and concepts in xenobiotic metabolism surfaced from the study of juvenoids. The Hoffman *et al.* (1973) report of a Baeyer–Villiger-like oxidation has not been verified *in vitro*, but it presents interesting mechanistic considerations. The report of Quistad *et al.* (1974b) that a pesticide can be metabolized and then converted into natural products certainly illustrates that retention of tissue radioactivity following administration of a xenobiotic should not automatically be considered as a sign of biostability. The elucidation of epoxide hydration as a major route of insect metabolism of epoxide-containing juvenoids (Gill *et al* 1972) coincided with the analogous discovery of epoxide hydration and ester hydrolysis being major routes of natural juvenile hormone metabolism (Slade and Zibitt, 1971, 1972). The significance of epoxide reduction initially detected in rats (Hoffman et al., 1973), and clearly shown to occur in the rumen contents of steers and sheep (Ivie et al., 1976; Ivie, 1976), seems to occur in the gut contents of a variety of mammals, including man. This route of metabolism, known from other organisms (Yamamoto and Higashi, 1978), surely warrants further investigation and may be important in the inactivation of toxins, mutagens, and carcinogens in the gut (Ivie, 1976; Callen, 1978). In spite of the early reports of Gill et al. (1972, 1974) to the contrary, it had been widely assumed that all epoxide hydrolase activity was membrane bound. Subsequent reports certainly demonstrate that for terpenoid epoxides high hydrolase activity is in the cytosolic or soluble fraction (Hammock et al., 1976; Mumby and Hammock, 1979a, b, c) as well as the mitochondrial fraction (Gill and Hammock, unpublished). They explain how the activity has been overlooked in many laboratories for a decade (Hammock et al., 1980a, b; Ota and Hammock, 1980) and demonstrate that the cytosolic enzyme hydrates a wide variety of substrates including mutagens and suspect carcinogens (El-Tantawy and Hammock, 1980; Hasegawa and Hammock, unpublished). The investigation of precocene metabolism may help to explain the mechanism of action of these interesting compounds as well as the intracacies of JH biosynthesis (Jennings and Ottridge, 1979; Soderlund et al., 1980).

The IGRs represent an exciting concept in the development of new insecticides. The investigation of their metabolism and action should continue to provide insights useful for improving IGR structures. Hopefully, such improved

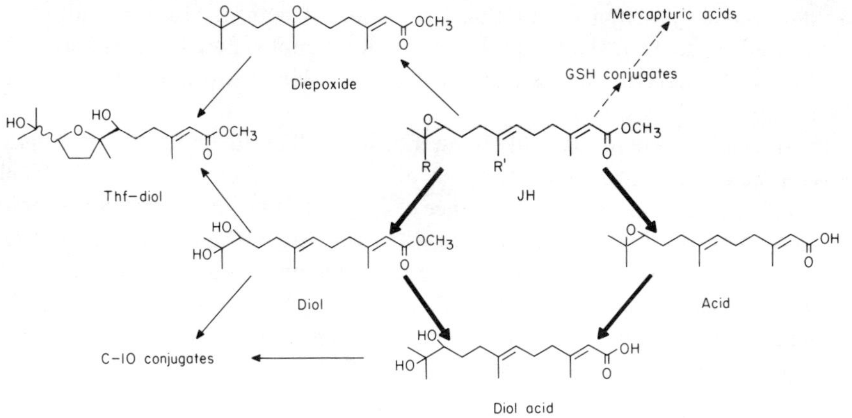

Figure 1.1 Metabolism of juvenile hormone (JH). Metabolites are shown for JH III ($R = R' = CH_3$). The structures of JH I ($R = R' = (C_2H_5)$) and JH II ($R = C_2H_5$, $R' = CH_3$) are similar. Heavy lines indicate the major pathways of metabolism shown to occur in most insects. GSH conjugation (dashed line) has not yet been demonstrated in insects

compounds will be even more compatible with environmentally sound integrated pest management programmes.

JUVENOIDS—MODE OF ACTION

Numerous publications have appeared over the last decade concerning the mode of action of insect juvenile hormones. Such studies can be conveniently divided into the following chronological categories.

(1) Synthesis (including its initiation and regulation) in the corpora allata.
(2) Transport of JH in haemolymph (binding proteins).
(3) Degradation.
(4) Cellular targets for expression (epidermis, fat body, prothoracic gland, etc.).
(5) Macromolecular events (synthesis of protein, RNA, or DNA, and enzyme regulation).
(6) Overt morphological effects.

The synthesis and transport of juvenile hormones in two insect species have been compared by Kramer and Law (1980) while the morphological results of juvenile hormone activity in insects have also been summarized (Staal, 1975; Sehnal, 1976). A comprehensive compilation of data relating to the mode of action of juvenile hormones was represented by Gilbert (1976) and Gilbert and King (1973). In analysing information concerning the mode of action of juvenile hormones, it is readily apparent that many uncertainties complicate the issue. A unified explanation of insect JH mode of action is still non-existent, but rather in its place are numerous, often unrelated facts from a multitude of different researchers and different insects. In this review we will address the question of *juvenoid mode of action* rather than JH mode of action. However, although considered separately, the modes of action of juvenoids and JH are likely to be synonymous. Indeed, there are many examples where juvenoids can serve as direct replacements for JH.

In overview, there are two likely methods by which juvenoids disrupt insect development. In one method juvenoids may act as perfect mimics of JH. Their disruptive effect stems from the presence of a relatively large amount of juvenoid, overpowering the homeostatic mechanisms present in the insect or from their presence at inopportune times. A second mechanism may involve juvenoids acting as imperfect mimics of JH. In this case, the juvenoid may be a potent JH mimic at some sites but a poor mimic or even antagonist at other sites leading to a disruption of insect development.

Effects on JH Degradation

Slade and Wilkinson (1973) proposed a rather controversial explanation of the mode of action of juvenoids in insects. They claimed that rather than being intrinsically hormonal, juvenoids were synergistic in preventing the degradation of natural JHs. This proposal was supported by data showing the apparent stabilization of JH I *in vitro* when midgut preparations from *Spodoptera eridania* (southern armyworm) were exposed to several juvenoids (including methoprene and hydroprene). Slade *et al.* (1975) and Brooks (1973) showed that hydration of an epoxide of a cyclodiene (HEOM) was also inhibited by several juvenoids (including R-20458 and methoprene). These authors suggest that inhibitors of cyclodiene epoxide hydrolase may also inhibit JH epoxide hydrolase. Hence, these juvenoids are potential synergists of natural JHs. In conflict with the above conclusions were data from Terriere and Yu (1973, 1974) using *M. domestica* which showed not inhibition of cyclodiene epoxide hydrolase (using heptachlor epoxide), but rather either no effect (hydroprene) or slight enhancement of epoxide hydrolase activity (methoprene). Downer *et al.* (1975) offered further support for the Slade–Wilkinson hypothesis of a synergistic mode of action by showing that esterases of *Aedes* were inhibited by methoprene. However, the conclusions of Downer *et al.* (1975) are suspect since they used α-naphthylacetate as a substrate to assay for JH esterase and α-naphthyl esters may only be reflective of general esterase activity, rather than JH-specific esterase (see Hooper, 1976).

Although under certain conditions juvenoids may exert their action (at least in part) through synergism of endogenous hormone, considerable data refute the notion of juvenoids solely as alternative substrates or inhibitors of the enzymes of JH degradation (Solomon and Walker, 1974). In the Colorado potato beetle (*Leptinotarsa decemlineata*) it has been established that JH is metabolized mainly by JH esterases (Kramer *et al.*, 1977), but two biologically active juvenoids (methoprene and hydoprene) are not substrates for this insect's JH esterase (Kramer and deKort, 1976b) and actually increase *in vivo* esterase activity for JH I (Kramer, 1978). Although pupal haemolymph from *Tenebrio molitor* contained the highest esterase activity for JH I of five insect species tested by Weirich and Wren (1976), methoprene was not hydrolysed. Methoprene is also not hydrolysed by esterases of *Culex pipiens* (Hooper, 1976), *Blaberus giganteus* (cockroach; Hammock *et al.*, 1977b), and *Manduca sexta* (Weirich and Wren, 1973) even though the compound is morphologically active on these insects. Larval esterases of *Diatraea grandiosella* (southwestern corn borer) bound neither hydroprene nor methoprene, but did bind JH I; hence, in this insect also, these juvenoids do not appear to synergize JH by inhibiting its ester hydrolysis (Brown *et al.*, 1977). Methoprene also failed to inhibit the *in vitro* hydrolysis of JH by esterases of *Trichoplusia ni* (Sparks and Hammock, 1980b), and as discussed later some juvenoids may even stimulate JH metabolism.

Effects on JH Synthesis

The actual modulation of JH III synthesis by hydroprene in *Diploptera punctata* (a viviparous cockroach) was shown elegantly by Tobe and Stay (1979).

By determination of the titre of JH III these authors demonstrated that hydroprene stimulated JH synthesis at low doses and suppressed hormone biosynthesis at high doses. It is likely that the juvenoid is acting on a normal JH receptor by feedback regulation of JH biosynthesis. Of course, since hydroprene is intrinsically gonadotrophic in this cockroach, even though natural JH III synthesis has been suppressed at high doses of juvenoid, the overt morphological response (i.e. oocyte elongation) could mistakenly suggest a high JH III titre.

These studies and those of Schooley and Bergot (1979) illustrate the interesting concept that many known juvenoids are likely to have some JH antagonist action at the physiological, if not the biochemical level. However, such antagonistic effects may be difficult to distinguish in the intact insect, but they may enhance the disruption of insect development by some juvenoids.

Effects on Site of Action

Since JHs are both morphogenetic and gonadotrophic, it is not surprising that the tissues associated with such effects would be regulated by juvenoids. Using the cabbage armyworm (*Mamestra brassicae*) Hiruma *et al.* (1978b) showed that methoprene inhibits release of neurosecretory material (i.e. prothoracicotropic hormone, PTTH) from cerebral cells, but may not affect the synthesis of new PTTH in the same cells. These results suggest that a high JH titre late in the last larval instar of *M. brassicae* inhibits the release of PTTH from the brain and the subsequent lack of PTTH decreases α-ecdysone secretion from the prothoracic gland. Hence, the cerebral neurosecretory cells appear to be regulated by both β-ecdysone and JH. In the absence of PTTH the prothoracic gland in *M. brassicae* is activated by methoprene, but only in the last part of the final instar and in the pupal stage (Hiruma *et al.*, 1978a). By means of ligature experiments Cymborowski and Stolarz (1979) demonstrated that methoprene inhibits the prothoracic glands of *Spodoptera littoralis* at the beginning of the final instar and then stimulates the glands shortly before pupation. Activation of pupal prothoracic glands by juvenoids seems to be a rather general response in lepidopterous insects since Krishnakumaran and Schneiderman (1965) showed increased prothoracotrophic activity upon treatment of brainless, diapausing pupae with farnesol, farnesyl methyl ether, farnesyl diethylamine, nerolidol, and dodecyl methyl ether. As already mentioned, Tobe and Stay (1979), using hydroprene, have shown that the biosynthetic activity of the corpus allatum itself is regulated by JH.

After secretion from the glands responsible for morphogenetic hormone production, JH is transported (in part) to the epidermis, a primary target site. By using (2E,6E)-10,11-epoxyfarnesyl propenyl ether Schmialek *et al.* (1973) reported a JH receptor in the epidermis of *Tenebrio molitor* pupae which appeared to accumulate the juvenoid against a concentration gradient. Once in the epidermis it appears (at least for *M. sexta*) that JH is hydrolysed primarily by esterases (Mitsui et al., 1979). Specific effects of methoprene on the fat body (Couble et al., 1979) and ovary (Postlethwait and Gray, 1975) will be discussed subsequently.

Effect on Macromolecular Biosynthesis

Krypsin-Sorensen et al. (1977) found that the abnormally increased somatic growth of *Spodoptera littoralis* treated with methoprene was not associated with hypermetabolic activity (e.g. unusually large O_2 consumption and CO_2 release). Work in the same laboratory had previously revealed that juvenoid treatment of dermestid beetle larvae caused an enormous rise in total metabolic rate, but without formation of extra-larval instars (Sláma and Hodkova, 1975). The glycogen and lipid reserves were considerably depleted in *Aedes aegypti* pupae resulting from fourth instar larval treatment with methoprene (Downer et al., 1976). Thus, reduced energy supplies may contribute to pupal mortality.

Since it has been suggested that insect hormones may act primarily at the genetic level, a number of investigations of juvenoid effects on nucleic acid synthesis have been reported. With imaginal wing discs of *Calliphora*, farnesol inhibits incorporation of thymidine into DNA (Vijverberg and Ginsel, 1976), while methoprene has no effect on thymidine incorporation, but does decrease inclusion of uridine into DNA (Scheller et al., 1978). In the *Calliphora* wing disc assay methoprene increases rRNA synthesis and inhibits production of mRNA (Scheller et al., 1978). Likewise, de novo synthesis of rRNA from [2-^{14}C]glycine is accelerated by methoprene in *Musca domestica* (Miller and Collins, 1975). Himeno et al. (1979) reported that DNA and particularly RNA synthesis was inhibited by methoprene in *Culex molestus*. Methoprene was a stronger inhibitor of RNA synthesis than either puromycin or actinomycin D, but the RNA inhibition was reversible. Both methoprene and hydroprene reduce uridine incorporation into the RNA of larval *Drosophila* (Breccia et al., 1976) while the RNA of adult *D. melanogaster* treated with methoprene was 6% enriched in poly(A) sequences (Gavin and Williamson, 1978). Thus, juvenoids regulate nucleic acid synthesis with the observed effect (i.e. stimulation or repression) dependent on stage of development, type of insect, method of assay, etc.

Since genetic manipulation is often expressed as protein synthesis, it follows that protein production would also be regulated by juvenoids. Methoprene is an inhibitor of protein synthesis in larval *Drosophila* homogenates (Breccia et al., 1976) and cultured *Culex* cells (Himeno et al., 1979), but had no effect in a *Calliphora* wing disc assay (Scheller et al., 1978). JH itself inhibits protein synthesis in imaginal discs of *D. melanogaster* (Fristrom et al., 1976).

A particularly important protein is vitellogenin which is essential to insect egg maturation. Methoprene is effective in converting the fat body of adult female *Locust migratoria* from a nutrient storage depot to a site of vitellogenin synthesis and secretion (Chen et al., 1976; Couble et al., 1979). Methoprene activates the fat bodies in isolated *D. melanogaster* abdomens to promote vitellogenesis, thereby promoting the maturation of oocytes (Postlethwait *et al.*, 1976; Handler and Postlethwait, 1978). Methoprene not only enhances the synthesis of vitellogenin, but also increases its sequestration from the haemolymph into the oocyte (Handler and Postlethwait, 1978). The stimulation of yolk protein synthesis in *D. melanogaster* has been attributed to production of poly(A)-containing RNA (Gavin and Williamson, 1978). Methoprene also regulates the acid phosphatase activity in the ovary of *D. melanogaster* and this enzyme is necessary for yolk metabolism (Postlethwait and Gray, 1975).

Thus, in summary, juvenoids regulate the activity of insect glands associated with growth hormone production. The type of effect on these glands (neurosecretory cells, prothoracic gland, corpus allatum) is largely dependent on the timing of application relative to the stage of metamorphosis. The stimulation and inhibition of nucleic acid or protein synthesis are equally a function of the exact timing of juvenoid application. Hence, observations of gland activity or macromolecular biosynthesis are intimately controlled by metamorphosis.

Several facets of the mode of action of juvenoids in non-insect species have been examined. Both methoprene and hydroprene decrease protein synthesis in larval shrimp (Breccia *et al.*, 1977). Although JH I, II, and III are uncouplers of oxidative phosphorylation in mouse liver mitochondria, methoprene and epofenonane were 10-fold less active (Chefurka, 1978). JH I, methoprene, and triprene (ZR-619) all depress the synthesis of DNA, RNA, and protein in mouse cells, but both juvenoids were less inhibitory than JH itself (Chmurzyńska *et al.*, 1979).

RADIOSYNTHESIS OF JH AND JUVENOIDS

Several pathways have been used for the radiosynthesis of juvenoids. As with any pesticide, a ^{14}C label is useful for research on mammalian metabolism and environmental degradation because it unambiguously traces at least one carbon in the molecule. However, the extraordinary effectiveness of juvenoids leads to very low effective doses in insects. Thus, ^{14}C and even many ^{3}H labels have too low a specific activity for many mechanism of action and metabolism studies at reasonable doses.

Racemic juvenile hormone I (JH I, Figure 1.1) (25 mCi mmol^{-1}) was first labelled in the 2-position by an Emmons modification of the Wittig reaction using trimethyl [2-^{14}C]phosphonoacetate prepared from methyl [2-^{14}C]bromoacetate (Hafferl *et al.*, 1971). The first commercially available JH I was prepared by New England Nuclear Corporation by selective reduction of an alkyne to an ethyl branch at C-7 of the *E,E,cis* precursor. In later years, this product has been replaced by JH I and now JH III labelled at C-10 by sodium borotritide reduction of the corresponding haloketone to the *erythro-* and *threo-* diastereomers of the resulting halohydrins. Following chromatographic separation, the *threo* isomer was cyclized in base to yield E,E,cis-JH I or E,E-JH III (~ 10 Ci mmol^{-1}) (Ahern and Schooley, unpublished).

The lability of the methyl ester of juvenile hormone limits the usefulness of such methyl-labelled preparations; however, the production of methanol from the hydrolysis of JH forms the basis of a number of rapid methods for the assay of JH esterase activity (Sanburg *et al.*, 1975a; Nowock *et al.*, 1976; Hammock *et al.*, 1977b; Vince and Gilbert, 1977). The availabiity of potent JH esterase inhibitors may also extend the usefulness of such preparations (Hammock *et al.*, 1977b; Sparks and Hammock, 1980a, b). Sanburg *et al.* (1975a) prepared methyl-labelled JH I by sodium methoxide catalysed transesterification with [^{3}H]methanol to yield a product with a specific activity of ~ 8.3 mCi mol^{-1}. Trautmann *et al.* (1974) and Hammock *et al.* (1977b) prepared labels of 4.3 and 2.5 mCi mol^{-1}, respectively, by exchanging the acid proton of JH acid with T_2O and exposing the acid to diazomethane. This method was based on the earlier work by Trautmann (1972) on JH III and a dichloro analogue labelled at 0.49 and 3.7 Ci mmol^{-1}, respectively. Still higher activities could theoretically be obtained by the alternate pathway of using tritiated diazomethane (Denmore and Davidson, 1959). Hammock *et al.* (1977b) reported that diazomethane may have reacted with the terpenoid chain of JH and that this byproduct(s) was only detected by high pressure liquid chromatography. Such side reactions should not present problems if lower specific activities for metabolism studies are required. Peter *et al.* (1979a) demonstrated a biochemical method for preparing [*methyl*-^{14}C]JH I based on the methylation of JH acid catalysed by homogenates of the sex accessory glands of male *Hyalophora cecropia* (Shirk *et al.*, 1976; Weirich and Culver, 1979). The methyl donor, carrier-free S-[*methyl*-^{3}H]-adenosyl-L-methionine, is commercially available and remarkable yields of $>90\%$ ^{3}H incorporation into JH I have been reported. The method uses expensive starting materials, but an optically active product with very high specific activity can be obtained in small batches. Modifications of this procedure could yield labelled JH at a reasonable price to biological laboratories.

For conjugated systems such as the juvenile hormones, metal-catalysed proton exchange is seemingly an attractive route of radiosynthesis. When E,E-farnesoic acid was heated in a dimethylformamide/D_2O mixture with an activated-platinum catalyst, high deuterium incorporation was obtained. G.l.c.-m.s. and n.m.r. analysis indicated that $>35\%$ of the material was the E,E isomer following the reaction and that the other mass could be largely accounted for by isomerization at the $2E$ and $6E$ positions. The majority of the deuterium was at the C-3 methyl group with moderate exchange at other allylic positions and some exchange of the olefinic C-2 proton itself. An analogous reaction with carrier free T_2O resulted in a product of the correct polarity, but much of the product apparently lacked the 10,11-olefin. Following reaction with diazomethane, only a low yield of E,E methylfarnesoate (8.4 Ci mmol^{-1}) was obtained (Hafferl, Schooley, and Hammock, unpublished; Hammock, 1975). D_2O and low specific activity T_2O reactions with methoprene and methoprene acid indicate even more facile incorporation in these dienoate molecules which also lack the labile 10,11-olefin.

It is very likely that JH mimics of high specific activity could be useful, perhaps even more useful than the natural hormone, in mechanism of action studies. The utility of such compounds can even be further enhanced if they are refractory to metabolism. Such an approach was demonstrated by Schmialek et al. (1976) by the synthesis of (2E, 6E)-10,11-epoxyfarnesyl[2,3-^{3}H]propenyl ether from the corresponding propargyl ether by reduction with tritium gas catalysed by quinoline (chinoline) poisoned Lindlar catalyst.

The radiosynthesis of [10-^{3}H]methoprene involved methoxymercuration of the olefinic precursor with mercuric acetate in methanol (Brown and Geoghegan, 1970; Henrick et al., 1973) followed by conversion of the resulting organomercurial acetate to its chloride. The organomercurial chloride was reduced with sodium borotritide to yield an isomer mixture of [10-^{3}H]methoprene with a specific activity of >1.9 Ci mmol^{-1} (Schooley et al., 1975a) (Figure 1.2).

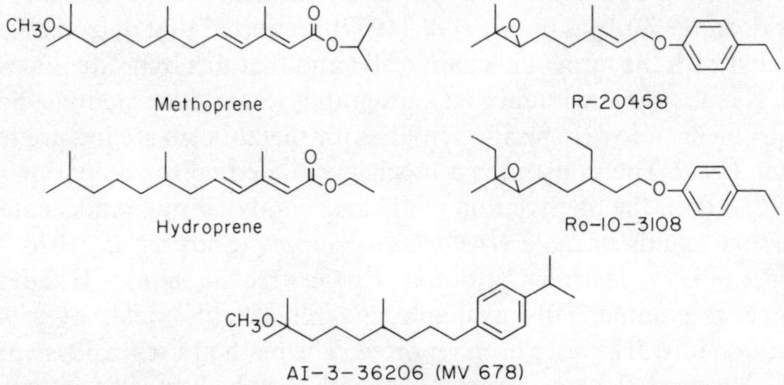

Figure 1.2 Representative juvenoid structures. Methoprene and hydroprene are referred to as dodecadienoates (Henrick et al., 1973), R-20458 (Pallos et al., 1971) and Ro-10-3108 are referred to as terpenoid phenoxy ethers, and AI-3-36206 (Hangartner et al., 1976) is referred to as an arylterpenoid (Schwarz et al., 1974)

Synthesis of [5-^{14}C]methoprene involved the carbonation of the Grignard of 1-bromo-2,6-dimethyl-5-heptene with $^{14}CO_2$ and subsequent conversion to [1-^{14}C]citronellal. Condensation with di-isopropyl 3-isopropoxycarbonyl-2-methyl-2-propenyl phosphonate (Henrick et al., 1973) yielded [5-^{14}C]isopropyl (4E)-3,7,11-trimethyl-2,4,10-dodecatrienoate which could be reduced to hydroprene (Bergot and Schooley, unpublished) (Figure 1.2) or subjected to methoxymercuration-reductive demercuration to yield [5-^{14}C]methoprene with a specific activity of 58 Ci mmol^{-1} (Schooley et al., 1975a). Hydroprene, methyl epoxyfarnesoate, ethyl dichlorofarnesoate, and R-20458 were prepared with specific activities in the 50 mCi mmol^{-1} range by tritiation (T_2O) of intermediate aldehydes (Ajami and Crouse, 1975).

Several syntheses of the juvenoid R-20458 (Figure 1.2) have been reported with a label in both the geranyl and ethylphenoxy portions of the molecule. The reduction of citral (a geranial/neral or $2E/2Z$ mixture) with sodium borotritide to $(2E/2Z)$ [1-^{3}H]geraniol, bromination of the alcohol and a Williamson ether synthesis with 4-ethylphenol followed by epoxidation of the 6,7-olefin resulted in [1-^{3}H]geranyl R-20458 at 33 mCi mmol^{-1} (Kamimura et al., 1972). Alternatively, the aromatic protons on 4-ethylphenol were exchanged by heating with sulphuric acid in tritium water. The resulting tritiated phenol was used as above in the synthesis of R-20458 labelled at >650 mCi mmol^{-1} in the phenyl ring (Kamimura et al., 1972). Kalbfeld et al. (1973) synthesized [phenyl-^{14}C]R-20458 at 17 mCi mmol^{-1} by a similar procedure using a Williamson ether synthesis to couple the labelled phenol with 6,7-epoxygeranyl bromide. All three radiolabels proved useful in metabolism studies with several limitations. Ether cleavage immediately resulted in the loss of half of the [geranyl-1-^{14}C] label and oxidation of the resulting aldehyde, undoubtedly, resulted in total loss of the label. The ^{3}H-ring label was suitable for *in vitro* work, but apparently acidic conditions in the stomach caused some exchange of the labile aromatic protons. The ^{14}C label has proven very useful for metabolism studies, but following ether cleavage, it fails to trace the interesting geranyl portion of the molecule, and its specific activity is too low for most biological research.

The closely related juvenoid, Ro-10-3108 (Figure 1.2), was labelled in the 2,3-position by exposure to tritium gas in the presence of platinum. A product of ~2.3 Ci mmol^{-1} was obtained with reduction of the epoxide apparently not a problem (Dorn et al. 1976). The 7-methoxy and 7-ethoxy analogues (Figure 1.2) of R-20458 were prepared from the corresponding ^{3}H-ring labelled diene described above by solvomercuration–demercuration in methanol or ethanol to yield products labelled at >600 mCi mmol^{-1} (Hammock et al., 1975a).

Future work in the area of radiosynthesis will probably lead to the very high specific activities usually needed for hormone receptor studies. In this research, juvenoids which are biologically stable and easier to label are, in some cases, likely to be used in lieu of natural JH.

JUVENILE HORMONE METABOLISM

Major Pathways of JH Metabolism

The basic pathways of juvenile hormone metabolism were first illustrated by Slade and Zibitt (1971, 1972) and include ester hydrolysis and epoxide hydration followed by conjugation (Figure 1.1). These observations were expanded to other insects in surveys by White (1972), Ajami and Riddiford (1971, 1973), and a host of subsequent workers. The literature through 1975 on JH metabolism was reviewed by Hammock and Quistad (1976) and some subsequent work has dealt with expanding early observations to a wider variety of insects, correlations of

JH metabolism with developmental changes, and a biochemical characterization of the proteins involved. Hopefully, such work will continue to expand yielding data on the regulation of JH metabolism and allowing development of a model of JH *in vivo* kinetics. General aspects of JH metabolism in several insect orders will first be covered followed by a discussion of insect epoxide hydrolases (EH). JH esterases (JHE) will be covered in more detail including assay methods, changes during development, inhibition, and regulation. Finally, the influence of JH carriers on metabolism will be considered. For more detailed information on aspects of JH chemistry, biosynthesis, and action, numerous reviews are available (Schneider and Aubert, 1971; Menn and Beroza, 1972; Akamatsu *et al.*, 1975; Gilbert, 1972, 1974, 1976; Gilbert *et al.*, 1976, 1977, 1978; Riddiford and Truman, 1978; Kramer and Law, 1980; deKort, 1981). The metabolism of JH in relation to insecticide resistance has been reviewed by Sparks and Hammock (1980a).

Although numerous studies on JH degradation have been published, most studies examine a single aspect of metabolism. Very few investigations have been complete enough to determine the relative importance of alternate pathways of metabolism or JH clearance. Most research has been in the Lepidoptera and no information is available on many orders of comparative interest. Hopefully, such information will soon be available.

Erley *et al.* (1975) extended the studies of Slade and Zibitt (1972) on the vagrant grasshopper, *Schistocera nitens* (syn *vaga*) and White (1972) on *Schistocera gregaria* to *Locusta migratoria*. JH I moved rapidly from an oil droplet into circulation and was rapidly excreted as unchanged JH, acid, diol, diol acid, and conjugates. Excretion rates were similar in adults of both sexes at various ages. This work has been continued at the biochemical level (Peter *et al.*, 1979b). Sams *et al.* (1978) evaluated JH metabolism in cultured fat bodies and ovary from the cockroach, *Periplaneta americana*. Other studies with the Orthoptera include those of Pratt (1975) and Hammock *et al.* (1977b).

Ajami and Riddiford (1973) found JH diol, acid, diol acid, and conjugates in the yellow mealworm, *Tenebrio molitor*. Subsequent workers have found high JH hydrolase as well as esterase at various times during development (Weirich and Wren, 1976; Mumby and Hammock, 1979a; McCaleb *et al.*, 1980; Reddy and Kumaran, 1980; Sparks and Hammock, 1980b). JH I is metabolized in the flour beetle, *Tribolium castaneum*, byester hydrolysis, epoxide hydration, and apparent conjugation (Edwards and Rowlands, 1977). The Colorado potato beetle, *Leptinotarsa decemlineata*, has been the subject of many endocrine investigations and an impressive amount of information is accumulating on JH metabolism in this species (Kramer and deKort, 1976a, b; deKort *et al.*, 1978; Kramer, 1978). Metabolism is again dependent upon esterases, hydrolases and conjugating enzymes with little oxidative metabolism (Kramer *et al.*, 1977).

The Diptera often seem unlike other insects in their response to juvenoids, the amount of JH present and other factors. Their metabolism of JH also seems somewhat unique. Slade and Zibitt (1972) found the conjugated diol ester to be a major metabolite in third instar larvae of the flesh fly, *Sarcophaga bullata*. Ajami and Riddiford (1973) failed to find JH diol in any of the four species of Diptera examined. They report evidence for oxidative metabolism although the very high R_f 'tetraol' metabolite probably indicates that it was actually a mixture consisting largely of tetrahydrofuran diols (Thf-diol, Figure 1.1). The diol acid does not appear as important a metabolite in the house fly, *Musca domestica*, as in many insects. JH diol and JH acid were found to be major metabolites in susceptible *M. domestica* while oxidative metabolism predominated in insecticide resistant strains (Hammock *et al.*, 1977a; Yu and Terriere, 1978a, b). Both cytosolic and microsomal JH esterases are present in homogenates of *M. domestica* larvae and these enzymes are quite unstable (Mumby *et al.*, 1979; Sparks and Hammock, 1980b). Following topical application to three stages of *Drosophila melanogaster*, Wilson and Gilbert (1978) found apparent conjugates, JH diol, and diol acid to be major metabolites with a trace of JH acid present. JH was rather stable in *D. melanogaster* haemolymph as earlier reported for haemolymph of *Sarcophaga bullata* (Weirich and Wren, 1976). Klages and Emmerich (1979a) also found no haemolymph JHE in larval *Drosophila hydei*, but esterase activity appeared in the pupal body fluid. Ester hydrolysis appeared to be the dominant metabolic pathway in tissue homogenates. Haemolymph JH esterase levels are very low in the honey bee, *Apis mellifera*, and metabolism seems to largely be due to tissue esterases (deKort *et al.*, 1977; Mane and Rembold, 1977). In two other species of Hymenoptera, Ajami and Riddiford (1973) also found the JH acid to predominate. Continued work on this order may help to answer interesting questions on the relationship of JH metabolism to social behaviour and host–parasite interaction as well as to facilitate more extensive comparisons of dipterous and hymenopterous insects.

Slade and Zibitt (1971, 1972) reported that the tobacco hornworm, *Manduca sexta*, hydrolyses the methyl ester of JH I and subsequently hydrates the epoxide. The diol ester is detected as an additional metabolite in *Hyalophora cecropia*. Ajami and Riddiford (1973) similarly found the diol to be a minor metabolite in pupae of each of three Saturniid species examined, but not in *M. sexta*. Conjugation pathways have not received adequate attention. Ajami and Riddiford (1973) and White (1972) present evidence for glucuronide and glucoside as well as sulphate formation. The most thorough study on insect conjugation of JH, to date, has been that of Slade and Wilkinson (1974) who report only sulphate formation in the southern armyworm, *Spodoptera eridania* (syn. *Prodenia*). Most subsequent work has been with *M. sexta*. These studies and similar work on several other lepidopterous species indicate that ester cleavage is the primary route of metabolism.

Information is lacking on the metabolism of JH in other arthropods. JH appears to have very low toxicity to mammals (Siddall and Slade, 1971; Slade and Zibitt, 1972), and it is rapidly metabolized in isolated rat hepatocytes

(Morello and Agosin, 1979). Very high levels of JH do interfere with macromolecular biosynthesis in cultured mouse cells (Chmurzyńska et al., 1979) and bovine lymphocytes (Kensler and Mueller, 1978; Laskowska-Bożek and Zielińska, 1978; Zielińska et al., 1978) as well as uncoupling oxidative phosphorylation (Chefurka, 1978). The high levels of hormone sometimes used suggest physical disruption of membranes.

Insect Epoxide Hydrolases

Epoxide hydrolases (E.C. 3.3.2.3, formerly E.C. 4.2.1.63 and known as hydrases or hydratases) add water to three-membered epoxide rings to yield 1,2-diols or glycols (Figure 1.1). Knowledge of insect epoxide hydrolases lags far behind knowledge of mammalian epoxide hydrolases (Oesch, 1973; Hammock et al., 1980; Lu and Miwa, 1980). The high level of interest in mammalian epoxide hydrolases has stemmed from interest in the mutagenicity and carcinogenicity of some highly reactive epoxides such as the arene oxides and other natural and man-made xenobiotics. It is interesting, however, that the first demonstration of epoxide formation in mammals emerged from the study of an exceedingly non-reactive epoxide, the cyclodiene insecticide heptachlor (Davidow and Radomski, 1953). Initial work on insect epoxide hydrolases involved the investigation of the metabolism of these cyclodiene epoxides and only later involved the study of JH and juvenoid metabolism. In an early and very comprehensive study, Brooks et al. (1970) demonstrated that dieldrin and several related cyclodiene epoxides are metabolized very slowly by microsomal epoxide hydrolases in both insects and mammals. The relative importance of microsomal metabolism and epoxide hydration was nicely shown by comparing metabolism and toxicity of the dieldrin analogues HEOM and HCE (which are degraded largely by hydration and hydroxylation, respectively) in several insects and vertebrates (Brooks et al., 1970; El Zorgani et al., 1970; Nelson and Matsumura, 1973; Walker and El Zorgani, 1974; Brooks, 1977, included references). The readily hydrated HEOM has been used as a model substrate for several subsequent studies of insect epoxide hydrolases (Brooks, 1973, 1974; Slade et al., 1975, 1976; Craven et al., 1976). Although no highly active and specific inhibitors were found in these studies, a wide variety of compounds including JH, juvenoids methylene dioxyphenyl, and benzothiozole mixed function oxidase (MFO) inhibitors, and organophosphates were found to cause some inhibition. Trichloropropene oxide and tetrahydronaphthalene 1,2-epoxide, which are inhibitors of styrene oxide epoxide hydrolase in mammals, also inhibited HEOM hydrolases in insects, although kinetics of inhibition were different. A very interesting observation was that some glycidyl ethers proved to be effective inhibitors of the insect enzymes. For a more comprehensive review of hydration of cyclodiene epoxides see Brooks (1977) and included references.

Compared with published work on insect esterases, investigation of the involvement of epoxide hydrolases in JH metabolism has been very limited. The tedious nature of the assays involved and the lack of effective hydrolase inhibitors have certainly limited the progress of research in this area. The most rapid assays involve either incomplete partitioning or the use of cellulose prelayer t.l.c. plates (Mumby and Hammock, 1979a; Mullin and Hammock, 1980). Although a continuous assay has been developed for epoxide hydrolases using *trans*-stilbene oxide, this substrate may be hydrated by enzymes not involved in JH metabolism. As illustrated by the early surveys of Slade and Zibitt (1971, 1972), White (1972), and Ajami and Riddiford (1973), the relative importance of ester hydrolysis and epoxide hydration in JH metabolism is variable in different insects, and the careful *in vivo* and *in vitro* kinetic studies to delineate the role of epoxide hydration in JH action are still lacking.

Using the juvenoid R-20458 (Figure 1.2), epoxide hydrolase activity was shown to be membrane bound and largely in the microsomal fraction of house fly, *M. domestica,* heads, abdomens, and thoraces, while little activity was detected in the soluble subcellular fraction. The hydration was fastest at neutral pH in *M. domestica* and distinct pH optima were observed, suggesting the involvement of several enzymes. A variety of compounds failed to cause strong inhibition of hydrolase activity, but the poor inhibition caused by trichloropropene oxide was especially notable (Hammock *et al.,* 1974a). Yu and Terriere (1978a, b) demonstrated that most hydrolase activity on JH I was microsomal in the house fly *M. domestica,* flesh fly *Sarcophaga bullata,* and black blow fly, *Phormia regina.* As reported for the metabolism of JH I and R-20458 for three strains of *M. domestica* (Hammock *et al.,* 1977a), it was reassuring that Yu and Terriere noted no major differences in hydrolase levels in insecticide susceptible and resistant strains. Possibly, epoxidized juvenoids may thus offer some advantages for insect control (Siddall, 1976; Zurflüh, 1976; Sparks and Hammock, 1980a). Yu and Terriere (1978b) report induction of JH I hydrolase activity by phenobarbital, and Mullin and Wilkinson (1980a) similarly observed induction in *Spodoptera eridania* midguts following exposure to pentamethylbenzene. When expressed in terms of metabolism per *M. domestica* equivalent, Yu and Terriere (1978a, b) observed high hydrolase activity towards the end of the last larval instar, a drop in the pupa, and a subsequent increase during the first two weeks of adult life. Several juvenoids inhibited JH I hydration, suggesting hydration by the same enzyme, while cyclodiene insecticides, several synergists, styrene oxide, and cyclohexene oxide caused no inhibition.

Slade and Wilkinson (1974) found JH epoxide hydrolase activity absent from the haemolymph, but widely distributed in other tissues from *Spodoptera eridania, H. cecropia, M. sexta,* and the cockroach, *Gromphadorhina portentosa.* The activity was largely microsomal. Hydrolase activity was similarly reported from the cockroach, *Periplaneta americana; T. molitor;* cabbage looper, *Trichoplusia ni;* and *M. sexta* (Hammock *et al.,* 1974a). Yawetz and Agosin

rt epoxide hydrolase in epimastigotes of *Trypanosoma crazi* and
at it may be involved in degrading the hosts' JH. Slade *et al.* (1975)
emonstrated that HEOM hydrolase activity was largely membrane
bound in *S. eridania* and *G. portentosa* midgut homogenates. The hydrolase pH
optima for these two insects as well as for the blow fly, *Calliphora erythrocephala*,
was quite basic. The lower pH optimum for JH hydration compared to HEOM
and the weak inhibitory action of trichloropropene oxide on JH hydration
suggest that different hydrolases may be involved with the metabolism of these
two substrates. However, studies with the mammalian hydrolases have demonstrated that apparently the same enzyme can show greatly differing pH
optima with different substrates and hydration of substrates with low apparent
K_ms is poorly inhibited while substrates with high K_ms are effectively inhibited by
trichloropropene oxide. Mumby and Hammock (1979a) found a relatively low
apparent K_m (7.1 μM) for the hydration of R-20458 by *M. domestica* larval
microsomes.

Slade *et al.* (1976) monitored epoxide hydrolase activity during the development of *S. eridania* using HEOM and JH I as substrates. As later verified on a
separate study with the same species (Mullin and Wilkinson, 1980a), hydrolase
levels were maximal in midgut homogenates during the middle of the last instar.
Similar results were obtained by Wing *et al.* (1980) as they found the highest fat
body hydrolase activity on R-20458 between the two major esterase peaks in the
last larval instar (Figure 1.3).

Mullin and Wilkinson (1980a, b) reported the purification of a Cutsum[R]-solubilized epoxide hydrolase from midgut microsomes of *S. eridania*. Five
purification steps yielded the most active hydrolase reported from any eukaryote
when styrene oxide or 1,2-epoxyoctane were used as substrates. The lack of
activity on either HEOM or JH demonstrates that there are multiple forms of
insect hydrolases and illustrates the need for further biochemical studies on these
enzymes.

JH Esterases

JHE assay methods

Numerous methods have been developed for monitoring JHE activity since
Weirich *et al.* (1973) first used a thin layer chromatography (t.l.c.) assay. Weirich
and Wren (1973) extended this technique to a substrate specificity study in *M.
sexta* haemolymph where they found that the esterase was specific for methyl
esters and 2E geometry. Chromatographic assays using radiolabelled JH are
certainly the most definitive methods since they can potentially discern other
pathways of metabolism; however, they are laborious and the relative R_fs of JH
and JH diol are variable depending upon the acidity of the chromatographic
media and other factors (Hammock and Sparks, 1977). The use of cellulose

prelayer plates has greatly increased analytical speed in this laboratory; however, the reaction must be carefully terminated and chemiluminescence may create radio-counting problems. Higher resolution can be obtained through the use of high performance liquid chromatography (h.p.l.c.) (Morello and Agosin, 1979). Yu and Terriere (1975a) used a gas–liquid chromatographic assay for juvenoids, while Pratt (1975) used an electrophoretic assay. Sanburg *et al.* (1975a) report a rapid method using methoxy-labelled JH, and this method has been slightly modified for use in several laboratories (Nowock *et al.*, 1976; Hammock *et al.*, 1977b; Vince and Gilbert, 1977; Hwang-Hsu *et al.*, 1979; Klages and Emmerich, 1979a; Sparks *et al.*, 1979a). Unfortunately, the methoxy-labelled JHs are not commercially available. Hammock and Sparks (1977) report a rapid partition method using a commercially available JH I which can be modifed to monitor EH activity (Mumby and Hammock, 1979a). The method is also applicable to JH III analysis (Jones *et al.*, 1980; Wing *et al.*, 1980). Continuous assay for JHE and assays suitable for JHE staining on gels are still lacking. Analysis by pH stat and u.v. shift of the conjugated carbonyl proved too insensitive with JH and several model substrates. Coupling the hydrolysis of ethyl esters of JH and some analogues to alcohol dehydrogenase was only successful in the case of *M. domestica* because the JHE of many insects will not hydrolyse ethyl esters (Hammock *et al.*, 1977b, and unpublished work); however, such an assay may prove useful in some insects such as *T. molitor* and *Samia cynthia* (Weirich and Wren, 1976). The hydrolysis of β-methylumbelliferone esters was effectively used to monitor esterase activity during partial purification of JHE from *G. mellonella* haemolymph, but these compounds have not yet been shown to be specifically hydrolysed by JHE (Rudnicka *et al.*, 1979; Sláma and Jarolím, 1980). Several studies have demonstrated that α-naphthyl acetate (α-NA) is a poor marker for JHE activity in several insects (Weirich *et al.*, 1973; Hammock *et al.*, 1977b; Sparks and Hammock, 1979a); thus, one must be cautious about drawing conclusions regarding JH metabolism from model substrates.

JHE changes during development

The first developmental study of JH metabolism was that of Weirich *et al.* (1973) in *M. sexta*. Very little JHE activity was found in haemolymph from early instars or pupae of either sex. A major prewandering peak (Figure 1.3A) and a minor prepupation peak (C) of JHE activity were found which did not correlate with α-NA esterase levels. Subsequent studies in the Lepidoptera verified the report of Weirich *et al.* (1973). Numerous studies have dealt with monitoring α-NA hydrolysis during insect development. Since JH has often been shown to be hydrolysed by esterases different from those hydrolysing α-NA, the relevance of such work to JH metabolism is questionable. Brown *et al.* (1977) extended such an electrophoretic comparison of *Diatraea grandiosella* (southwestern corn borer) esterases to define those esterase bands inhibited by JH and several juvenoids.

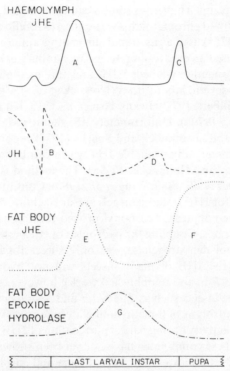

Figure 1.3 Relative levels of JH and enzymes possibly influencing JH titre during the last instar of a hypothetical lepidopterous larva. Juvenile hormone esterase (JHE) activity in the haemolymph (——) demonstrates a prewandering (A) and a prepupation (C) burst of activity. JH titre (– – –) drops at the moult, increases in the early last instar (B) and increases again before pupation (D). The increase in fat body JHE (·····) in the prewandering last instar (E) coincides with the haemolymph increase (A), but the levels again increase in the prepupa (F) to remain high during most of the pupal stage. Epoxide hydrolase levels (–·–·–·) are highest in the mid–last instar (G)

Sanburg et al. (1975a) monitored DFP sensitive and resistant esterase activity in fourth and early fifth instar M. sexta haemolymph. Vince and Gilbert (1977) reported occurrence of the major JHE peak just before M. sexta clears its gut and correlated the prepupation peak with the formation of tanned, sclerotized bands in the prepupa. Similar patterns of JHE activity were found in black mutant and allatectomized M. sexta except that the first JHE peak was smaller in both cases and the second JHE peak was greatly reduced in allatectomized larvae (Riddiford and Hammock, unpublished; Jones et al., unpublished). Mitsui et al. (1979) found M. sexta epidermis degrades JH largely by ester hydrolysis and the half-life of JH in epidermal cultures is lowest when haemolymph JHE activity peaks. Weirich and Wren (1976) report a large prewandering JHE peak in the Saturniid *Samia cynthia* (Figure 1.3A), while Slade *et al.* (1976) monitored JH hydrolysing enzymes during the development of *Spodoptera eridania*.

Sparks *et al.* (1979a) studied hydrolysis of JH I and JH III in *T. ni* larvae. No difference was noted in male and female larvae and the second JHE peak was much larger than that reported in *M. sexta*. The second peak occurred early in the morning of the day of pupation regardless of what gate larvae were used (Sparks and Hammock, 1979b). Wing *et al.* (1980) monitored haemolymph and tissue JHE and EH as well as haemolymph JH binding during *T. ni* development. Peak EH levels lag behind JHE levels, and there is a correlation between fat body JHE and haemolymph JHE until pupation when the fat body JHE remains high (Figure 1.3). Hwang-Hsu *et al.* (1979) monitored JHE and α-NA esterase activity during development of *G. mellonella* as well as ecdysone titre and sensitivity to JH. The sharp peaks of JHE observed in *M. sexta* and *T. ni* were not reported, but there is JHE activity in the prewandering stage and activity remains reasonably high until pupation. In a survey of haemolymph JHE from several Lepidoptera, both the prewandering and prepupation JHE peaks generally occurred with JHE levels usually lower in butterfly larvae than moth larvae (Wing *et al.*, Jones *et al.*, unpublished). As indicated by Akamatsu *et al.* (1975), most workers who have monitored JHE have hypothesized an *in vivo* role for the enzyme based on correlations with expected decreases in JH titre. Indeed, in several insects the appearance of the first JHE peak correlates with the drop in haemolymph titre of JH, and haemolymph JH has again been detected in postwandering larvae prior to the prepupal JHE peak. Nijhout (1975), however, was quite justified in questioning the role of JHE in clearing JH from *M. sexta* haemolymph. By applying EPPAT (*O*-ethyl-*S*-phenyl phosphoramidothiolate) which inhibits JHE but not EH and other esterase inhibitors, Sparks and Hammock (1980b) demonstrated that inhibition of JHEs stabilized radioactive JH *in vivo*, and caused developmental aberrations consistent with stabilizing intrinsic JH. These studies provided direct evidence for the hypothesis that JHE actually play a role in normal JH metabolism *in vivo* in at least one insect.

Studies on JH metabolism in *L. decemlineata* have proved very useful because of the large background of biological and endocrinological literature on this beetle. Kramer and deKort (1976a) found low haemolymph JHE in third instar, but high JHE in fourth instar larvae which would correlate with a presumed decline in JH before pupation. High JHE was found in beetles just before diapause, again correlating with decreasing JH titres. Unlike the Lepidoptera examined, Kramer and deKort (1976a) report that developmental profiles of JHE and α-NA esterase were similar in *L. decemlineata* and speculate that JHE activity may comprise a substantial portion of the observed esterase activity on α-NA. Injections of Triton X-100 increased the half-life of radiolabelled JH in short-day beetles and inhibited JHE and EH. These studies indicate an *in vivo* role for JH metabolism in clearing at least exogenously administered JH (Kramer *et al.,* 1977). Using *T. molitor,* Weirich and Wren (1976) report very high levels of JHE in pupal and pharate adult haemolymph, lower levels in larvae, and very low levels in adults. Reddy and Kumaran (1980) examined α-NA esterase and JHE during the development of carefully timed *T. molitor* and found a rapid increase of JHE in prepupae, slowly decreasing levels of JHE during the pupal stage and low levels in the adult. Sparks and Hammock (unpublished) found very high levels of JHE in the 100,000 g soluble fraction and high epoxide hydrolase levels in the 100,000 g pellet of *T. molitor* pupal homogenates during development, but no difference in enzyme activity was noted between male and female insects. Edwards and Rowlands (1977) monitored *in vivo* JH I metabolism in the flour beetle, *Tribolium castaneum,* at eight times during development. Metabolism by both hydrolase and esterase pathways was highest in pupal and last larval instars and was lower in adults and early larvae.

JH metabolism in the Diptera seems to vary greatly from species to species and to be quite different from other insects. Yu and Terriere (1978a, b) followed esterase, oxidase, and hydrolase action on JH during the development of *M. domestica.* Activities were high during the larval stages, low in the early pupae, and increased prior to eclosion. These workers also compared the activities in three fly species and four strains of *M. domestica.* Differences were noted, especially in oxidase activity, which raised the question of how JH is regulated in insecticide resistant strains which rapidly metabolize it (Sparks and Hammock, 1980a). Wilson and Gilbert (1978) conclude that JH metabolism may not be a major mechanism of JH titre regulation in *D. melanogaster.* Klages and Emmerich (1979a) report that JHE in haemolymph and body fluid of *Drosophila hydei* larvae is very low, but that it increases in prepupae and pupae. Metabolism of JH is much higher in the fat body and body wall, due largely to a cytosolic esterase.

Mane and Rembold (1977) followed JH metabolism in the 20,000 g supernatant of homogenates of queen and worker honey bee, *Apis mellifera,* larvae and pupae. As reported earlier for adults (Ajami and Riddiford, 1973), ester hydrolysis accounted for the majority of the metabolism. The activity was soluble at 100,000 g and was highest in late larval instars and pupae. Interestingly, the activity was much higher in queen than in worker larvae.

The metabolism of JH at various times during development has been examined in several insects. With techniques now well established for monitoring *in vivo* metabolism of JH as well as measuring the major enzymes involved in its degradation, careful metabolism studies need to be extended to a variety of precisely timed insects from different orders.

JHE inhibition

The first indication that JH esterases in many insects were not 'typical' carboxylesterases stemmed from studies with inhibitors. Although α-NA esterase activity which co-migrates with some JHE from *Hyalophora gloveri* pupae on electrophoresis can apparently be inhibited by O,O-di-isopropyl phosphorofluoridate (Whitmore *et al.*, 1972), JHEs are often much more resistant to inhibition than most serine esterases and proteases (Kramer *et al.*, 1974). Pratt (1975) in a screen of four compounds found paraoxon to be a good inhibitor of haemolymph JHE from the desert locust, *Schistocera gregaria*, and that inhibition was irreversible. Hammock *et al.* (1977b) demonstrated multiple forms of JH esterases in the cockroach *Blaberus giganteus* and screened 42 potential inhibitors using partially purified JHE fractions. Phosphoramidothiolates and *S*-phenylphosphates were found to be superb inhibitors; several classical esterase inhibitors were found to be very poor inhibitors, and most α-NA hydrolysis could be inhibited separately from JH hydrolysis. The studies with *S. gregaria* and *B. giganteus* also indicated that esterase inhibitors could be useful in biosynthesis studies by blocking subsequent JH metabolism (Hammock, 1975; Pratt, 1975; Hammock and Mumby, 1978). The use of such inhibitors can possibly be extended to studying *in vitro* action and receptor binding of JH as well as to *in vivo* kinetic studies.

Hooper (1976) reported a very extensive study of esterase action on malathion, α-NA, methoprene, hydroprene, and JH in *Culex pipiens pipiens*. He concluded that esterases hydrolysing malathion and α-NA are similar to each other and distinct from those hydrolysing JH. The enzymes could be classed as B- and A-esterases, respectively (Aldridge, 1953), and Hooper's (1976) studies illustrate the difficulty in forcing insect esterases into an artificial classification system designed for mammalian or avian esterases. JH esterases in the Diptera do appear to be quite different from the JH esterases in most other insects examined. In *M. domestica* larvae, activity was found in both the 100,000 g soluble and pellet (microsomal) subcellular fractions, and the activity in each fraction could be differentially inhibited. The soluble and especially the microsomal activity was quite unstable when compared to haemolymph or tissue JHE from Lepidoptera. *M. domestica* JHE activity was not inhibited by many of the compounds active on *T. ni*, *T. molitor*, or *B. giganteus*, but it was inhibited by a paraoxon and some *N*-ethyl carbamates (Yu and Terriere, 1978a; Mumby *et al.*, 1979; Sparks and Hammock, 1980b). These carbamates caused little or no inhibition of JHE from *B. giganteus*, *T. ni*, and *T. molitor*. The JHEs from *D. hydei* are also inhibited by several classical esterase inhibitors inactive on JHE from lepidopterous larvae, and Chang (unpublished) found that *N*-ethyl carbamates could greatly stabilize JH in *D. melanogaster* cell lines.

Kramer and deKort (1976b) report that haemolymph JHE of the Colorado potato beetle, *Leptinotarsa decemlineata*, are resistant to DFP but are very susceptible to inhibition by Triton X-100, and this compound has been shown to stabilize JH *in vivo* (Kramer *et al.*, 1977). The mechanism of Triton X-100 inhibition has not been elucidated, but there are several possibilities. One possibility is direct interaction with the enzyme, but another possibility is simply sequestration of the JH substrate in micelles. JH not only forms micelles, but it is, undoubtedly, quite soluble in detergent micelles. The stability of JH in such micelles has been previously noted as a possible explanation for observations by Schmialek *et al.* (1975) on JH binding (Law, personal communication). It is known that some enzymes will not recognize substrate in micelle form (Mumby and Hammock, 1979b,c; Hammock *et al.*, 1980), and Armstrong *et al.* (1980) demonstrated that substrate sequestration is the mechanism by which non-ionic detergents inhibit the activity of a mammalian epoxide hydrolase. The affinity of such detergent micelles for lipophilic molecules may be quite high: even approaching the affinity of the *M. sexta* carrier protein for JH. Thus, detergent micelles may provide a nice model for the study of JH carrier–enzyme interaction, and the lack of Triton X-100 inhibition of JHE in the Lepidoptera may be of biological significance.

Triton X-100 also selectively inhibits haemolymph JHE in *T. molitor* while DFP and paraoxon are poor inhibitors (McCaleb *et al.*, 1980; Reddy and Kumaran, 1980); EPPAT emerged as a useful inhibitor of JH hydrolysis in soluble and microsomal fractions of *T. molitor* (Sparks and Hammock, 1980b). Mane and Rembold (1977) report that high levels of Triton X-100 inhibit *A. mellifera* JHE. Such differential inhibitors of JHE as Triton X-100, EPPAT, paraoxon, and DFP may be useful to investigate the apparent dichotomy of JHE in various insect groups as well as the physiological role of esterases. Ajami (1975) screened 16 potential esterase inhibitors using *M. sexta* pupal haemolymph. The haemolymph, if free of fat body, is rich in carboxylesterase, but JH hydrolysis is slow. Ajami (1975) was able to stabilize JH *in vivo* and demonstrate a synergistic effect of some esterase inhibitors. Slade and Wilkinson (1974, and unpublished) took an analogous approach and demonstrated *in vivo* stabilization of JH in the southern armyworm, *Spodoptera eridania*, using juvenoids and a range of EH and MFO inhibitors. In a comparison of 27 potential esterase inhibitors in three insects, Sparks and Hammock (1980b) report that EPPAT is a potent inhibitor in *T. ni* and *T. molitor*, but not in *M. domestica*. EPPAT will delay pupation by inhibiting the first JH esterase peak in the last larval instar of *T. ni* thus stabilizing JH and apparently delaying PTTH release. This study, like the Triton X-100 study in *L. decemlineata* (Kramer *et al.*, 1977), demonstrated an *in vivo* role for JHE as discussed earlier. The trifluoromethylketones represent a second novel series of inhibitors where the polarized ketone is hypothesized to mimic the tetrahedral transition state involved in ester cleavage. The inhibition caused by these compounds is reversible and quite specific for JHE in *T. ni, M. sexta,* and *G. mellonella* (Sparks and Hammock, 1980b, and unpublished).

JHE inhibitors have proved useful for testing the similarity of esterases from subcellular fractions, from different developmental stages, from different tissues, and acting on different substrates (Hooper, 1976; Hammock *et al.*, 1977b; Sparks and Hammock, 1979b; Jones *et al.*, 1980; Wing *et al.*, 1980). There was early hope that JHE inhibitors might act as synergists for JH or juvenoids (Solomon and

Metcalf, 1974; Ajami, 1975), but several factors make such action unlikely. The problems involved in the registration of a synergized formulation and the cost effectiveness of such a formulation are major economic considerations. JH will certainly not be used as an insecticide, and the most promising juvenoids either lack an ester function or possess an ester stable to most esterases so that the need for synergists is limited to isolated cases (Bigley and Vinson, 1979b). There is the possibility that inhibition of JHE may lead to disruption of insect development due to intrinsic JH, and since JHE of several insects appear to be so specific, there is hope that highly selective chemicals can be developed. Sparks and Hammock (1980a) conclude that such JHE inhibition by some classical organophosphates might have led to sublethal effects observed with insects in the field. However, they further conclude that, at least in *T. ni*, the inhibition of JHE does not appear a promising course for the development of control agents for several reasons:

(1) JHE appears to be present in large excess so that almost total inhibition is needed for an *in vivo* effect.
(2) JH production as well as metabolism appear to be precisely regulated.
(3) the inhibitors would only be effective during narrow periods of development; and
(4) the *in vivo* effects would include prelongation of a potentially destructive instar.

However, as tools for studying insect development, specific inhibitors are vital, and inhibitors such as the trifluoromethyl ketone moiety could be used for active site studies, purification via affinity chromatography, and possibly to aid in distribution of a JH mimic by binding with the JHE. For direct use in insect control, the disruption of JH metabolism by inhibition of enzyme production or precocious production of the enzyme is more promising.

JHE regulation

Since JHEs appear to be involved in JH regulation, understanding the regulation of these and other metabolic enzymes is a logical extension of our present knowledge. Retnakaran and Joly (1976) demonstrated that cautery of the A and B neurosecretory cells of *Locusta migratoria* reduces JHE activity. Hopefully, more research will soon emerge with the adult Hemimetabola. Kramer (1978) illustrated a complex response of *L. decemlineata* JHE to exogenous JH, which may possibly be correlated with corpus allatum activity (Schooneveld *et al.*, 1979). Treatment of diapausing beetles with JH I or several juvenoids caused a JHE increase, and this increase could be blocked by the use of puromycin or actinomycin D. This apparent induction was largely blocked by neck ligations suggesting that the JH effect on the fat body was indirect. A series of experiments using long- and short-day beetles with and without allatectomy suggest that the fat body's history of exposure to JH may influence its ability to produce JHE. Reddy and Kumaran (1980) report that in *T. molitor* pupae JH and ETB have no effect on haemolymph JHE while precocene II slightly depresses JHE.

Terriere and Yu (1973) demonstrated that JH, juvenoids, and β-ecdysone stimulate MFO activity in *M. domestica*. Reddy and Krishnakumaran (1974) followed MFO levels during the development of *G. mellonella* and found that JH reduced MFO activity *in vivo* and *in vitro* but that it was increased in JH induced superlarvae. Although these results have not been clearly tied to hormone metabolism, JH can be degraded by MFO. Such changes in MFO activity during development or in response to hormones or even dietary inducers may have an effect on JH titres (Terriere, 1980; Wilkinson, 1980). Downer *et al.* (1975) found that the juvenoid methoprene could depress the activity of a non-specific esterase in *Aedes aegypti*.

JHE regulation in the Lepidoptera has received the greatest attention. Evidence that JH could directly 'induce' its own metabolism was first provided by Whitmore *et al.* (1972). In a very meticulous study using *H. gloveri* and *H. cecropia*, Whitmore *et al.* (1974) subsequently demonstrated that the apparent induction was dose dependent; prevented by puromycin, cycloheximide, and actinomycin D; that it could be duplicated with fat body *in vitro*; and that the esterases produced *in vitro* were immunochemically similar to the ones produced *in vivo*. Attempts to specifically label the enzyme(s) were not successful. These studies suffered from two problems. One problem was interpreting the role of 'inducible' esterases in pupae theoretically devoid of JH. The second problem stemmed from the logical but probably incorrect assumption that α-NA hydrolysis could be used as a marker for JHE activity. In spite of these limitations, later studies have generally supported the conclusions of Whitmore *et al.* (1972, 1974), and no subsequent studies using more biochemically defined JHE have been so thorough. Reddy *et al.* (1979) using *G. mellonella* and Wing *et al.* (1980) using *T. ni* also demonstrated that JH application to pupae resulted in increased haemolymph JHE. Since pupal fat body has high JHE, this effect could result only from release, but the data of Wing *et al.* (1980) suggest that some esterase production is also involved (it is very difficult to prove unequivocally true induction).

Data of Hammock *et al.* (1975b) indicate that *M. sexta* fat bodies and imaginal wing discs, held in short term culture, release JHE activity. Nowock *et al.* (1976) report that the apparent release of JHE activity by fat body *in vitro* correlates with peak haemolymph JHE activity (Figure 1.3E) and that α-NA esterase activity appears to be under separate control. Nowock and Gilbert (1976) further demonstrated in *M. sexta* the JHE activity of the $100,000\,g$ supernatant of fat body homogenate paralleled that of the haemolymph for the first half of the instar. Inhibitor studies indicated that JHE release was an active process and not cell leakage. Similarly, in *G. mellonella* the release of JHE by fat body *in vitro* could be correlated with the haemolymph JHE (Reddy *et al.*, 1979). Wing *et al.* (1980) monitored fat body, midgut, and haemolymph JHE levels during *T. ni* development. The fat body and haemolymph JHE levels correlated well during the last larval instar (Figure 1.3E), but haemolymph JHE was low and fat body JHE high during the pupal stage (F). Isoelectric focusing and inhibition data indicated that the fat body and haemolymph JHE activity appeared to be due to the same enzyme, but that much of the lower midgut activity was due to enzymes of different isoelectric points. Such data support but still do not conclusively prove the fat body to be the source of haemolymph JHE, and it indicates that production and release of JHE may be under different control mechanisms.

Studies from several laboratories serve to build a theory of JHE regulation in Lepidoptera. The data have been largely obtained from *Hyalophora* sp., *G. mellonella*, *M. sexta*, and *T. ni*, so the following generalizations should be treated as only a working hypothesis. There are two peaks of haemolymph JHE activity in *M. sexta* and *T. ni* and the first peak in prewandering larvae (A) (Figure 1.3) correlates with the decline of JH in the early instar (B). The second peak (C) follows a presumed second burst of JH (D). In *G. mellonella*, ligation of early larvae or starvation partially blocks the subsequent appearance of JHE (Reddy *et al.*, 1979). This partial block may indicate that tissues other than those in the head are involved in stimulating JHE production or that these factors are released prior to ligation and are either relatively stable or have a delayed effect. McCaleb and Kumaran (1978) and McCaleb *et al.* (1980) subsequently demonstrated that JHE production is closely coupled to the cues that initiate the larval–pupal transformation by studying JHE levels in injured, chilled, and/or starved insects. The effects of ligation or starvation in *T. ni* (Sparks and Hammock, 1979a, b) are more dramatic, totally blocking the appearance of the first JHE peak (A) or quickly reducing haemolymph JHE levels (F) to the trace activity seen in earlier instars. The moisture content of the food and several nutritive factors apparently play a role in maintaining high haemolymph JHE with dietary protein having a major role in *T. ni*. Starvation causes an immediate increase in haemolymph JH titre in *M. sexta* (Cymborowski *et al.*, 1979) so it is of possible physiological significance that starvation either prevents JHE appearance or later causes a precipitous decline in haemolymph JHE titres in last instar *M. sexta* just as it does in *T. ni* (Riddiford and Hammock, unpublished). High levels of JH, ETB, or precocene II will reduce haemolymph JHE levels in early last instar *G. mellonella* and will also reduce the release of JHE by fat body held in short term culture (Reddy *et al.*, 1979). In *T. ni* JH will not increase haemolymph JHE in the penultimate instar larvae, but it will increase the activity in the first JHE peak (A). This JH effect is either not directly on the fat body or it must be in concert with other head factors, since JH application will not increase haemolymph JHE in isolated abdomens. Implantation of the brain or suboesophageal ganglion stimulates JHE production in chilled or ligated *G.*

mellonella larvae with the brain giving the highest activity (McCaleb *et al.*, 1980). Similar effects have been obtained with *T. ni* (and apparently *M. sexta*: Vince and Gilbert, 1977), but the suboesophageal ganglion appears more active than the brain. Jones *et al.* (1980) further found that the ability of brain and suboesophageal ganglion homogenates to induce JHE in isolated abdomens as well as the fat body's responsiveness to the homogenates was correlated with haemolymph JHE titre in *T. ni*. Perhaps a peak of brain and suboesophageal ganglion activity in the penultimate larval instar serves to prime the unresponsive fat body and there is some evidence that a head factor may be involved in actively turning off JHE production. In *T. ni* implantation of the brain-suboesophageal ganglion complex was much more effective in restoring JHE activity than either tissue alone. The data from *G. mellonella*, *M. sexta*, and *T. ni* are thus consistent with a neurohormone from the brain and/or suboesophageal ganglion controlling the first JHE peak.

Control of the second JHE peak appears quite different from the first peak. Sparks and Hammock (1979a) reported that after the wandering stage, JH and juvenoids could induce JHE in a dose dependent fashion in both normal *T. ni* larvae and isolated abdomens. This induction could be inhibited by application of the antihormone ETB (Sparks *et al.*, 1979b). Wing *et al.* (1980) found that JHE in the fat body of postwandering *T. ni* also increases following juvenoid application, JH application also increases haemolymph JHE levels in *G. mellonella* larvae just before pupation (McCaleb *et al.*, 1980). Several lines of evidence suggest that the second JHE peak is due, at least in part, to a response to natural JH (Sparks and Hammock, 1979a). Injection of the MFO inhibitor *O*-bromophenoxymethyl imidazole which is known to block JH biosynthesis in corpora allata of *Blaberus giganteus* (Hammock and Mumby, 1978) will reduce the level of the natural JHE peak, but will not block the response to exogenous JH in *T. ni*. Since such imidazole compounds have a variety of actions, these data must be treated cautiously. Allatectomy of *M. sexta* larvae causes a large reduction in, if not elimination of, the second JHE peak, although allatectomized larvae readily respond to JH or juvenoid application by increased JHE (Riddiford and Hammock, unpublished). Preliminary data from *M. sexta* and rather conclusive data from *T. ni* indicate that the two JHE peaks are caused by the same enzyme (Sparks and Hammock, 1979a). If the regulatory mechanisms outlined above are valid, this could be one of the rare examples in biology where the production of the same protein is under two completely different regulatory mechanisms at different times during development (Sparks and Hammock, 1979a, b).

A teleological interpretation can be inflicted on the above data using *T. ni* as a model. If adequate food is available, the mature larva will halt or reduce JH biosynthesis and increased JHE will begin to clear JH from the haemolymph preparing the insect for PTTH release and subsequent pupation. If adequate food is not available for development, the insect must quickly respond in *M.*

sexta and *T. ni* by increasing JH production and reducing JHE, thus extending the feeding stage. It is clearly not advantageous for JH to lead directly to a large increase in JHE at this stage of development, and neurosecretory control of JHE facilitates a rapid response to environmental factors. Once JH has been cleared and PTTH released, it seems important to reduce the high JHE levels and, in fact, this process may be active. The JHE activity must be reduced for a second prepupal increase in JH, and this increase in JH seems important to prevent precocious adult development of some tissues (Kiguchi and Riddiford, 1978; Cymborowski and Stolarz, 1979). It is critical for normal development that the JH titre is again reduced before pupation, so that the ability of JH to stimulate JHE directly, possibly in concert with other factors, is clearly of survival value. The role of high fat body JHE in the pupa remains unclear, but Reddy *et al.* (1979) and Wing *et al.* (1980) hypothesize that it may ensure JH removal before adult development. Although direct inhibition of the enzymes involved in JH metabolism does not appear promising for insect control, the stimulation or inhibition of their production at inappropriate times could clearly disrupt development.

Influence of JH Carriers on Metabolism

Juvenile hormone is not only susceptible to metabolism in insect haemolymph, it is also quite lipophilic and should tend to partition out of an aqueous compartment. The mechanism by which JH is transported in insect haemolymph was a subject of speculation for many years. The answer came from the discovery of JH carrier proteins in the Leptidoptera (Whitmore and Gilbert, 1972; Kramer *et al.*, 1974) and the most detailed subsequent work has been with members of this order. When present in high concentration, JH will bind to high capacity, low affinity lipoproteins in insect haemolymph. Whitmore and Gilbert (1972, 1974) found that six different lipoproteins from the saturniids *H. cecropia, H. gloveri,* and *Antheraea polyphemus* will bind JH *in vitro*, but only two will bind JH *in vivo*. Although a high affinity binding protein is also present in *H. cecropia* (Gilbert and Hammock, unpublished), the tremendous capacity of the lipoprotein component overshadows its presence. In most other Lepidoptera examined, the low capacity, high affinity binding protein probably accounts for most JH binding (Goodman and Gilbert, 1979; Kramer *et al.*, 1974; Kramer and Childs, 1977), and this protein has been most extensively studied in *M. sexta*.

The *M. sexta* carrier protein has been biochemically characterized as having an acidic pI, a molecular weight of 28,000 daltons, and a K_a for JH I of $\sim 10^{-7}$ M with one binding site (Kramer *et al.*, 1974, 1976a, b; Akamatsu *et al.*, 1975; Goodman *et al.*, 1978a) and its major site of biosynthesis appears to be the fat body (Nowock *et al.*, 1975, 1976; Nowock and Gilbert, 1976). The carrier protein demonstrates a high specificity for JH I, and even juvenoids which are quite active in *M. sexta* fail to have a high affinity for the carrier (Goodman *et al.*, 1976; Kramer *et al.*, 1976b). The carrier does not significantly bind JH metabolites, shows much higher affinity for JH I than JH III, and distinguishes the correct optical and geometrical isomer (Kramer *et al.*, 1974, 1976b; Goodman *et al.*, 1976, 1978b; Peterson *et al.*, 1977; Law, 1978; Gilbert *et al.*, 1978; Schooley *et al.*, 1978).

It is likely that the carrier protein influences JH stability and distribution in several ways. Sanburg *et al.* (1975a, b) demonstrated that esterases from *M. sexta* could be classified as general or JH specific enzymes based, in part, on the inability of general esterases to degrade JH bound to the carrier. Hammock *et al.* (1975b) similarly found that the carrier protein stabilized JH when added to *M. sexta* fat bodies in short term cultures. These two studies clearly indicate that the carrier protein reduces unwanted metabolism of JH (Gilbert *et al.*, 1976). It is not yet clear if JH 'specific esterases' actually metabolize bound JH or simply rely on a high affinity for the hormone and mass action. The carrier protein also tends to keep JH in solution. Although JH is readily water soluble at far above physiological concentrations, it tends to partition into lipophilic depots. Whether the depot is fat body or simply a lipid droplet, binding protein retards the uptake of and shifts the equilibrium away from lipophilic compartments and into aqueous compartments (Hammock *et al.*, 1975b; Nowock *et al.*, 1976; Mitsui *et al.*, 1979). Thus, the carrier protein helps to ensure an equal distribution of JH through the insect. It is likely that JH will behave as a surfactant and JH I has been shown to form micelles with a critical micelle concentration of about 10^{-5} M when determined by several independent methods (Kramer *et al.*, 1974, 1976b; Mayer and Burke, 1976; Hammock *et al.*, 1977b). Since such surfactants may accumulate near cell surfaces, the carrier may be important in preventing such surface excess of JH (Akamatsu *et al.*, 1975). Although the carrier seems to have a JH protective role for much of the insect's life, it also probably aids in the rapid clearing of JH from the insects' body by keeping it in circulation and accessible to JH 'specific esterases' rather than sequestered in depots refractory to metabolism (Gilbert *et al.*, 1978; Goodman and Gilbert, 1978; Sparks *et al.*, 1979a). There was some early indication that the carrier might additionally influence JH action, but subsequent work indicates that JH probably acts as a free molecule (Sanburg *et al.*, 1975b; Mitsui *et al.*, 1979). Kramer and Law (1980), in a nice comparison of JH biosynthesis and metabolism in *M. sexta* and *L. decemlineata*, suggest that the specific carrier molecule present in the Lepidoptera allows more precise regulation of JH and, thus, lower synthesis rates. Although such roles for the carrier protein appear obvious, and they have been demonstrated in numerous *in vitro* systems, *in vivo* demonstrations are still lacking.

The carrier protein titre fluctuates much as total haemolymph protein does in the penultimate larval instar of *M. sexta* (Goodman and Gilbert, 1978). Fluctuations during the last larval instar when changes in JH titre are most dramatic are of high interest, but this study has been complicated by high JHE titres. This problem was solved in *T. ni* using EPPAT to inhibit JHE. *T. ni* JH binding is largely due to a single protein with a similar titre and affinity for JH I and III as the *M. sexta* protein (Hammock *et al.*, 1977a; Sparks and Hammock, 1979b). JH I binding activity in *T. ni* haemolymph fluctuates only slightly during the last larval instar, being highest in mid instar (Wing *et al.*, 1980). Hopefully, similar information will shortly be available for *M. sexta*.

Similar studies with JH carriers have been done in the Indian meal moth, *Plodia interpunctella*, and *T. ni*, with results analogous to those in *M. sexta* (Ferkovich *et al.*, 1975, 1976, 1977; Ferkovich and Rutter, 1976; Hammock *et al.*, 1977b; Sparks *et al.*, 1979a; Sparks and Hammock, 1979b; Wing *et al.*, 1980). Surveys of several species of Lepidoptera indicate that low molecular weight, high affinity JH haemolymph carriers are present (Kramer *et al.*, 1976a; Kramer and Childs, 1977; Wing *et al.*, unpublished).

In several insects, from a variety of orders other than the Lepidoptera, low molecular weight carrier proteins have not been detected. In these insects, high molecular weight lipoproteins with low affinity for JH probably assume the role as JH carrier (Trautmann, 1972; Emmerich and Hartmann, 1973; Emmerich, 1976; Kramer and deKort, 1976b, 1978; Bassi *et al.*, 1977; Hammock *et al.*, 1977b; deKort *et al.*, 1977). An interesting exception was the report by Hartmann (1978) of a very high affinity lipoprotein separated from a large amount of low affinity lipoproteins in the haemolymph of the grasshopper, *Gomphocerus rufus*. Peter *et al.* (1979b) report a complex situation in *Locusta migratoria* in which both a large capacity, low specificity diglyceride carrier lipoprotein and a high affinity carrier protein exist. The high affinity carrier binds the natural (10R) isomer of JH III preferentially over the unnatural (10S) isomer and JH I. Unbound hormone is then metabolized by haemolymph esterases. Klages and Emmerich (1979b) report low affinity JH binding to a large molecular weight protein isolated from the haemolymph of third instar larvae of *Drosophila hydei*. Interestingly, they also report a small high affinity protein which is quite unstable. This discovery suggests that re-examination of insect haemolymph from orders other than Lepidoptera is warranted.

A demonstration of JH binding does not necessarily imply a physiological role. For instance, JH will even bind to some mammalian proteins (Mayer and Burke, 1976). Kramer and deKort (1978) demonstrated that lipoprotein-bound JH was poorly protected from *L. decemlineata* esterases so these proteins may serve largely for transport. The indication that lipoproteins from the haemolymph of the cockroach, *Periplaneta americana*, protect JH from metabolism in short term organ culture may indicate another physiological role for such proteins (Sams *et al.*, 1978) although they probably also are involved in transport of other insect lipids (Chino and Gilbert, 1971; Gilbert, 1974; Whitmore and Gilbert, 1974). Research is now needed which will define the *in vivo* roles of binding proteins in the dynamics of JH action.

DODECADIENOATES

Methoprene

Insect metabolism

Hammock and Quistad (1976) have reviewed the metabolism of methoprene (Figure 1.2) in insects by comparing its degradation to that of other juvenoids. Weirich and Wren (1973) using *M. sexta* first demonstrated that haemolymph esterases were unable to effectively hydrolyse methoprene. The refractory character of the isopropyl ester function was substantiated in *M. domestica* (Yu and Terriere, 1975a, 1977b), *P. regina,* and *S. bullata* by the same workers (Terriere and Yu, 1977). Yu and Terriere have also amply documented the importance of microsomal oxidases in methoprene degradation *in vitro* by flies. Quistad *et al.* (1975d) studied the *in vivo* degradation of methoprene in *Culex, Aedes,* and *M. domestica* larvae. A number of parameters (including olefinic isomerization, larval age, penetration, and synergists) were examined in an attempt to rationalize susceptibility as related to metabolism. Solomon and Metcalf (1974) reported the metabolic fate of methoprene applied to *Oncopeltus fasciatus* (milkweed bug) and *T. molitor.* With the aid of synergists these workers demonstrated metabolic activation via *O*-demethylation to the hydroxy ester which had fourfold greater biological activity in *O. fasciatus.*

Hammock *et al.* (1977a) explored the possible metabolic basis for resistance to methoprene in *M. domestica.* Larvae resistant to methoprene and cross-resistant larvae (R-dimethoate) both metabolized methoprene faster than susceptible larvae. Methoprene also penetrated more slowly into larvae of the resistant strain. Esterases, epoxide hydrolases, and olefinic isomerization of the 2E bond were inconsequential to the development of house fly resistance. Mixed function oxidase activity was considerably enhanced in resistant larvae which agrees with the work of Yu and Terriere (1977b) using another cross-resistant strain of *M. domestica* (R-diazinon).

Brown and Hooper (1979) compared the metabolism of methoprene in susceptible and methoprene-resistant larvae of *Culex pipiens.* Increased detoxication was an important factor in the high resistance developed after 30 generations of laboratory selection. Highly resistant larvae contained 11% less methoprene but more significantly, those larvae produced 40% more polar conjugates than susceptible larvae.

Bigley and Vinson (1979a, b) observed the degradation of methoprene by the imported fire ant (*Solenopsis invicta*). Adults and pupae metabolized methoprene primarily by *O*-demethylation to yield the hydroxy ester while larvae and pharate pupae produced mainly methoxy acid by esterase action. Although methoprene and its primary metabolites were all shown by bioassay to have juvenile hormone activity against *Solenopsis*, the hydroxy ester was particularly effective and alleged to be an activated metabolite (see Solomon and Metcalf, 1974). Adults not only produce principally the activated hydroxy ester, but by meticulous waste management, trophallaxis, and intimate social contact they distribute the insect growth regulator and its metabolites throughout the colony thereby providing a potential reservoir of insecticidal compounds. Bigley and Vinson (1979b) also modulated the degradation of methoprene by concurrent application of synergists (PB and DEF) and they suggest that synergists may be useful in extending the useful lifetime of methoprene in *S. invicta* bait formulations.

Mammalian metabolism

The fate of methoprene has been studied in several rodents, namely rats (Tokiwa *et al.*, 1975; Hawkins *et al.*, 1977), mice (Cline *et al.*, 1974), and a guinea pig (Chamberlain *et al.*, 1975). These studies were designed largely to follow the balance and distribution of radiolabel from methoprene and/or its metabolites; hence, there was minimal structure elucidation of degradation products. It became readily evident from this rodent work that although most radiolabel was readily excreted (Table 1.1), significant tissue residues remained upon termination of the animals. There was an apparent dichotomy of facile metabolism (considerable $^{14}CO_2$ evolution) and refractory elimination (high tissue residues). Whole-body autoradiographs of rats (Hawkins *et al.*, 1977) showed a very general distribution of radioactivity with a particularly high concentration in the adrenal cortex. Although the exact identity of tissue metabolites was not

Table 1.1 Mammalian metabolism of methoprene—distribution of radiolabel

	Per cent applied dose					
	Steer (14 day)	Cow (7 day)	Rats (5 day)		Mice (4 day)	Guinea pig (1 day)
Urine	22	20	20[a]	18[b]	68	24
Faeces	39	30	18	35	14	9
$^{14}CO_2$	3	15	39	25	NA[c]	17
Tissues	13	20	17	13	0.1	NE[d]
Milk	—	8	—	—	—	—
Total recovery	77	93	94	91	82	50

[a] Hawkins *et al.* (1977).
[b] Tokiwa *et al.* (1975).
[c] [10-^{3}H]methoprene dosed at $0.9\,mg\,kg^{-1}$, Cline *et al.* (1974).
[d] Not examined.

pursued, in light of other concurrent work (Quistad et al., 1974b, 1975b, c) the ^{14}C-residues were likely natural products (e.g. steroids) produced by extensive metabolism to anabolic precursors (e.g. [^{14}C]acetate).

The copious evolution of $^{14}CO_2$ from [5-^{14}C]methoprene was suggestive of extensive metabic degradation by mammals, particularly since the C-5 carbon was expected to be relatively inaccessible. A comprehensive investigation of the ^{14}C-tissue residues in bovines (Quistad et al., 1974b, 1975b, c) confirmed that methoprene was exhaustively metabolized to common precursors of intermediary metabolism (e.g. acetate) which were then incorporated into natural products

Table 1.2   Bovine metabolism of [5-^{14}C]methoprene to natural products

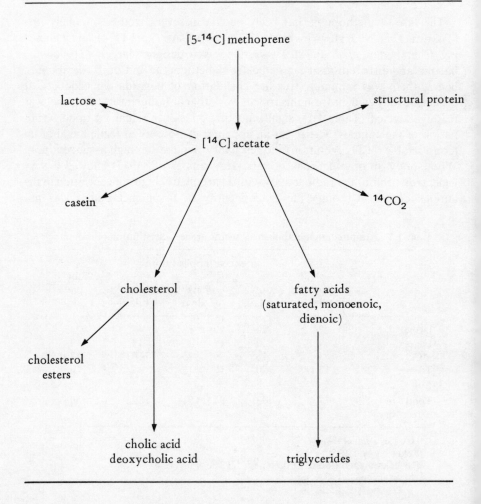

(Table 1.2). The bovine metabolism of methoprene to primary metabolites in excrement (Chamberlain et al., 1975) suggests partial degradation by the usual pathways of xenobiotic metabolism, but the quantitative abundance of acetate-derived ^{14}C-natural products and $^{14}CO_2$ implies that methoprene is also metabolized as a methyl-branched fatty acid (hence, as a 'food').

Environmental fate

Much of the environmental degradation of methoprene has already been reviewed (Quistad et al., 1975e; Schooley and Quistad, 1979). As a conjugated dodecadienoate the ester is relatively stable to chemical hydrolysis at pH 5, 7, and 9 (Schooley et al., 1975a). However, photochemical degradation is both rapid and extensive (Schaefer and Dupras, 1973; Quistad et al., 1975a). An important practical photochemical reaction is the isomerization of 2E,4E-methoprene to the biologically inactive 2Z,4E isomer. In addition to the photoproducts given in Table 1.3, a plethora of products remains uncharacterized. Microbial degradation of methoprene in pond water (Schooley et al., 1975a) and soil (Schooley et al., 1975b) gave only primary metabolites rather than radiolabelled natural products which could be readily characterized. However, the facile biodegradability of methoprene was again evidenced by copious evolution of $^{14}CO_2$. Degradation of [5-^{14}C]methoprene by bluegill fish (Quistad et al., 1976a) in an aquatic ecosystem gave a spectrum of radiolabelled natural products similar to that found in mammals (Table 1.2). Methoprene is also environmentally labile in the Metcalf ecosystem (Metcalf and Sanborn, 1975).

Methoprene is readily degraded by alfalfa and rice (Quistad et al., 1974a). Although several primary metabolites were identified after enzymatic cleavage of conjugates, the majority of the 'metabolite' fraction consisted of a diverse array of products. Gel permeation chromatography strongly suggested incorporation of radiolabel from extensively degraded [5-^{14}C]methoprene into higher molecular weight plant constituents such as carotenoids and chlorophylls. The principal non-polar metabolite was 7-methoxycitronellal which was isolated from vapours transpired from plants (13% applied dose from rice).

Methoprene is considerably more stable to degradation under conditions necessary for stored products pest control. Rowlands (1976) found a residual half-life of 2–3 weeks for methoprene on freshly harvested wheat grain. Gel filtration chromatography revealed negligible inclusion of radiolabel into high molecular weight natural products (*vide supra*) and the methoxy acid represented 20–40% of the total degradation products. The stability of methoprene in stored tobacco is even more impressive. After 31 months of storage for [5-^{14}C]methoprene on Bright-leaf tobacco in a mini-hogshead (6 × 6 × 60 cm) to simulate natural storage conditions, 69% of the applied dose was recovered as intact methoprene (Staiger et al., 1980).

Table 1.3 Environmental degradation products from methoprene

	methoprene	
Degradation product	Produced by (maximum % applied dose, including conjugates)	Reference
methoxy acid	Mosquito (7)	Quistad et al., 1975d
	Meal worm (1)	Solomon and Metcalf, 1974
	Milkweed bug (0.4)	Solomon and Metcalf, 1974
	Fire ant (8)	Bigley and Vinson, 1979a
	Aquatic microbes (6)	Schooley et al., 1975a
	Alfalfa (2)	Quistad et al., 1974a
	Wheat grain	Rowlands, 1976
	Chicken (3)	Quistad et al., 1976b
	Guinea pig	Chamberlain et al., 1975
	Steer	Chamberlain et al., 1975
hydroxy acid	Mosquito (8)	Quistad et al., 1975d
	House fly (10)	Quistad et al., 1975d
	Fire ant (10)	Bigley and Vinson, 1979a
	Milkweed bug (0.3)	Solomon and Metcalf, 1974
	Mealworm (1)	Solomon and Metcalf, 1974
	Aquatic microbes (3)	Schooley et al., 1975a
	Alfalfa (7)	Quistad et al., 1974a
	Chicken (2)	Quistad et al., 1976b
	Guinea pig	Chamberlain et al., 1975
	Steer	Chamberlain et al., 1975
saturated methoxy acid	Chicken (6)	Quistad et al., 1976b
saturated hydroxy acid	Chicken (0.5)	Quistad et al., 1976b
hydroxy ester	Mosquito (9)	Quistad et al., 1975d
	House fly (17)	Hammock et al., 1977
	Fire ant (10)	Bigley and Vinson, 1979a
	Milkweed bug (0.4)	Solomon and Metcalf, 1974

Table 1.3 (*cont.*)

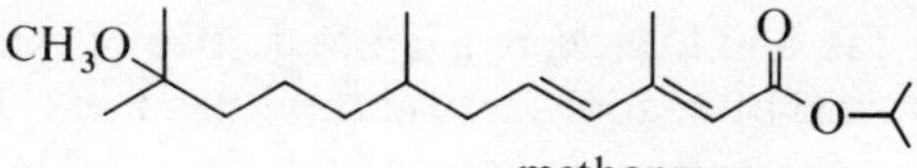

methoprene

Degradation product	Produced by (maximum % applied dose, including conjugates)	Reference
	Mealworm (2)	Solomon and Metcalf, 1974
	Aquatic microbes (7)	Schooley et al., 1975a
	Soil (0.7)	Schooley et al., 1975b
	Alfalfa (0.6)	Quistad et al., 1974a
	Rice (0.9)	Quistad et al., 1974a
	Blue gill fish	Quistad et al., 1976a
	Guinea pig	Chamberlain et al., 1975
	Steer	Chamberlain et al., 1975
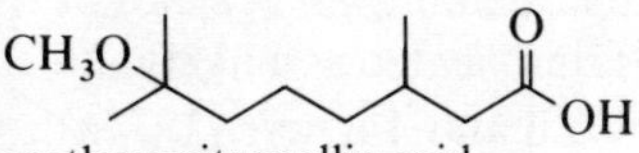 methoxycitronellic acid	Mosquito (0.3)	Quistad et al., 1975d
	House fly (0.3)	Quistad et al., 1975d
	Aquatic microbes (29)	Schooley et al., 1975a
	Alfalfa (0.8)	Quistad et al., 1974a
	Rice (1)	Quistad et al., 1974a
	Photochemical (7)	Quistad et al., 1975a
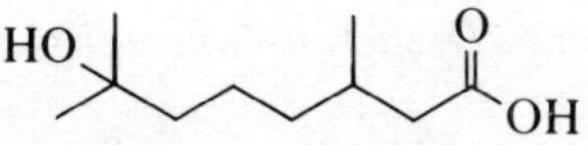 hydroxycitronellic acid	Mosquito (7)	Quistad et al., 1975d
	Alfalfa (3)	Quistad et al., 1974a
	Rice (0.3)	Quistad et al., 1974a
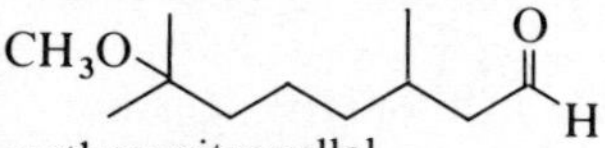 methoxycitronellal	House fly (0.6)	Quistad et al., 1975d
	Alfalfa (2)	Quistad et al., 1974a
	Rice (13)	Quistad et al., 1974a
	Photochemical (13)	Quistad et al., 1975a
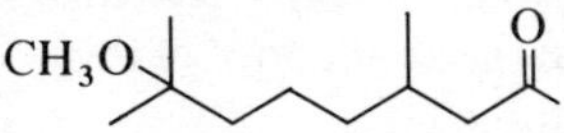	Photochemical (4)	Quistad et al., 1975a
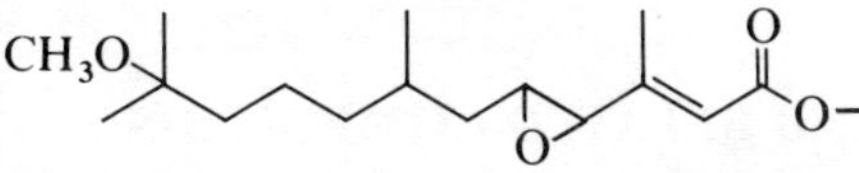	House fly (3)	Hammock et al., 1977
	Photochemical (6)	Quistad et al., 1975a

Chickens metabolized methoprene by pathways similar to those in mammals (Davison, 1976; Quistad *et al.*, 1976b). Radiolabelled triglycerides and cholesterol in egg yolks, fat, and liver were again reflective of exhaustive degradation to [^{14}C]acetate followed by anabolism to natural products. A unique pathway in chickens involved *reductive* metabolism to hydrogenated hydroxy and methoxy acids. These reduced primary metabolites were important constituents of triglycerides (egg yolk, fat, liver) and the reduced methoxy acid was even esterified with cholesterol (liver). Metabolic saturation of the dienoate was dose-dependent, decreasing with a reduction in dose level.

Hydroprene

Insect metabolism

Terrier and Yu have studied in detail the *in vitro* metabolism of hydroprene by esterases and microsomal oxidases of house flies (Terriere and Yu, 1973; Yu and Terriere, 1975a) and also of blow flies and flesh flies (Terriere and Yu, 1977). Both esterolytic and hydrolytic cleavage were important pathways for hydroprene degradation by flies. Since the hydroprene molecule exhibits an absence of multiple functionality, the 2,4-dienoate chromophore is implicated as a likely site of microsomal oxidation. Indeed, subsequent work by Yu and Terriere (1977a) using the same dipterans revealed several metabolites resulting from epoxidation of double bonds in [5-^{14}C]hydroprene (Table 1.4). Several additional epoxidized products (Table 1.4) were characterized by mass spectral fragmentation patterns of metabolites from hydroprene in the red cotton bug (*Dysdercus koenigii*) (Tungikar *et al.*, 1978).

Environmental fate

Since hydroprene is not as commercially developed as methoprene, it is not surprising that there is a paucity of detail concerning environmental degradation. Henrick *et al.* (1975) identified a 3-hydroxy-2-pyrone (Table 1.4) from the photosensitized oxygenation of hydroprene in ethanol, but the abundance of this photoproduct under environmental conditions is unexplored.

TERPENOID PHENOXY ETHERS AND RELATED COMPOUNDS

Insect Metabolism

The terpenoid phenoxy ether juvenoids will be treated together because the metabolic pathways involved appear common to most of the structures and the majority of the published research has dealt with a single compound, 1-(4-ethylphenoxy)-6,7-epoxy-3,7-dimethyl-2-octene (Figure 1.2) (R-20458 of

Table 1.4 Degradation products from hydroprene

Degradation product	hydroprene	Produced by	Reference
(epoxide, ethyl ester)		House fly, flesh fly, blow fly	Yu and Terriere, 1977a
		Red cotton bug	Tungikar et al., 1978
(epoxide, carboxylic acid)		Flies	Yu and Terriere, 1977a
		Red cotton bug	Tungikar et al., 1978
(diepoxide, ethyl ester)		Red cotton bug	Tungikar et al., 1978
(diepoxide, carboxylic acid)		Red cotton bug	Tungikar et al., 1978
(dienoic acid)		Flies	Yu and Terriere, 1977a
		Red cotton bug	Tungikar et al., 1978
(diol acid)		Flies	Yu and Terriere, 1977a
(pyranone)		Photochemical	Henrick et al., 1975

Stauffer Chemical Company). The first report on its metabolism in insects was a cursory study which showed the corresponding diol (K, Figure 1.4) as the only metabolite identified from the faeces of the American locust, *Schistocerca americana* (Gill et al., 1972). R-20458 (0.1–1 µg/insect) will cause normal adult coloration, mating, and reproduction in allatectomized male and female *S. americana c.* 2 weeks after a single injection. This juvenile hormone-like effect is apparently similar to the juvenile hormone-dependent maturation in male and female *S. gregaria* (Loher, 1960). At the effective doses of either 0.25 or 1 µg/insect of R-20458, >90% of the radioactivity is excreted into the faeces with much of the remaining dose localized at the injection site (Loher and Hammock, unpublished; Hammock, 1973).

Subsequent studies with the juvenoid R-20458 indicated that insects could be divided into two groups based on the relative contribution of hydrolases and oxidases to the metabolism. For instance, the diol is a major metabolite in larvae and pupae of the yellow meal worm, *T. molitor*, while numerous metabolites are detected in the faeces of the cockroach, *P. americana* (Hammock et al., 1974a). Similar metabolite distributions are found in tissue homogenates of a variety of insects including *T. molitor*, *P. americana*, *M. domestica*, *S. calcitrans*, *S. bullata*, *M. sexta*, and *T. ni* (Singh, 1973; Hammock et al., 1974a, 1975a). Since the majority of reported research has been with the house fly, *M. domestica*, it will be used as an example. Ether cleavage does not appear important in insects, with the possible exception of *P. americana*. The major routes of metabolism involve hydroxylation at the alpha (benzylic) and beta positions of the ethylphenoxy moiety and, in some cases, oxidation to the corresponding ketone (Figure 1.5). In a strain of house flies which had been previously selected with Baygon and which was known to have high MFO activity, hydroxylation at the alpha position on the ethyl side chain was greatly enhanced. Of much less importance was epoxidation at the 2,3-position. In all insects examined *in vivo* or *in vitro*, epoxide hydration was an important route of metabolism as discussed in more detail above.

An interesting series of metabolites may be formed from the diepoxide (Figure 1.4). Since treatment of the epoxide of JH or R-20458 (J) with aqueous acid gives a diol (K), one might expect a tetraol (H) (Figure 1.4) to result following similar treatment of a diepoxide (F) (Gill et al., 1972; Ajami and Riddiford, 1973). However, the 6,7-epoxide is much more labile to acid and, apparently more labile to enzyme catalysed hydration than the 2,3-epoxide (Mumby and Hammock, 1979b, c), so a diol-epoxide (G) results which cyclizes to a variety of 5-, 6-, and 7-membered heterocyclics, the most common of which are tetrahydrofuran diols (C,D) (Figure 1.4). Evidence for this cyclization was initially based on literature comparisons, chemical identification, synthesis of the tetrahydrofuran diols by alternative pathways, and n.m.r. detection of the ephemeral 6,7-diol-2,3-epoxide (G) following reaction of the corresponding olefin (K) with peracid (Hammock, 1973; Hammock et al., 1974b). More recently, the 6,7-diol-2,3-epoxide was

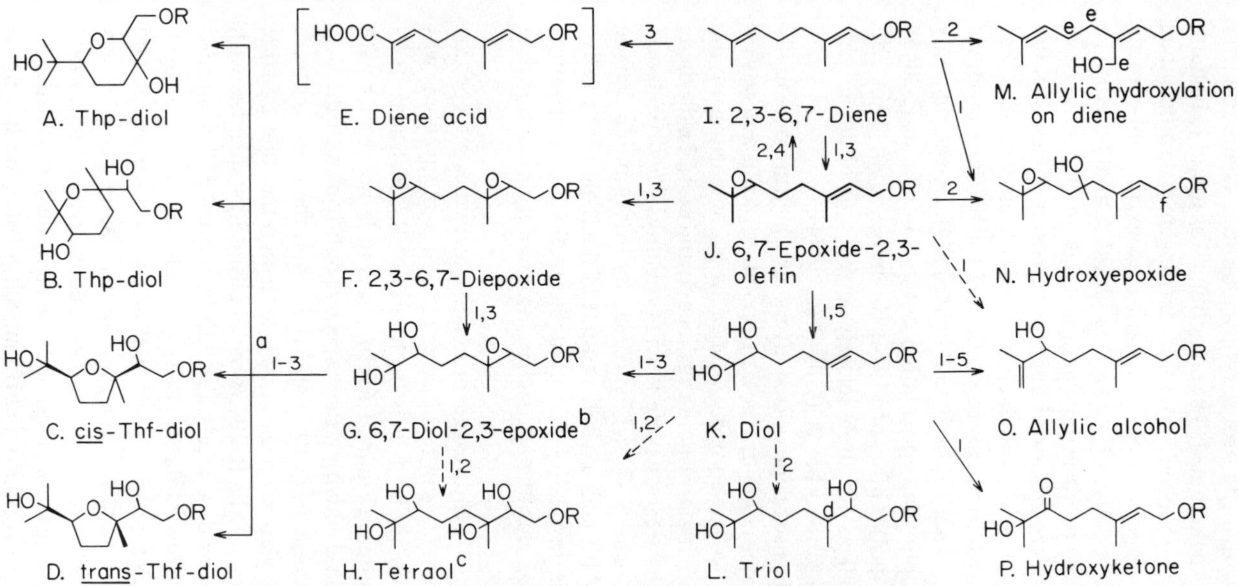

Figure 1.4 Pathways of metabolism and environmental degradation of the epoxygeranyl moiety of the juvenoid R-20458. Reference 5 concerns the related compound Ro-10-3108 which lacks the 2,3-olefin and has ethyl branches at C-3 and C-7. R = H or a phenoxy moiety (Figure 1.5). Numbers refer to literature references for the pathways: 1, Gill et al., 1972, 1974: 2, Hoffman et al., 1973; 3, Hammock et al., 1974a, 1975a, 1977a; 4, Ivie, 1976; Ivie et al., 1976; 5, Dorn et al., 1976; Hangartner et al., 1976. [a]The diol-epoxide rearranges to a variety of products including oxepane diols and bicyclic ethers, only the major products are shown (Hammock, 1973). [b]This unstable intermediate described in reference 2 had a m.s. appropriate for compounds C and D. [c]The tetraol does not form from either oxidation of the diol or hydration of the diepoxide *in vitro*, see text. [d]Hydroxylation also occurs at C-3. [e]Allylic hydroxylation occurs at several positions, and allylic hydroxylation followed by two dehydrogenation steps produces metabolite E. [f]Allylic hydroxylation alpha to the ether results in cleavage of the aryl alkyl ether

isolated following peracid oxidation of the corresponding olefin in a modified biphasic system. The diol-epoxide was chemically characterized, and its reactivity as an alkylating agent with 4-(*p*-nitrobenzyl)pyridine and its mutagenicity in the Ames' *Salmonella* assay determined. The compound was found to be quite reactive, but most reactions involved internal rearrangement rather than alkylation. For this reason, and possibly also for steric reasons, the diol-epoxide showed no detectable mutagenic activity (Ota and Hammock, unpublished). The proportions of furan, pyran and oxepane diols obtained following incubation of the diepoxide with acid, rat microsomes or house fly microsomes vary slightly, but isotope dilution techniques using authentic tetraol (H) demonstrated that less than 0.1% of the total hydration can be accounted for by these products (Hammock, 1973; Hammock *et al.*, 1974a, b; Gill *et al.*, 1974). Essentially, only tetrahydrofuran diols result following epoxidation of the 2,3-olefin of R-20458 diol (or the 6,7-olefin of JH diol) (Gill *et al.*, 1974; Hammock *et al.*, 1974a, b). However, the tetraol, as well as thf diols, clearly appears to be an *in vivo* metabolite in mammals as discussed below (Gill *et al.*, 1974).

As an approach to avoid the use of epoxidized juvenoids which are potentially unstable in the environment, one could use the corresponding olefins and rely on the insect to activate them to the epoxide *in vivo*. In a comparison of the metabolism of R-20458 (J) with its corresponding diene (I) and diol (K), the diene was metabolized at the highest rate to an olefinic acid probably by hydroxylation and subsequent oxidation of a C-7 methyl since hydroxy and aldehyde intermediates were formed. Although a small amount of the biologically active 6,7-epoxide was formed, the instability of the allylic methyl to microsomal oxidases limits the potential of such an approach in insect control (Hammock *et al.*, 1974b, 1975a).

The potential application of knowledge of relative activities of epoxide hydrolases and microsomal oxidases in test insects for design of juvenoids was illustrated by a comparison of the half-lives (in hours) of R-20458 and its 7-methoxy and 7-ethoxy analogues (Figure 1.6). In susceptible house flies with moderate hydrolase and oxidase activities, the respective half-lives were 0.5, 1.0, and 1.5 h while in resistant house flies with high MFO levels the half-lives were 0.3, 0.1, and 0.3 h with hydroxylation on the ethyl side chain rather than epoxide hydration accounting for most of the R-20458 metabolism. In stable flies with low hydrolase and moderate oxidase levels, the half-lives are 4.5, 3.5, and 6.5 while in *T. molitor* pupae with low oxidase activity the half-lives are 8.0, 57, 63 h (Hammock *et al.*, 1975a). The relative biological activities of epoxides and methoxides (ethoxides appear quite selective) in various insects largely reflect relative hydrolase and oxidase levels. Thus, for maximum activity on many field populations of resistant insects, juvenoids should be chosen which are refractory to attack by microsomal oxidases (Sparks and Hammock, 1980a). Studies on the metabolism of the geranyl phenyl ether alkoxides may facilitate interpretation

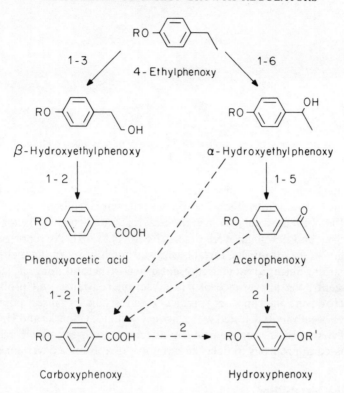

Figure 1.5 Pathways of metabolism and environmental degradation of the ethylphenoxy moiety of the juvenoids R-20458 and Ro-10-3108. R = a terpenoid moiety; R' = H, SO$_3$H, or glucuronide. Solid arrows indicate established pathways and dashed arrows indicate possible routes leading to metabolite formation. Numbers refer to literature references for the pathways: 1, Gill *et al.*, 1972, 1974; 2, Hoffman *et al.*, 1973; 3, Hammock *et al.*, 1974a, 1975a; 4, Ivie *et al.*, 1976; 5, Dorn *et al.*, 1976; Hangartner *et al.*, 1976; 6, Agosin *et al.*, 1979

of stability data on various arylterpenoid derivatives (Schwarz *et al.*, 1974) (Figure 1.2).

Most of the metabolites of R-20458 and the 7-methoxy and ethoxy analogues were devoid of morphorgenetic activity in the *T. molitor* pupal bioassay (Hammock *et al.*, 1974a, b, 1975a). The 7-hydroxy (*O*-dealkylation) products (Figure 1.6) were active in some insects, and benzylic oxidation of R-20458 actually resulted in an active juvenoid reported earlier by Bowers (1969). Possibly derivatives of acetophenoxy or α-hydroxyethylphenoxy juvenoids could be used for insect control (Sláma *et al.*, 1978).

Wright and Spates (1975) investigated the penetration and persistence of R-20458 in the stable fly, *S. calcitrans*, while Hammock *et al.* (1977a) compared the

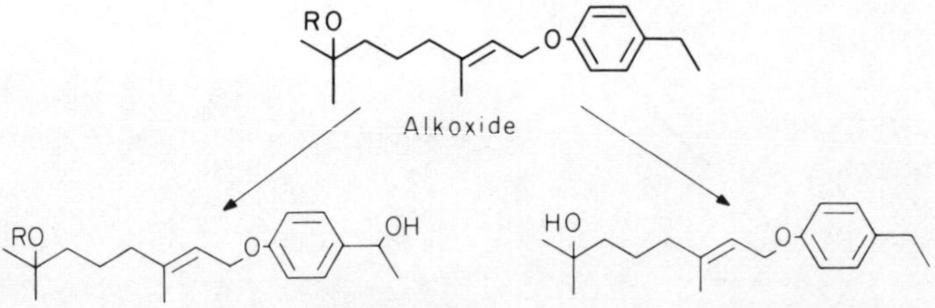

Figure 1.6 Major routes of metabolism of a 7 alkoxy geranylphenyl ether. R = CH_3 or C_2H_5

in vitro metabolism of methoprene, JH I and R-20458 in susceptible and methoprene resistant house flies. The diol was the only metabolite in the absence of NADPH, but a variety of metabolites resulting from oxidation of the ethylphenoxy and 2,3-olefin were detected with NADPH. As reported earlier with Baygon-resistant insects, metabolism was much faster in the methoprene-resistant strain. Since juvenoids represent a new structural approach to insect control agents, the appearance of site of action resistance will probably be delayed. However, one can safely predict cross-resistance in insect populations which have been heavily selected with classical pesticides. Sparks and Hammock (1980a) reviewed the resistance problem as related to insect growth regulators and discussed approaches to delay or circumvent resistance development.

Mammalian Metabolism

Several workers have examined the *in vivo* metabolism of R-20458. Following intraperitoneal administration to rats, Gill *et al.* (1972) reported 100, 96, and 73% recovery of radioactivity in excreta using the ^{14}C-phenyl-, ^{3}H-phenyl-, and geranyl labels respectively. The lower recoveries with ^{3}H-phenyl and geranyl labels were probably due to exchange and metabolic release, respectively. Hoffman *et al.* (1973) report quantitative recovery of radiolabel from rats following either oral or intraperitoneal administration of the ^{14}C-phenyl compound. Approximately equal amounts of radioactivity were recovered in the urine and faeces, and no radioactivity was detected as expired $^{14}CO_2$ or as tissue residue. Either 1 mg or 100 mg kg^{-1} doses of ^{14}C-phenyl R-20458 delivered orally to mice or intraperitoneally to rats resulted in almost quantitative recovery of the radioactivity in the urine and faeces within 96 h (Singh, 1973; Gill *et al.*, 1974). Following oral administration to rats at 1 mg kg^{-1}, maximum blood and liver levels of 11 and 26 p.p.m. relative to dry tissue weight, respectively, were reached at 0.5 h, and levels of ^{14}C decreased rapidly thereafter (Singh, 1973). Following oral administration of R-20458 to a steer, Ivie *et al.* (1976) reported >98% recovery with >84% of the dose in the urine and >14% in the faeces. Only trace (low p.p.b. or p.p.t.) levels remained in the tissues 7 days after oral administration. Seven days after dermal application of R-20458 to a steer, approximately

30% of the material had been absorbed and was rapidly eliminated in the urine and faeces.

As would be expected from a compound with so many sites labile to metabolism, a plethora of metabolites was observed in all cases. Much of the metabolism can, however, be explained on the basis of products resulting from a combination of several primary steps in metabolism. For simplicity, the metabolism of the phenoxy and geranyl portions of the molecule will be treated separately. As with insects, the major route of oxidative metabolism involves hydroxylation in the benzylic position of the ethylphenoxy moiety resulting in the α-hydroxyethylphenoxy moiety which can be dehydrogenated to acetophenoxy compounds (Figure 1.5). The less abundant β-hydroxyethylphenoxy compounds are presumably oxidized to phenoxyacetic acid derivatives and possibly decarboxylated to the carboxyphenoxy derivatives. Hoffman *et al.* (1973) reported the *p*-hydroxy compounds as rat metabolites and speculated that they might arise from a Baeyer–Villiger-like oxidation of the acetophenoxy moiety. Aromatic hydroxylation has not been confirmed, but one could safely predict that such pathways play, at least, a limited role in metabolism further increasing the number of possible metabolites. These phenoxy derivatives are present with the geranyl phenoxy ether intact with various modifications on the geranyl portions (Figure 1.4), as free phenols, and as several conjugates. Phenols and conjugates of phenols constitute a major portion of the metabolites from all mammals examined; for instance, in steers ∼37% of the total radiocarbon existed as conjugates of either ethylphenol or acetophenol (Ivie *et al.*, 1976). Hoffman *et al.* (1973) speculate that formation of the acetophenoxy derivative facilitates subsequent cleavage of the geranyl phenoxy ether linkage.

Since an adequate label was lacking, metabolism of the epoxygeranyl portion of the R-20458 molecule was only followed while the ether linkage was intact. One might expect ether cleavage to occur by hydroxylation alpha to the allylic ether (N) (Figure 1.4). The resulting product would be expected to rearrange to the corresponding geranyl derivative and be subsequently dehydrogenated to the acid. Once a free geranyl derivative is formed, one would predict rapid metabolism to acetate and other natural products as shown for geraniol in *Pseudomonas citronellois* by Seubert and Fass (1964) and possibly incorporation into natural products as described for methoprene by Quistad *et al.* (1974b). A major discrepancy between *in vivo* and *in vitro* data is that ether cleavage is a major *in vivo* metabolic route which has not been detected in tissue homogenates or reconstituted P450 (Gill *et al.*, 1972, 1974; Agosin *et al.*, 1979). Possibly, such ether cleavage is catalysed by gut microorganisms.

The *in vivo* metabolism of the geranyl portion of the molecule involves epoxide hydration, epoxidation of the 2,3-olefin (possibly with subsequent cyclization), allylic hydroxylation, and possibly either reduction of the 2,3-olefin followed by hydroxylation or some form of reductive hydroxylation (Figure 1.4) (Gill *et al.*, 1972, 1974; Hoffman *et al.*, 1973; Ivie *et al.*, 1976). A search for metabolites of the epoxide other than the corresponding diol was not fruitful since several possible rearrangement products were not detected (Gill *et al.*, 1974; Hammock *et al.*, 1974a). Hoffman *et al.* (1973) reported fairly large amounts of the corresponding 6,7-olefin (I) (Figure 1.4) in rat faeces following administration of R-20458 which they attributed to olefinic impurities in material administered to the rat. Ivie *et al.* (1976) also found the 6,7-olefin as a metabolite from the steer, but traced it to an epoxide reduction occurring in the rumen fluid (Ivie, 1976). Subsequent work has shown that the contents of the large intestine of a variety of mammals, including man, when incubated under carbon dioxide will all reduce epoxides very cleanly, to the corresponding olefin (Hammock *et al.*, 1980b).

Hoffman *et al.* (1973) report metabolites resulting from hydroxylation of the terpenoid chain of R-20458 (N) (Figure 1.4) and the corresponding diene (M). Apparently, the 2,3-olefin can be reduced and hydroxylated resulting ultimately in a pair of triols (L). Epoxidation of the 2,3-olefin of either the epoxide or the diol occurs (F,G) but biological and chemical epoxidation of the 2,3-olefin is much slower than the 6,7-olefin (Gill *et al.*, 1972, 1974; Hammock *et al.*, 1974a, b, 1975a). The 6,7-epoxide is similarly much more labile to acid catalysed and enzymatic hydration than the 2,3-epoxide probably resulting in the ephemeral diol-epoxide intermediate (G) (Mumby and Hammock, 1979b, c). The diol-epoxide rapidly rearranges to a variety of cyclic and bicyclic products unless held under basic conditions (Ota and Hammock, unpublished). The most abundant of these products are shown in Figure 1.4 (A–D), but oxepane diols and bicyclic ethers also result from the rearrangement (Hammock, 1973; Gill *et al.*, 1974; Hammock *et al.*, 1974a, b). The mass spectrum of the diol-epoxide (G) reported as a metabolite in rats is consistent with the mass spectra of the tetrahydrofuran diols (C and D). The tetraol (H) clearly does not arise from hydration of either the diepoxide or the diol-epoxide *in vitro*, but the support for the *in vivo* formation of the acetophenoxytetraol is quite good (Hoffman *et al.*, 1973; Gill *et al.*, 1974). Possibly a very rapid oxidation–hydration occurs before cyclization in some tissues or in the gut. The acetophenoxy moiety might also encourage hydration over cyclization.

Although many metabolites are present, most of the dose of R-20458 given to mammals can be accounted for by hydroxylation at the benzylic position of the ethylphenoxy moiety, epoxide hydration, and ether cleavage. Subsequent conjugation of the resulting phenols, presumably as sulphates and glucuronides, appears to be extensive (Hoffman *et al.*, 1973; Gill *et al.*, 1974; Ivie *et al.*, 1976).

With the exception of ether cleavage, metabolism of R-20458 by tissue homogenates illustrates the primary sites of metabolism which lead to the plethora of metabolites observed *in vivo*. The *in vitro* metabolism of R-20458 was approached by Gill *et al.* (1972, 1974) by basing identification of metabolites on cochromatography with a series of synthetic standards and on a series of reactions diagnostic for various functionalities (Hammock *et al.*, 1974b). The **metabolism** of the $2Z$ and the $2E$ isomers of R-20458 as well as further metabolism of primary metabolites was investigated in addition to the olefinic precursor of R-20458 and juvenoids with p-nitrophenyl and methylenedioxyphenyl substituents in subcellular fractions of a variety of tissues from the rat, mouse, and rabbit with and without induction (Singh, 1973). The 6,7-epoxide, 2,3-olefin, and ethylphenoxy moieties were established as being labile to rapid hydrolytic and oxidative metabolism (Gill *et al.*, 1972, 1974). The 6,7-olefin of the diene (I) is much more rapidly epoxidized than the 2,3-olefin group. Epoxide hydration was observed in all hepatic subcellular fractions, but in contrast to investigations with other substrates (Jerina *et al.*, 1968; Oesch, 1973), much of the hydrolase activity was found in the 100,000 g soluble or cytosolic fraction (Gill *et al.*, 1972, 1974). Hammock *et al.* (1976) compared the metabolism of R-20458, JH I, and Ro-8-4314 in the microsomal and cytosolic fractions of mouse liver and kidney and found that the cytosolic fraction made a significant contribution to hydration and the relative rates of hydration were R-20458 > JH I > Ro-8-4314 (Figures 1.1 and 1.2). Mumby and Hammock (1979a, b, c) synthesized and monitored the relative initial rates of hydration of a number of juvenoids by the cytosolic fraction. Gill and Hammock (1979, 1980) further investigated the properties of the enzyme in several tissues of four mammals using R-20458 and *cis*- and *trans*-methylepoxystearates where substantial activity was additionally found in the mitochondrial fraction. Subsequently, the effect of this enzyme system has been studied with a variety of substrates (Hammock *et al.*, 1980a, b; Ota and Hammock, 1980). With the related methoxy and ethoxy juvenoids, *O*-dealkylation replaces epoxide hydration as a major route of metabolism in mouse liver microsomes (Figure 1.6) (Hammock *et al.*, 1975a).

Using isolated rat hepatocytes and two forms of non-induced rat hepatic cytochrome P450 (P451) in a reconstituted system, Agosin *et al.* (1979) and Morello and Agosin (1979) studied the metabolism of R-20458, methoprene, hydroprene, and JH I. They found rapid conversion of R-20458 to the α-hydroxyethylphenoxy derivative by both forms of P450. These workers also demonstrate reasonable proof for glutathione-*S*-epoxide transferase activity in rat hepatocytes using JH. Although one would expect such a reaction to occur, Gill *et al.* (1974) did not find extensive evidence for glutathione conjugation using R-20458 as could be predicted for trisubstituted epoxides (Chasseaud, 1979). However, the conjugated 2,3-olefin groups of JH, methoprene, and hydroprene will chemically react with glutathione (Sparks and Hammock, 1980a) as expected since an analogous reaction with thiophenol is used in the synthesis of methoprene and hydroprene (Siddall, 1976).

Toxicology

Juvenile hormone (Siddall and Slade, 1971) and two early epoxy amide and dichloro amide juvenoids (Cruickshank, 1971) demonstrated very low toxicity to rodents. Furthermore, it appears that recent toxicological studies with the terpenoid phenoxy ether juvenoids have been generally favourable. Pallos *et al.* (1971) report negligible acute toxicity for R-20458. A series of studies was carried out on the toxicology of R-20458 to cattle, sheep, and swine following dermal spray or oral administration; blood chemistry and numerous other parameters were monitored in the test animals. Even at very high doses no effects of R-20458 were noted in swine or cattle, but leucopenia in sheep and testicular atrophy in rams were noted. A more highly purified technical sample failed to demonstrate the above symptoms when retested. No teratogenicity or toxicity was observed following intraperitoneal injection of R-20458 to hamsters on day 8 of gestation (Smalley *et al.*, 1974; Wright, 1976; Wright and Smalley, 1977). From a biochemical perspective, it was found that R-20458, the corresponding diene (I) and diol (K) at 0.1–2 mM, exhibit a concentration dependent inhibition of O_2 uptake by rat liver mitochondria (Hammock, 1973), and a similar observation was made with JH using mitochondria from the Indian meal moth, *Plodia interpunctella* (Firstenberg and Silhacek, 1973). It is unlikely that effects at such high concentrations represent a toxicological risk to man or a JH mode of action in insects, and the effects are probably related to the hydrophobicity of the compounds involved.

A number of juvenoids and, of course natural JH, contain the epoxide functionality. As highly strained ethers, epoxides are electrophilically reactive and some epoxides may alkylate biological materials including proteins and nucleic acids (Miller and Miller, 1977). In order to be a potent mutagen, such a compound must (1) readily alkylate biological material; (2) have an affinity for nucleic acids; and (3) either be stable enough to reach a critical site or be formed at the site. The extraordinary stability of the trisubstituted alkyl epoxides present in JH and juvenoids to acid hydration was first demonstrated by the rigorous conditions necessary to form the corresponding diols (Gill *et al.*, 1972; Slade and Zibitt, 1972). Although such epoxides can be detected in solution or on paper or thin layer chromatograms with 4-(*p*-nitrobenzyl) pyridine (selective for alkylating agents) (Hammock *et al.*, 1974c), juvenoids are much less reactive with this reagent than epoxides known to mutagenic (Hammock, unpublished). In general, monosubstituted alkyl epoxides are much less mutagenic than the corresponding disubstituted epoxides (El-Tantawy and Hammock, 1980, and included references), and Ivie *et al.* (1980b) demonstrated that trisubstituted psoralen glycidyl ethers were, at best, very weak mutagens, while the corresponding monosubstituted compound demonstrated the highest mutagenicity reported for any alkyl epoxide in the Ames' *Salmonella* system (Ames *et al.*, 1975). The same paper again illustrated the importance in mutagenicity of a group with high affinity for DNA or perhaps a group that can intercalate with it. In a screen of JH, geranyl phenoxy ether juvenoids, and a series of metabolites including the diepoxide (F), the 2,3-epoxide, and the diol-epoxide (G), no mutagenicity was detected in the *Salmonella* system (Hammock, unpublished). One possible risk from such juvenoids is that they appear to inhibit the hydration of some epoxides by the mammalian cytosolic hydrolase (Hasegawa and Hammock, unpublished), and another possibility is that esters of the α-hydroxyethylphenoxy derivatives (Figure 1.5) might be alkylating agents (Miller and Miller, 1977), but there is no evidence for a problem with these metabolites. However, even the most stable juvenoids are so biodegradable that such inhibition is probably a laboratory curiosity. Although no compound can be proven safe, the available data indicate that mutagenic risk from those juvenoids tested is minimal.

R-20458 and, to a lesser extent, other phenyl ethers are toxic to algae and retard algal growth. However, these effects are seen at such high concentrations that it is unlikely that the use of these compounds in reasonable insect-control programmes would present a significant risk (Gill *et al.*, 1972, 1974; Hammock, 1973). The juvenoids so far examined seem to present minimal acute risk to aquatic invertebrates. For instance, R-20458, to the limit of its solubility, caused no 48 h toxicity to the marine isopod *Sphaeroma quoyanum* or its commensal *Iais californica*, but it has profound effects on sexual dimorphism in the terrestrial isopod, *Porcellio scaber* (Rotramel and Hammock, unpublished). Since juvenoids may disrupt the development of arthropods, careful studies are needed, but there is no indication, to date, that juvenoids present a substantially greater risk to most non-target arthropods than do classical pesticides (Breaud *et al.*, 1977; Costlow, 1977).

Norris *et al.* (1974) report on some rodent studies with 4-(4,8-dimethyldecyloxy)-1,2-(methylenedioxy)benzene. The acute oral LD_{50} was $>4\,g\,kg^{-1}$ and long term feeding studies were carried out at $1\text{--}100\,mg\,kg^{-1}\,day^{-1}$ while monitoring numerous parameters. Toxicological problems appeared minimal with neonate survival and growth, and litter size decreased at $100\,mg\,kg^{-1}\,day^{-1}$ but not at lower levels. Unsworth *et al.* (1974) reported that a methylenedioxyphenoxy juvenoid was teratogenic to mice at $1\,mg\,kg^{-1}$ while several very closely related compounds (Figure 1.2) showed no effects. The juvenoid Ro-10-3108 and its major metabolites show little if any acute toxicity to mice ($LD_{50} >5\,g\,kg^{-1}$), and following a 96 h exposure Ro-10-3108 demonstrated LC_{50}s of >5000 p.p.m. to three fish species (Dorn *et al.*, 1976). A tabular summary of the toxicology of Ro-10-3108 has been given by Zurflüh (1976).

Environmental Fate

The first reports on the environmental degradation of juvenoids were those of Gill *et al.* (1972) on R-20458 (Figure 1.2) and Pawson *et al.* (1972) on JH I and a methylenedioxyphenoxy juvenoid. Gill *et al.* (1972) reported slow decomposition on silica gel plates exposed to sunlight unless photosensitizers were present. Rapid decomposition then occurred especially at the benzylic position yielding α-hydroxyethylphenoxy and acetophenoxy derivatives (Figure 1.5). Pawson *et al.*

(1972) report rapid degradation of both compounds (and loss of biological activity) when exposed to ultraviolet light as thin films on glass. Numerous photodegradation products were formed including the corresponding diepoxides (F).

Much of the subsequent work on environmental degradation of such compounds has been reviewed by Schooley and Quistad (1979). Gill et al. extended their preliminary study on R-20458 in a more comprehensive 1974 report. Photodecomposition of the epoxide R-20458 (J), diene (I), and diol (K) were compared on silica gel chromatoplates exposed to natural sunlight, a sunlamp, or u.v. lamp. R-20458 is very stable on silica gel in the dark, it decomposes at relatively low identical rates which are similar with sunlight or sunlamp exposure, but at a high rate with u.v. exposure. The rate curves are definitely phasic with all three light sources. U.v. irradiation resulted in highly polar products which failed to move from the origin of a t.l.c. plate in several systems, while sunlight led to products of intermediate polarity. Twenty-two candidate triplet sensitizers were screened, most of which enhanced the photochemical degradation of R-20458. No correlation could be drawn between the triplet energy (39–74 kcal mol^{-1}) and the rate of R-20458 degradation. The most active sensitizer was anthracene which, on exposure to sunlight, converted R-20458 to very polar products reminiscent of u.v. exposure. Dyes generally failed to sensitize decomposition, and rotenone, which is a very potent sensitizer for dieldrin (Ivie and Casida, 1971) failed to sensitize R-20458 degradation (Singh, 1973).

A comparison of the stabilities of R-20458 and the corresponding methoxide and ethoxide (Figure 1.6) when exposed to sunlight on silica gel demonstrated that the epoxide was only slightly less stable than the alkoxides. Since epoxide hydration and rearrangement is much less important than other photodecomposition pathways, such a result could be anticipated (Hammock et al., 1975a).

On silica gel, benzylic oxidation of R-20458 to α-hydroxyphenoxy and acetophenoxy derivatives was a major route of degradation. Rates of epoxide hydration to the diol (K, Figure 1.4) were somewhat variable and, as expected, they increased under moist conditions. Epoxidation at the 2,3-olefin yields the diepoxide (F) and a variety of cyclic (A–D) and bicyclic ethers (not shown). The diol was subsequently oxidized to the hydroxyketone (P), and either the epoxide, diene, or diol, depending on conditions, may yield the allylic alcohol (O). The diene (I) was converted to the epoxide (J) on silica gel in the dark and at a much higher rate upon exposure to light, while the diepoxide (F) yielded largely rearrangement products (A–D) discussed earlier. In all compounds, ether cleavage was evidenced by the presence of free phenols and, in all cases, several unknown photoproducts were detected. As discussed earlier, photodegradation products had greatly reduced biological activity (Singh, 1973; Gill et al., 1974; Hammock et al., 1974a).

Photoalteration of the diene (I), R-20458 (J), and diol (K) at 0.5 p.p.m. in distilled water was also studied following exposure to sunlight or a sunlamp. As on silica gel, the diene was much less stable than the epoxide or diol, giving a variety of photoproducts including the allylic alcohol (O). Peroxides may be involved in this pathway (Hammock, 1973). Benzylic oxidation, 2,3-epoxidation and cyclization, epoxide hydration, and ether cleavage again occurred (Singh, 1973; Gill et al., 1974).

The photodecomposition of the diene (I), R-20458, and diol (K) was also studied in monoaxenic cultures of living or dead *Chlorella* sp. and *Chlamydomonas teinhardii* and compared with decomposition in chlorophyll solutions. Epoxide hydration was a major route of metabolism in *C. teinhardii* while *Chlorella* sp. demonstrated a high rate of benzylic oxidation. With culture medium, chlorophyll solutions, or dead algae, the diepoxide (F) was a predominate metabolite. With the algae, especially *Chlorella,* surprisingly large amounts of phenolic and acidic metabolites formed (Hammock, 1973; Gill et al., 1974).

Hangartner et al. (1976) reported a comparison of the stability of several juvenoids on silica gel irradiated with a high intensity sunlamp, in pukka water under field conditions, in water at pH 4, and on bean leaves. Ro-10-3108 (Figure 1.2) was significantly more stable than a variety of other juvenoids. The methylenedioxyphenyl group of a related juvenoid was very unstable photochemically and hydrolytically. Elimination of the 2,3-olefin of R-20458 definitely increased the photochemical stability of the resulting compound. Surprisingly, methylethyl rather than dimethyl substituents on the epoxide also increased the stability. This observation was further investigated by Mumby and Hammock (1979b) by comparing the stability of 17 juvenoids to aqueous acid. The influence of electronic substituent effects could be predicted from the classic studies of Pritchard and Long (1956; Long and Pritchard, 1956), but an increase in only the hydrophobicity of the epoxide substituents (dimethyl to methylpropyl) could increase the stability c. 9 times and the biological activity on *T. molitor* c. 50 times. The enhanced stability of Ro-10-3108 over R-20458 and methoprene could clearly be advantageous for some applications (Hargartner et al., 1976) and this approach could possibly still be taken further with compounds such as those described by Schwarz et al. (1974) (Figure 1.2). The stability predicted in the laboratory for Ro-10-3108 was demonstrated under field conditions on the summer fruit tortrix moth (*Adoxophyes orana*), the San José scale (*Quadraspidiotus perniciosus*), and the citrus snow scale (*Unaspis citri*).

Dorn et al. (1976) investigated the degradation of Ro-10-3108 in polluted water held under field conditions. The major routes of degradation included benzylic oxidation (Figure 1.5), ether cleavage, epoxide hydration (K) (Figure 1.4), and conversion of the resulting diol to two isomeric allylic alcohols analogous to product O (Figure 1.4). The metabolites all had reduced biological activity, although the α-hydroxyphenoxy and acetoxyphenoxy analogues of

Ro-10-3108 are still active juvenoids. Dorn et al. (1976) attribute the limited number of degradation products in Ro-10-3108 relative to R-20458 to the absence of the biodegradable 2,3-olefin.

ANTI-JUVENILE HORMONES

The theoretical attraction of anti-juvenile hormones (anti-JHs) as insect control agents has stimulated work in numerous laboratories for several years. It was Bowers (1976) who transformed speculation into reality with the announcement of the structures of the precocenes. One could obtain anti-JH effects by inhibiting JH production or release, disrupting JH transport, stimulating JH degradation, blocking JH action at a target site, or subtly disrupting insect regulation. The search for such compounds has involved screening of natural and man-made products, investigating insect–plant or insect–insect interaction, or attempting to disrupt known biosynthetic pathways. To date only two series of compounds have been described in the literature which demonstrate clear anti-JH activity (i.e. precocenes and ETB).

Ethyl 4-[2-(*tert*-butylcarbonyloxy)]butoxybenzoate (ETB) causes black pigmentation in *Manduca sexta* larvae, an effect which is alleged to indicate JH deficiency at a level not severe enough to produce premature metamorphosis (Staal, 1977). The mode of action of ETB is largely unknown, but it is a weak JH esterase inducer in *T. ni* and it appears to act as a JH agonist/antagonist (Sparks et al., 1979b).

Precocene—Mode of Action

Much of the information concerning the mode of action of precocene II comes from work with *Oncopeltus fasciatus*. Precocene II inactivates the corpora allata (Masner et al., 1979) which lose the ability to secrete JH (Müller et al., 1979). The effectiveness of precocene II is dependent on the timing of application and maximal reponses are obtained if treatment occurs when the corpora allata are active (Masner et al., 1979; Unnithan and Nair, 1979). When precocene II is applied to certain aged larvae or 1–7 day old adults, the insects are rendered sterile (Müller et al., 1979). Ultrastructural examination of the corpora allata of *O. fasciatus* treated with precocene II as young adults revealed that the glands consisted of immature cells (Liechty and Sedlak, 1978) which implied lack of differentiation of tissue.

Schooneveld (1979a, b) has shown that precocene II induces collapse of the corpora allata of nymphal *Locusta migratoria*. Necrosis of cells is followed by phagocytosis of cell fragments by haemocytes. Hence, *precocene II is cytotoxic*. Brooks et al. (1979a, b) have postulated that the cytotoxicity of precocene II in *L. migratoria* results from an activated alkylating agent such as the 3,4-epoxide of precocene II. Support for this hypothesis is based partly on the known high chemical reactivity of this 3,4-epoxide (Jennings and Ottridge, 1979) and also on the results of Brooks *et al.* (1979a, b) which demonstrate that the corpora allata metabolize precocene II cleanly to the 3,4-dihydrodiol (presumably via the 3,4-epoxide).

Precocene II is also known to inhibit vitellogenesis in *O. fasciatus* (Masner *et al.*, 1979) and *D. melanogaster* (Landers and Happ, 1979), but it is not antigonadotropic in the adult female *A. aegypti*, a morphogenetically insensitive insect (Kelly and Fuchs, 1978). Precocene II also interferes with sex attractant production in the brown cockroach (Burt *et al.*, 1979).

Insect Metabolism of Precocene II

Nine insect species showed a 37-fold variation in the rate of precocene II metabolism *in vivo* (Ohta *et al.*, 1977). The principal metabolite in each insect was a 3,4-diol (Figure 1.7) which allegedly arose from hydration of a 3,4-epoxide. This 3,4-diol is also the major aglycone from metabolism of precocene II by the brown cockroach (*Periplaneta brunnea*) (Burt *et al.*, 1979). Although Ohta *et al.* (1977) claimed to have synthesized the 3,4-epoxide as an authentic standard, the structure of their product was subsequently disputed (Bergot *et al.*, 1980; Soderlund *et al.*, 1980). More recently using carefully controlled conditions the preparation of authentic 3,4-epoxide has been reported, but it is readily evident that this epoxide is chemically quite reactive and unstable (Jennings and Ottridge, 1979; Soderlund *et al.*, 1980). Most workers agree that the 3,4-epoxide metabolite probably plays a key role in solving questions about the precocene mode of action, but thus far the 3,4-epoxide metabolite itself has not been conclusively identified.

A comparison of different tissues showed that the fat bodies of both *Trichoplusia ni* and *Ostrinia nubilalis* fifth instar larvae possessed high *in vitro* metabolic activity toward precocene II (Burt *et al.*, 1978). Although these lepidopterans are insensitive to the morphogenetic effects of precocene, Burt *et al.* (1978) studied the effects of suspected epoxide hydrolase inhibitors on precocene metabolism in fat body homogenates from these insects. Inhibition of precocene metabolism could be shown with all five inhibitors used, but accumulation of intermediate metabolites (e.g. 3,4-epoxide) could not be shown. A 3-hydroxy hydration metabolite was shown as a major product in *T. ni* homogenates (Ohta *et al.*, 1977).

Bergot *et al.* (1980) compared the metabolism of precocene II in several insect species which showed varying response to its morphogenetic effects. By using a treatment method (topical) and dose rate appropriate for inducing the desired morphological reponse in sensitive insects, Bergot *et al.* (1980) hoped to detect an activated metabolite, if such a compound existed. The metabolites in insect haemolymph were strikingly different from previous results reported for fat body

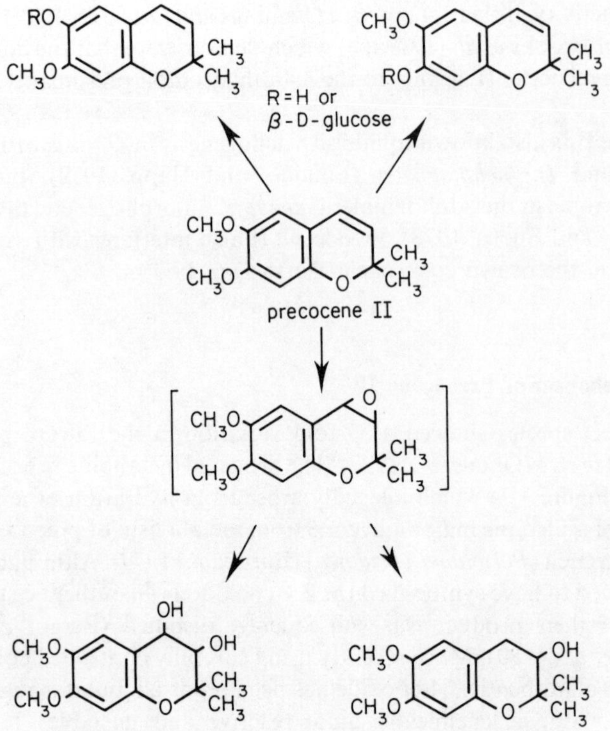

Figure 1.7 Insect metabolism of precocene II

homogenates. The primary metabolites in haemolymph were glucosides of O-demethylated precocene (Figure 1.7). Since these O-β-glucosides of 6- and 7-monomethylated precocene II were demonstrated in both sensitive and insensitive species, no evidence was found for a haemolymph-borne, biologically effective 'activated metabolite'. If such a biologically active product exists, it is likely produced *in situ* at the target tissue (Bergot *et al.*, 1980). Although Bergot *et al.* (1980) were unable to detect free phenolic metabolites in haemolymph, Soderlund *et al.* (1980) using *T. ni* fat body homogenates were able to detect the corresponding 6- and 7-desmethylated precocenes as free metabolites. They also recovered the 3,4-diol metabolites as a mixture of *cis* and *trans* isomers. A summary of the insect metabolites of precocene is given in Figure 1.7.

BENZOYLPHENYL UREAS—MODE OF ACTION

History

The mode of action of benzoylphenyl ureas has been intensely investigated over the last decade (for reviews see Post and Mulder, 1974; Marx, 1977; Verloop and Ferrell, 1977; Ker, 1978). The insecticidal activity of benzoylphenyl ureas was discovered at Philips-Duphar (The Netherlands) originally with DU-19111 (Figure 1.8) which was synthesized as an analogue of dichlobenil, a pre-emergence herbicide. It became visibly evident that insect larval death was invariably connected with the moulting process (Post and Mulder, 1974).

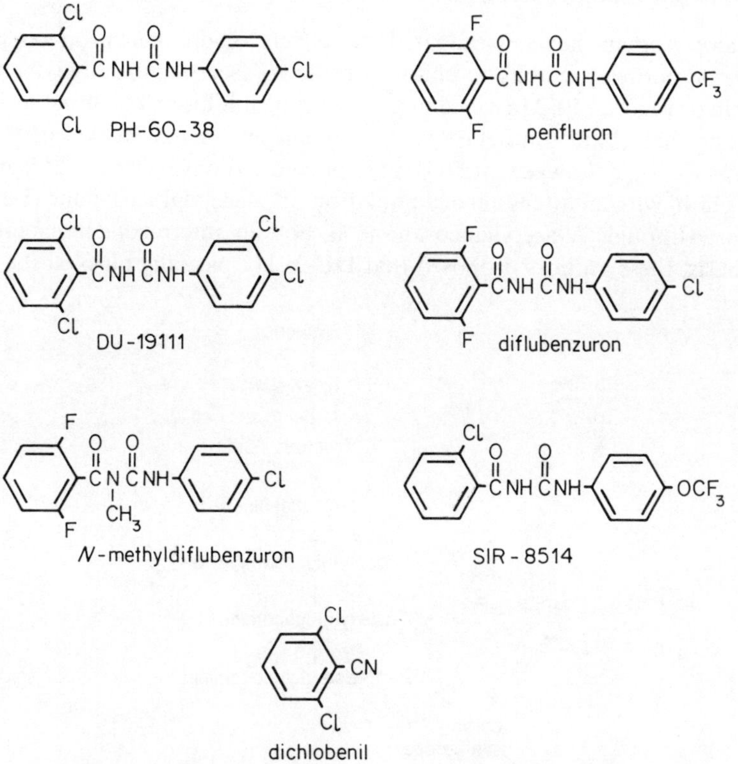

Figure 1.8 Benzoylphenyl ureas and related compounds

Although several alternative theories of the mode of action for this new class of compounds arose subsequently, the initial report implicating chitin inhibition as a prelude to insect mortality also came from Philips-Duphar (Post and Vincent, 1973). Using DU-19111, Post and Vincent (1973) showed the benzoylphenyl urea prevented incorporation of [^{14}C]glucose into the cuticular chitin of *Pieris* larvae. In the process of optimizing the insecticidal activity of DU-19111 by analogue synthesis, diflubenzuron arose as the leading candidate for commercialization. Diflubenzuron has been shown to inhibit chitin biosynthesis in a number of insects: *Lymantria dispar* (Salama *et al.*, 1976), *Leptinotarsa* (Post and Vincent, 1973; Grosscurt, 1977, 1978), *Oncopeltus fasciatus* (Hajjar and Casida, 1978, 1979), *Pieris brassicae* (Mulder and Gijswijt, 1973; Post *et al.*, 1974; Deul *et al.*,

1978). *Culex pipiens* (Hajjar, 1979), locusts (Hunter and Vincent, 1974), *Locusta migratoria* (Clarke *et al.*, 1977), *Schistocerca gregaria* (Ker, 1977), *Chilo suppressalis* (Nishioka *et al.*, 1979), and *Musca domestica* (Ishaaya and Casida, 1974).

Inhibition of Chitin Biosynthesis

Many workers have investigated the effects of diflubenzuron on the biosynthetic pathway leading to chitin in order to ascertain exactly which step is inhibited (Figure 1.9). Most evidence concerning inhibition of chitin biosynthesis suggests that chitin synthetase is the key enzyme. Early work with larval *P. brassicae in vitro* showed that DU-19111 prevented conversion of [^{14}C]glucose to [^{14}C]chitin with an attendant accumulation of *N*-acetylglucosamine (Post *et al.*, 1974). Although *N*-acetylglucosamine is not an intermediate in chitin biosynthesis, these authors proposed that DU-19111 partially blocked the process

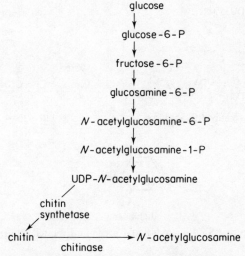

Figure 1.9 Biosynthetic production of chitin

by which *N*-acetylglucosamine units added to the growing chitin polymer. The same group (Post *et al.*, 1974) found that diflubenzuron completely blocked chitin formation and uridine 5′-diphospho-*N*-acetylglucsoamine (UDPAG) accumulated, suggesting inhibition of chitin synthetase. UDPAG also accumulates upon diflubenzuron inhibition of chitin biosynthesis in *O. fasciatus* (Hajjar and Casida, 1978, 1979) and *M. domestica* (van Eck, 1979). Hence, it appears that the most experimentally supported explanation is that the *primary mode of action of diflubenzuron in insects is prevention of chitin biosynthesis via chitin synthetase inhibition,* although some workers recommend caution in accepting this explanation (Mayer et al., 1980). Proof that chitin synthetase is the target for benzoylphenyl ureas awaits isolation of the enzyme from insect tissues for in vitro analysis, but enzyme purification heretofore has been elusive (van Eck, 1979; Hajjar and Casida, 1979).

The morphological manifestation of diflubenzuron treatment of insects is an imperfect cuticle which is particularly noticeable at the time of ecdysis. The endocuticle contains reduced levels of chitin while epicuticular and exocuticular tissues (relatively chitin-free) are unaffected (Mulder and Gijswijt, 1973). The impaired attachment of endocuticle to the epidermis (Mulder and Gijswijt, 1973; Salama et al., 1976) and lower chitin content confer reduced cuticular rigidity (Hunter and Vincent, 1974). Hence, insect mortality results from cuticular malformation.

A useful byproduct of investigations of the mode of action of benzoylphenyl ureas has been the development of a number of systems for the bioassay of potential chitin synthesis inhibitors (Table 1.5). A very sensitive in vitro system was developed using cockroach leg regenerates (Marks and Sowa, 1974; Sowa

Table 1.5 Assay of chitin biosynthesis

Precursor	Insect	Type	Reference
D-[6-^{14}C]glucose	Pieris, larvae	in vivo	Deul et al., 1978
[6-^{3}H]glucosamine	Musca, larval body walls	in vitro	van Eck, 1979
[acetyl-^{14}C]N-acetylglucosamine [U-^{14}C]glucose [U-^{14}C]glucosamine	Oncopeltus, adult abdomens	in vitro	Hajjar and Casida, 1979
[U-^{14}C]glucose	Oncopeltus, larvae	in vivo	Hajjar and Casida, 1979
[^{14}C]N-acetyl glucosamine [^{14}C]glucosamine	Leucophaea, nymphal leg regeneration	in vitro	Marks and Sowa, 1974 Sowa and Marks, 1975 Cohen and Marks, 1979
D-[6-^{3}H]glucose	Plodia, larval wing discs	in vitro	Oberlander and Leach, 1974
None, cuticle thickness measured	Chilo, diapausing larvae	in vitro	Nishioka et al., 1979
[^{14}C]acetylglucosamine D-[^{14}C]glucosamine D-[^{14}C]glucose D-[^{14}C]fructose	Stomoxys, imaginal pupal tissue	in vitro	Mayer et al., 1980

and Marks, 1975). In this system diflubenzuron inhibited chitin synthesis with an I_{50} of 6×10^{-10} M, but unfortunately the assay takes 2 weeks and requires β-ecdysone activation. By tissue culturing of integumentary cuticle of the rice stem borer (*Chilo suppressalis*) and then measuring cuticle thickness after exposure to potential inhibitors Nishioka *et al.* (1979) showed a respectable 6×10^{-7} M I_{50} for PH 60–38 (the dichlorobenzamide analogue of diflubenzuron, Figure 1.8). In the *in vivo* assay of Deul *et al.* (1978) virtually complete inhibition of chitin biosynthesis was demonstrated 15 min after application of diflubenzuron (1 μg/*Pieris* larva). One of the better *in vitro* bioassays was developed by Hajjar and Casida (1978, 1979) using the isolated abdomens of readily available *O. fasciatus*. This convenient chitin-synthetisizing system showed good structure activity correlations for 24 diflubenzuron analogues in comparing toxicity with chitin inhibitory activity. Diflubenzuron gave an I_{50} of 6×10^{-7} M (0.25 μg g^{-1} of abdomen) when [^{14}C]glucose was used as substrate. Using *M. domestica* larval body walls van Eck (1979) reported a 10-fold increase in sensitivity over the Hajjar–Casida system ($I_{50} = 0.03$ μg g^{-1} tissue). Hence, the target for diflubenzuron in house flies is apparently more sensitive than that of milkweed bugs.

Secondary Effects

An alternative proposal for the mode of action of diflubenzuron is that it impedes metabolism of ecdysone (Yu and Terriere, 1975b, 1977c). The observed morphological deformities after treatment with diflubenzuron could then result from increased levels of chitinase and phenol oxidase which has indeed been shown in *M. domestica* (Ishaaya and Casida, 1974). However, subsequent data refute this hypothesis as the mode of action for diflubenzuron. Deul *et al.* (1978) found that neither DU-19111 nor diflubenzuron has any effect on chitinase activity in *P. brassicae* either *in vitro* or *in vivo*. Hajjar and Casida (1979) provided the most conclusive evidence that diflubenzuron does not alter *in vivo* metabolism of either α- or β-ecdysone by fifth instar milkweed bug nymphs nor does it alter the endogenous titre of β-ecdysone in pharate *Stomoxys calcitrans* pupae (O'Neill *et al.*, 1977). The juvenoid R-20458 does not synergize diflubenzuron activity nor does β-ecdysone affect diflubenzuron activity in mosquitoes and milkweed bugs (Hajjar, 1978).

In general, diflubenzuron does not affect protein synthesis in insect cuticle. Protein synthesis is unaffected in the cuticle of *Leptinotarsa* (Post and Vincent, 1973), *P. brassicae* (Hunter and Vincent, 1974; Post *et al.*, 1974), *Anthonomus grandis* (Mitlin *et al.*, 1977), or *Leucophaea maderae* leg regenerates (Marks and Sowa, 1974). Ishaaya and Casida (1974) reported a dose dependent increase in the protein:chitin ratio when larval house fly cuticle was treated with diflubenzuron, an alteration which they suggest affects the elasticity and firmness of endocuticle. Contrary to this finding, Clarke *et al.* (1977) reported that diflubenzuron treatment of the peritrophic membrane of *Locusta* resulted in a constant ratio of protein:chitin, both of which were individually reduced. The constancy of the protein:chitin ratio may be limited to cuticles which are not covalently cross-linked since deposition of protein in sclerotized areas is largely unaffected (Clarke *et al.*, 1977). Although the protein:chitin ratio in the peritrophic membrane of adult *Calliphora* was not determined, diflubenzuron treatment decreased chitin levels as well as decreasing the mass and length of peritrophic membrane (Becker, 1978). Grosscurt (1978) showed that diflubenzuron modified the mechanical penetrability of adult *L. decemlineata* elytra, an effect which paralleled the kinetics of inhibition of chitin formation. Grosscurt concluded that effects on penetrability were due to interference of diflubenzuron with chitin–protein bonding in the elytra.

Several miscellaneous aspects of the mode of action of diflubenzuron have been explored. In locust species diflubenzuron has no effect on cuticular tanning (Hunter and Vincent, 1974; Ker, 1977). It also has no effect on transport of [^{14}C]glucose and its metabolites into the integument of *C. pipiens* (Hajjar, 1978; Hajjar and Casida, 1979). Although Salama *et al.* (1976) reported that diflubenzuron did not affect internal tissue or spermatogenesis in lepidopterous insects, Mitlin *et al.* (1977) found effects on lipoprotein synthesis and inhibition of testicular growth in male *A. grandis*. The diminishment of sexual function was attributed in part to inhibition of DNA synthesis by diflubenzuron. Interestingly, although diflubenzuron was originally the result of a probe for herbicidal activity, it has little effect on plants. Diflubenzuron has no effect on photosynthesis or leaf ultrastructure in soy beans and respiration is only stimulated in a transitory manner at high rates (Hatzios and Penner, 1978). Unlike Polyoxin D which inhibits chitin biosynthesis in both fungi and insects, benzoylphenyl ureas block this synthesis only in insects (van Eck, 1979). However, Gijswijt *et al.* (1979) have shown that for *P. brassicae* the inhibition of cuticle deposition and of chitin synthesis appears to be the same for both Polyoxin D and diflubenzuron.

METABOLISM OF DIFLUBENZURON AND RELATED COMPOUNDS

Insect Metabolism

In general, diflubenzuron is refractory to degradation by most insects studied to date. Minimal metabolism of diflubenzuron is reported for *Estigmene acrea* (salt marsh caterpillar; Metcalf *et al.*, 1975; Verloop and Ferrell, 1977), *Culex pipiens* (mosquito; Metcalf *et al.*, 1975), and *Pieris brassicae* (cabbage butterfly; Verloop and Ferrell, 1977). *Musca domestica* (house fly) degrades diflubenzuron more effectively than *Stomoxys calcitrans* (stable fly), but even *M. domestica* only metabolized about 10% of the applied dose after topical application (Ivie and Wright, 1978). Reduced penetration coupled with more efficient metabolism and rapid excretion are important factors in explaining resistance to diflubenzuron in

M. domestica (Pimprikar and Georghiou, 1979). An unnatural method of application (i.e. injection) resulted in only 21% degradation of diflubenzuron by *M. domestica* after 3 days (Chang, 1978).

The metabolism of diflubenzuron has been most extensively studied in the boll weevil (*Anthonomus grandis*). Although initial reports indicated essentially no degradation by this insect (Still and Leopold, 1975, 1978; Verloop and Ferrell, 1977), more recent data (Chang and Stokes, 1979) demonstrated up to 23% degradation after 4 days. Much interest has focused on the transfer of diflubenzuron from treated to untreated weevils (Moore *et al.*, 1978) and the secretion of unmetabolized compound into eggs which is responsible for inhibition of hatching (Bull and Ivie, 1980). The metabolites of diflubenzuron reported from insects are given in Table 1.6, and the metabolic fate of diflubenzuron has also been discussed by Sparks and Hammock (1980a).

Mammalian Metabolism and Toxicology

Hydrolysis and aromatic hydroxylation are the two primary metabolic pathways in rats (Verloop and Ferrell, 1977). About 20% of the diflubenzuron was hydrolysed by rats to 2,6-difluorobenzoic acid and 4-chlorophenylurea. The isolation of considerably less than stoichiometric amounts of 4-chlorophenylurea suggested that it is further degraded. Aromatic hydroxylation of diflubenzuron at both phenyl rings contributed 80% of the metabolic degradation. These hydroxylated metabolites represented almost all of the biliary ^{14}C products and about half of those in urine. Hydroxylated diflubenzurons are devoid of insecticidal activity on *A. Grandis* (Bull and Ivie, 1980) and are readily excreted by rats if given orally (Ivie, 1978).

The fate of diflubenzuron has been studied comprehensively in sheep and cattle (Ivie, 1977, 1978). Although sheep and cattle metabolized diflubenzuron in a qualitatively similar manner (Table 1.7), the major sheep metabolites arose from cleavage of the amide bond by hydrolysis whereas, in the lactating cow, metabolic transformation resulted primarily from hydroxylation at the 3-position of the 2,6-difluorobenzoyl moiety. The fate of the 4-chlorophenyl ring was largely undetermined since 4-chlorophenylurea was the single such metabolite identified and it was recovered in low yield. In both castrate male sheep and a lactating cow dosed orally, diflubenzuron was extensively metabolized and almost totally excreted. Cannulation of the bile duct in sheep demonstrated the importance of biliary excretion while minor levels of radiolabelled metabolites were secreted into milk.

Diflubenzuron has low acute mammalian toxicity (Ferrell and Verloop, 1975) with an acute LD_{50} in rats of 4600 mg kg^{-1} (Lewis and Tatken, 1979). Holstein bull calves can consume up to 1 mg kg^{-1} per day of diflubenzuron without affecting growth or organ histopathology (Miller *et al.*, 1979). *In vitro* studies with rat C6 glial cells demonstrated that diflubenzuron is neither cytotoxic nor

Table 1.6 Insect metabolites of diflubenzuron

diflubenzuron (structure: 2,6-difluorobenzoyl-CNHCNH-phenyl-Cl)

Metabolite	diflubenzuron Insect (% applied dose)	Reference
2,6-difluorobenzoyl-CNHCNH-(3-HO-phenyl)-Cl	Boll weevil[a] House fly (12)	Chang and Stokes, 1979 Chang, 1978
(3-HO-2,6-difluorobenzoyl)-CNHCNH-phenyl-Cl	Boll weevil[a]	Chang and Stokes, 1979
2,6-difluorobenzoyl-CNH$_2$	Boll weevil (2) Stable fly (0.8) House fly (0.3)	Chang and Stokes, 1979; Bull and Ivie, 1980 Ivie and Wright, 1978 Ivie and Wright, 1978 Pimprikar and Georghiou, 1979
Cl-phenyl-NHCCH$_3$	Stable fly (<0.2)	Ivie and Wright, 1978
Cl-phenyl-NHCNH$_2$	Stable fly (0.5) House fly (0.3)	Ivie and Wright, 1978 Ivie and Wright, 1978

[a] Hydroxylated diflubenzurons were detected only as conjugates, collectively representing up to 19% of the applied dose.

does it inhibit the synthesis of complex carbohydrates (glycosaminoglycans) in mammalian cells (Bishai and Stoolmiller, 1979). Using a *Salmonella* mutagenicity assay Seuferer *et al.* (1979) found that only 2,6-difluorobenzoic acid seemed mutagenic in bacteria (further testing showed this result to be a false positive) while diflubenzuron itself, 4-chlorophenylurea, 4-chlorophenol, and 4-chloroaniline were only borderline mutagens. Toxicology fears were somewhat ameliorated by recent data from the National Cancer Institute which partially absolve 4-chloroaniline from earlier allegations as a mammalian carcinogen

Table 1.8 Mammalian metabolites of diflubenzuron

	Per cent total ^{14}C in excrement				
	Cow		Sheep		Rat
	urine	faeces	urine	faeces	
Diflubenzuron structure (R₁, R₂, R₃ substituents on difluorophenyl–CO–NH–CO–NH–chlorophenyl)					
$R_1 = OH, R_2 = R_3 = H$	45	18	1	0.4	⎫
$R_1 = R_3 = H, R_2 = OH$	2	0.6	0.2	0.8	⎬ 80
$R_1 = R_2 = H, R_3 = OH$	4	0.7	0	0.4	⎭
Cl–C₆H₄–NHCONH₂	0.6	—	0	—	2
2,6-difluoro–C₆H₃–CO₂H	6	—	27	—	20
2,6-difluoro–C₆H₃–CONHCH₂CO₂H	7	—	22	—	—

(National Cancer Institute, 1979). Diflubenzuron is essentially non-toxic to fish (McKague and Pridmore, 1978) although 4-chloroaniline is relatively more toxic to bluegill (Julin and Sanders, 1978).

Environmental Fate

Schaefer and Dupras (1976, 1977) found that the environmental persistence of diflubenzuron in water appeared to be determined by the rate of hydrolysis and adsorption onto organic matter. High temperature and elevated pH enhanced instability. Schaefer and Dupras (1976) found minimal photodecomposition of diflubenzuron when a thin film on glass or a 0.1 p.p.m. aqueous solution was exposed to sunlight even though substantial photodegradation occurs when methanolic solutions are irradiated with artificial light sources (Ruzo *et al.*, 1974; Metcalf *et al.*, 1975). Ivie *et al.* (1979, 1980a) studied the fate of diflubenzuron in water and found a half-life of 56, 7 and <3 days for pHs 4, 6, and 10, respectively.

The major degradation products in water were quantitated as a function of pH and consisted of 4-chlorophenylurea, 2,6-difluorobenzoic acid, small amounts of 2,6-difluorobenzamide, and a unique quinazolinedione (4% yield after 56 days pH 10).

Quinazolinedione

The degradation of diflubenzuron by aquatic organisms has been reviewed by Schooley and Quistad (1979). Aquatic microbial metabolism has been reported by Metcalf et al. (1975), Schaefer and Dupras (1976, 1977), and Booth and Ferrell (1977). The degradation of diflubenzuron by fish, as well as the other components of an aquatic ecosystem, was detailed by Metcalf et al. (1975) and Booth and Ferrell (1977).

The degradation of benzoylphenyl ureas in soil was significant historically in the selection of diflubenzuron for commercial development (Verloop and Ferrell, 1977). Indeed an initial lead structure, the analogous 2,6-dichlorobenzamide (PH-60-38), was not pursued commercially because of its extreme persistence in agricultural soil (6–12 months). By replacing the two chlorines with fluorines an analogue (i.e. diflubenzuron) was found which was not only more susceptible to environmental degradation by hydrolysis, but also was unexpectedly more insecticidal. An important stability property of diflubenzuron was discovered from initial soil degradation studies, i.e. the rate of degradation was greatly dependent on particle size with smaller particles being degraded more rapidly. This axiom also extends to other degradation studies involving diflubenzuron including insect metabolism where topical application of organic solutions often results in deposits of large crystals which are more refractory to breakdown (Still and Leopold, 1978).

Diflubenzuron is still relatively persistent in soil. Metcalf et al. (1975) found virtually no degradation in soil after 4 weeks. When cotton plants previously sprayed with diflubenzuron were cultivated into soil, residues had not dissipated appreciably after 6 months (Bull and Ivie, 1978). Unmetabolized diflubenzuron represented 81% of the ^{14}C-residue with 2% as 4-chlorophenylurea the only identified metabolite. *The major metabolic pathway for diflubenzuron in soil involves hydrolysis to 2,6-difluorobenzoic acid and 4-chlorophenylurea* (Verloop and Ferrell, 1977; Mansager, et al., 1979). Verloop and Ferrell (1977) found that up to 70% of the aniline-labelled diflubenzuron was recovered as 4-chlorophenyl urea which was relatively persistent whereas benzoyl-labelled diflubenzuron gave only 20% as 2,6-difluorobenzoic acid, indicating greater environmental lability of this metabolite. By isolating 28% of the applied [^{14}C]diflubenzuron as $^{14}CO_2$ after 91 days Mansager et al. (1979) demonstrated that one or both of the aromatic rings can be exhaustively oxidized, albeit at a slow rate. These authors also found 4-chlorophenylurea as the major metabolite in soil (24% applied dose after 21 days).

Degradation of diflubenzuron in soil is primarily biological with little breakdown under sterile conditions (Verloop and Ferrell, 1977). Seuferer et al. (1979) have isolated four eucaryotic microorganisms from soil capable of degrading diflubenzuron. One species of *Fusarium* was capable of using diflubenzuron as a sole carbon source, producing 2,6-difluorobenzoic acid, 4-chlorophenylurea, 4-chloroaniline, 4-chloroacetanilide, acetanilide, and 4-chlorophenol as metabolites. Seuferer et al. (1979) concluded that fungi degraded diflubenzuron more rapidly than bacteria.

Diflubenzuron is essentially not metabolized by cotton (Bull and Ivie, 1978; Mansager et al., 1979) and other plants (including soy beans, apple, maize, and cabbage; Verloop and Ferrell, 1977). Absorption, translocation, and photodegradation were insignificant when diflubenzuron was applied to cotton foliage (Bull and Ivie, 1978; Mansager et al., 1979). Cultivation of diflubenzuron-treated cotton into soil followed by planting of wheat and collards resulted in minimal residues in these rotational crops (Bull and Ivie, 1978). Cotton planted in [^{14}C]diflubenzuron-treated soil acquired only 3% of the applied ^{14}C after 89 days and this small residue may be due in part to $^{14}CO_2$ fixation and incorporation into plant structural components (Mansager et al., 1979).

Metabolism of Benzoylphenyl Ureas

Although the metabolic fate of diflubenzuron has been adequately documented, the degradation of relatively few analogues has been reported. As previously discussed, the analogous 2,6-dichlorobenzamide (PH-60-38) received considerable attention as possibly the first benzoylphenyl urea candidate for commercial development (Verloop and Ferrell, 1977). In part because of extended persistence in soil, emphasis shifted to diflubenzuron. The metabolic fate of penfluron was studied in boll weevils (Chang and Woods, 1979b) and house flies (Chang and Woods, 1979a). For both insects penfluron was degraded qualitatively similar to diflubenzuron, but at a slower rate. Bull and Ivie (1980) have reported the metabolic fate of N-methyl diflubenzuron and several methoxylated derivatives in the boll weevil, but their results identified no metabolites and were restricted to observation of absorption, excretion, and secretion into eggs. Schaefer and Dupras (1979) examined the environmental stability of SIR-8514, including effects of pH, temperature, sunlight, microbes, and plants.

FUTURE OF INSECT GROWTH REGULATORS

Williams of Harvard popularized the concept of the third 'generation of pesticides' in 1967 with the thesis that the resulting compounds would solve numerous insect control problems. The resulting juvenile hormone mimics (juvenoids) and other insect growth regulators fall short of being panaceas, but they have resulted in marketable compounds. Of greater importance, they represent a concept for pesticide development which is still valid. As an example of the chitin synthesis inhibitors, diflubenzuron has proven to be useful in the control of major pests throughout the world. It should be noted that the term IGR does not confer *a priori* that a compound has only beneficial effects. The benzoylphenyl ureas being considered for development are generally broad spectrum, rather persistent compounds. It is thus necessary to weigh the benefits and risks associated with each compound. Hopefully, the future will see the development of more chitin synthesis inhibitors of varying specificity and selectivity.

Two juvenoids are marketed by Zoecon Corp. (methoprene and kinoprene). Because of their very high biological activity, the sale represents a much larger tonnage expressed in terms of classical pesticides. Methoprene is quite selective, and the majority of sales are confined to controlling insects of medical and veterinary importance. Methoprene will not see large volume use on row, field, or orchard crops. However, its uses are likely to expand in certain clearly defined areas. The prospect of other juvenoids being developed in the immediate future is slight unless there is a change in the philosophy of pesticide usage and in registration procedures. The advantages of juvenoids represent also their limitations. Juvenoids are generally quite selective, and there are very few markets which will bear the cost of registration of such selective compounds. Possible exceptions include compounds active on the boll weevil or specific for the Lepidoptera. There are compounds which show potential in this area. However, it is unlikely that even they will be developed unless several conditions are met. These conditions include development of high resistance to pyrethroids and other insecticides leaving a clear void in the market, economic incentives for the development of compounds which are very safe for humans and the environment, and rapid registration of compounds compatible with IPM programmes. Our knowledge of the environmental stability and metabolism of IGRs in target and non-target organisms provides a sound basis for synthesis of such third generation compounds. Although the technology is present, the economic incentives are regrettably absent.

The most exciting aspect of the IGR field is its future. The rate of discovery of new agents with novel sites of action on the insect integumental, endocrine, or other systems will probably increase. Such compounds will, hopefully, possess more properties of the ideal insecticide and novel structures are certain to present challenges to future pesticide chemists.

ACKNOWLEDGEMENTS

Original research presented in this manuscript was supported in part by NIEHS Grant 5 RO1 ES01260-04. B. D. Hammock was supported by NIEHS Research Career Development Award 1 KO4 ES00046-02.

REFERENCES

Agosin, M., Morello, A., White, R., Repetto, Y., and Pedemonte, J. (1979). 'Multiple forms of noninduced rat liver cytochrome P-450: metabolism of 1-(4'-ethylphenoxy)-3,7-dimethyl-6,7-epoxy-*trans*-2-octene by reconstituted preparations', *J. Biol. Chem.* **254**, 9915–9920.

Ajami, A. M. (1975). 'Inhibitors of ester hydrolysis as synergists for biological activity of cecropia juvenile hormone', *J. Insect Physiol.* **21**, 1017–1025.

Ajami, A. M., and Crouse, D. N. (1975). 'Synthesis of tritiated juvenile hormones', *J. Label. Compds.* **XI**, 117–126.

Ajami, A. M., and Riddiford, L. M. (1971). 'Comparative metabolism of the Cecropia juvenile hormone', *Amer. Zool.* **11**, 108–109.

Ajami, A. M., and Riddiford, L. M. (1973). 'Comparative metabolism of the Cecropia juvenile hormone', *J. Insect Physiol.* **19**, 635–645.

Akamatsu, Y., Dunn, P. E., Kézdy, F. J., Kramer, K. J., Law, J. H., Reibstein, D., and Sanburg, L. L. (1975). 'Biochemical aspects of juvenile hormone action in insects', in *Control Mechanisms in Development* (Eds. R. H. Meints and E. Davies), pp. 123–149, Plenum Press, New York.

Aldridge, W. N. (1953). 'Serum esterases: two types of esterase (A and B) hydrolysing *p*-nitrophenyl acetate, propionate and butyrate, and a method for their determination', *Biochem. J.* **53**, 110–117.

Ames, B. N., McCann, J., and Yamasaki, E. (1975). 'Methods for detecting carcinogens and mutagens with the *Salmonella*/mammalian-microsome mutagenicity test', *Mutation Res.* **31**, 347–364.

Armstrong, R. N., Levin, W., and Jerina, D. M. (1980). 'Hepatic microsomal epoxide hydrolase: mechanistic studies of the hydration of K-region arene oxides', *J. Biol. Chem.* **255**, 4698–4705.

Bassi, S. D., Goodman, W., Altenhofen, C., and Gilbert, L. I. (1977). 'The binding of exogenous juvenile hormone by the haemolymph of *Oncopeltus fasciatus*', *Insect Biochem.* **7**, 309–312.

Becker, B. (1978). 'Effects of 20-hydroxy-ecdysone, juvenile hormone, Dimilin, and captan on *in vitro* synthesis of peritrophic membranes in *Calliphora erythrocephala*', *J. Insect. Physiol.* **24**, 699–705.

Bergot, B. J., Judy, K. J., Schooley, D. A., and Tsai, L. W. (1980). 'Precocene II metabolism: comparative *in vivo* studies among several species of insects, and structure elucidation of two major metabolites', *Pestic. Biochem. Physiol.* **13**, 95–104.

Bigley, W. S., and Vinson, S. B. (1979a). 'Degradation of [^{14}C]methoprene in the imported fire ant, *Solenopsis invicta*', *Pestic. Biochem. Physiol.* **10**, 1–13.

Bigley, W. S., and Vinson, S. B. (1979b). 'Effects of piperonyl butoxide and DEF on metabolism of methoprene by the imported fire ant, *Solenopsis invicta* Buren', *Pestic. Biochem. Physiol.* **10**, 14–22.

Bishai, W. R., and Stoolmiller, A. C. (1979). 'Uptake of diflubenzuron (N-[[(4-chlorophenyl)amino]carbonyl]-2,6-difluorobenzamide) by rat glial cells *in vitro*', *Pestic. Biochem. Physiol.* **11**, 258–266.

Booth, G. M., and Ferrell, D. (1977). 'Degradation of Dimilin® by aquatic foodwebs', in *Pesticides in Aquatic Environments* (Ed. M. A. Q. Khan), pp. 221–243, Plenum Press, New York.

Bowers, W. S. (1969). 'Juvenile hormone: activity of terpenoid ethers', *Science* **164**. 323–325.

Bowers, W. S. (1976). 'Discovery of insect antiallatotropins', in *The Juvenile Hormones* (Ed. L. I. Gilbert), pp. 394–408, Plenum Press, New York.

Bowers, W. S., Thompson, M. J., and Uebel, E. C. (1965). 'Juvenile and gonadotropic hormone activity of 10,11-epoxyfarnesenic acid methyl ester', *Life Sci.* **4**, 2323–2331.

Breaud, T. P., Farlow, J. E., Steelman, C. D., and Schilling, P. E. (1977). 'Effects of the insect growth regulator methoprene on natural populations of aquatic organisms in Louisiana intermediate marsh habitats', *Mosq. News* **37**, 704–712.

Breccia, A., Gattavecchia, E., Albonetti, G., and DiPietra, A. M. (1976). 'Radiobiochemistry of phytodrugs: I, role of juvenile hormones and analogs in the biosynthesis of proteins and RNA in *Drosophila* larvae', *J. Environ. Sci. Hlth.-Pestic. Food Contam. Agr. Wastes* **B11**, 1–7.

Breccia, A., Gattavecchia, E., DiPietra, A. M., and Lumare, F. (1977). 'Radiobiochemistry of phytodrugs: II, activity of Altosid® and Altozar® in the biosynthesis of proteins and RNA in larvae of shrimps *in vivo* studied by leucine-U-^{14}C and uridine-2-^{14}C', *J. Environ. Sci. Health.* **B12**, 105–112.

Brooks, G. T. (1973). 'Insect epoxide hydrase inhibition by juvenile hormone analogues and metabolic inhibitors', *Nature, New Biol.* **245**, 382–384.

Brooks, G. T. (1974). 'Inhibitors of cyclodiene epoxide ring hydrating enzymes of the blowfly, *Calliphora erythrocephala*', *Pestic. Sci.* **5**, 177–183.

Brooks, G. T. (1977). 'Epoxide hydratase as a modifier of biotransformation and biological activity', *Gen. Pharmacol.* **8**, 221–226.

Brooks, G. T., Hamnett, A. F., Jennings, R. C., Ottridge, A. P., and Pratt, G. E. (1979a). 'Aspects of the mode of action of precocenes on milkweed bugs (*Oncopeltus fasciatus*) and locusts (*Locusta migratoria*)', in *Proceedings 1979 British Crop Protection Conference*, Vol. 1, pp. 273–279, British Crop Protection Council, London.

Brooks, G. T., Harrison, A., and Lewis, S. E. (1970). 'Cyclodiene epoxide ring hydration by microsomes from mammalian liver and houseflies', *Biochem. Pharmacol.* **19**, 255–273.

Brooks, G. T., Pratt, G. E., and Jennings, R. C. (1979b). 'The action of precocenes in milkweed bugs (*Oncopeltus fasciatus*) and locusts (*Locusta migratora*)', *Nature* **281**, 570–572.

Brown, H. C., and Geoghegan, P. J. Jr. (1970). 'Solvomercuration–demercuration I. The oxymercuration–demercuration of representative olefins in an aqueous system. A convenient mild procedure for the Markovnokov hydration of the carbon–carbon double bond', *J. Org. Chem.* **35**, 1844–1850.

Brown, J. J., Chippendale, G. M., and Turunen, S. (1977). 'Larval esterases of the southwestern corn borer, *Diatraea grandiosella*: temporal changes and specificity', *J. Insect Physiol.* **23**, 1255–1260.

Brown, T. M., and Hooper, G. H. S. (1979). 'Metabolic detoxication as a mechanism of methoprene resistance in *Culex pipiens pipiens*', *Pestic. Biochem. Physiol.* **12**, 79–86.

Bull, D. L., and Ivie, G. W. (1978). 'Fate of diflubenzuron in cotton, soil, and rotational crops', *J. Agr. Food Chem.* **26**, 515–520.

Bull, D. L., and Ivie, G. W. (1980). 'Activity and fate of diflubenzuron and certain derivatives in the boll weevil', *Pestic. Biochem. Physiol.* **13**, 41–52.
Burt, M. E., Kuhr, R. J., and Bowers, W. S. (1978). 'Metabolism of precocene II in the cabbage looper and European corn borer', *Pestic. Biochem. Physiol.* **9**, 300–303.
Burt, M. E., Kuhr, R. J., and Bowers, W. S. (1979). 'Distribution and metabolism of precocene II in the brown cockroach, *Periplaneta brunnea* Burmeister', *Bull. Environ. Contam. Toxicol.* **22**, 586–589.
Callen, D. F. (1978). 'A review of the metabolism of xenobiotics by microorganisms with relation to short-term test systems for environmental carcinogens', *Mutation Res.* **55**, 153–163.
Chamberlain, W. F., Hunt, L. M., Hopkins, D. E., Miller, J. A., Gingrich, A. R., and Gilbert, B. N. (1975). 'Absorption, excretion, and metabolism of methoprene by a guinea pig, a steer, and a cow', *J. Agr. Food Chem.* **23**, 736–742.
Chang, S. C. (1978). 'Conjugation: the major metabolic pathway of ^{14}C-diflubenzuron in the house fly', *J. Econ. Entomol.* **71**, 31–39.
Chang, S. C., and Stokes, J. B. (1979). 'Conjugation: the major metabolic pathway of ^{14}C-diflubenzuron in the boll weevil', *J. Econ. Entomol.* **72**, 15–19.
Chang, S. C., and Woods, C. W. (1979a). 'Metabolism of ^{14}C-penfluron in the house fly', *J. Econ. Entomol.* **72**, 482–485.
Chang, S. C., and Woods, C. W. (1979b). 'Metabolism of ^{14}C-penfluron in the boll weevil', *J. Econ. Entomol.* **72**, 781–784.
Chasseaud, L. F. (1979). 'The role of glutathione and glutathione-S-transferases in the metabolism of chemical carcinogens and other electrophilic agents', *Adv. Cancer Res.* **29**, 175–274.
Chefurka, W. (1978). 'Sesquiterpene juvenile hormones: novel uncouplers of oxidative phosphorylation', *Biochem. Biophys. Res. Commun.* **83**, 571–578.
Chen, T. T., Couble, P., DeLucca, F. L., and Wyatt, G. R. (1976). 'Juvenile hormone control of vitellogenin synthesis in *Locusta migratoria*', in *The Juvenile Hormones* (Ed. L. I. Gilbert), pp. 505–529, Plenum Press, New York.
Chino, H., and Gilbert, L. I. (1971). 'The uptake and transport of cholesterol by haemolymph lipoproteins', *Insect Biochem.* **1**, 337–347.
Chmurzyńska, W., Grzelakowska-Sztabert, B., and Zielinska, Z. M. (1979). 'Interference of a synthetic C_{18} juvenile hormone and related insect growth regulators with macromolecular biosynthesis in mammalian cells', *Toxicol. Appl. Pharmacol.* **49**, 517–523.
Clarke, L., Temple, G. H. R., and Vincent, J. F. V. (1977). 'The effects of a chitin inhibitor—Dimilin—on the production of peritrophic membrane in the locust, *Locusta migratoria*', *J. Insect Physiol.* **23**, 241–246.
Cline, N. L., Cohen, E. N., and Trudell, J. R. (1974). Personal communication.
Cohen, C. F., and Marks, E. P. (1979). 'Comparison of *in vivo* and *in vitro* activity of three chitin synthesis inhibitors', *Southwest. Entomol.* **4**, 294–297.
Costlow, J. D., Jr. (1977). 'The effect of juvenile hormone mimics on development of the mud crab, *Rhithropanopeus harrisii* (Gould)', in *Physiological Responses of Marine Biota to Pollutants* (Eds. F. J. Vernberg, A. Calabrese, F. P. Thurberg, and W. B. Vernberg), pp. 439–457, Academic Press, New York.
Couble, P., Chen, T. T., and Wyatt, G. R. (1979). 'Juvenile hormone-controlled vitellogenin synthesis in *Locusta migratoria* fat body: cytological development', *J. Insect Physiol.* **25**, 327–337.
Craven, A. C. C., Brooks, G. T., and Walker, C. H. (1976). 'The inhibition of HEOM epoxide hydrase in mammalian liver microsomes and insect pupal homogenates', *Pestic. Biochem. Physiol.* **6**, 132–141.

Cruickshank, P. A. (1971). 'Some juvenile hormone analogs. A critical appraisal', *Mitt. Schweiz. Ent. Ges.* **44**, 97–113.
Cymborowski, B., Riddiford, L. M., Williams, C. M., and Beckage, W. E. (1979). 'Endocrine control of starvation-induced supernumerary moulting in *Manduca sexta* larvae', Abstracts Western Regional Conference on Comparative Endocrinology, Division of Comparative Endocrinology, American Society of Zoologists, Corvalis, Oregon.
Cymborowski, B., and Stolarz, G. (1979). 'The role of juvenile hormone during larval–pupal transformation of *Spodoptera littoralis*: switchover in the sensitivity of the prothoracic gland to juvenile hormone', *J. Insect Physiol.* **25**, 939–942.
Davidow, B., and Radomski, J. L. (1953). 'Isolation of an epoxide metabolite from fat tissues of dog fed heptachlor', *J. Pharmacol. Exp. Ther.* **107**, 259–265.
Davison, K. L. (1976). 'Carbon-14 distribution and elimination in chickens given methoprene-^{14}C', *J. Agr. Food Chem.* **24**, 641–648.
Denmore, W. B., and Davidson, N. (1959). 'Photochemical experiments in rigid media at low temperatures. I. Nitrogen oxides and ozone', *J. Amer. Chem. Soc.* **81**, 5869–5874.
Deul, D. H., deJong, B. J., and Kortenbach, J. A. M. (1978). 'Inhibition of chitin synthesis by two 1-(2,6-disubstituted benzoyl)-3-phenylurea insecticides. II', *Pestic. Biochem. Physiol.* **8**, 98–105.
Dorn, S., Oesterhelt, G., Suchý, M., Trautmann, K. H., and Wipf, H.-K. (1976). 'Environmental degradation of the insect growth regulator 6,7-epoxy-1-(p-ethylphenoxy)-3-ethyl-7-methylnonane (Ro-10-3108) in polluted water', *J. Agr. Food Chem.* **24**, 637–640.
Downer, R. G. H., Spring, J. H., and Smith, S. M. (1976). 'Effect of an insect growth regulator on lipid and carbohydrate reserves of mosquito pupae (Diptera : Culicidae)', *Can. Entomol.* **108**, 627–630.
Downer, R. G. H., Wiegand, M., and Smith, S. M. (1975). 'Suppression of pupal esterase activity in *Aedes aegypti* (Diptera : Culicidae) by an insect growth regulator', *Experientia* **31**, 1239–1240.
Edwards, J. P., and Rowlands, D. G. (1977). 'Metabolism of a synthetic insect juvenile hormone (JH-I) during the development of *Tribolium castaneum* (Herbst) (Coleoptera, Tenebrionidae)', *Pestic. Biochem. Physiol.* **7**, 194–201.
El-Tantawy, M. A., and Hammock, B. D. (1980). 'The effect of hepatic microsomal and cytosolic subcellular fractions on the mutagenic activity of epoxide containing compounds in the *Salmonella* assay', *Mutation Res.* **79**, 55–71.
El Zorgani, G. A., Walker, C. H., and Hassall, K. A. (1970). 'Species differences in the *in vitro* metabolism of HEOM, a chlorinated cyclodiene epoxide', *Life Sci.* **9**, 415–420.
Emmerich, H. (1976). 'Summary of session IV. Juvenile hormone effects at the molecular level (binding and transport)', in *The Juvenile Hormones* (Ed. L. I. Gilbert), pp. 323–326, Plenum Press, New York.
Emmerich, H.,and Hartmann, R. (1973). 'A carrier lipoprotein for juvenile hormone in the haemolymph of *Locusta migratoria*', *J. Insect Physiol.* **19**, 1663–1675.
Erley, D., Southard, S., and Emmerich, H. (1975). 'Excretion of juvenile hormone and its metabolites in the locust, *Locusta migratoria*', *J. Insect Physiol.* **21**, 61–70.
Ferkovich, S. M., Oberlander, H., and Rutter, R. R. (1977). 'Release of a juvenile hormone binding protein by fat body of the Indian meal moth, *Plodia interpunctella*, in vitro', *J. Insect Physiol.* **23**, 297–302.
Ferkovich, S. M., and Rutter, R. R. (1976). 'Influence of a haemolymph protein fraction on the binding of juvenile hormone in homogenates of insect epidermis (*Plodia interpunctella* (Hübner))', *Roux's Arch. Develop. Biol.* **179**, 243–248.

Ferkovich, S. M., Silhacek, D. L., and Rutter, R. R. (1975). 'Juvenile hormone binding proteins in the haemolymph of the Indian meal moth', *Insect Biochem.* **5**, 141–150.

Ferkovich, S. M., Silhacek, D. L., and Rutter, R. R. (1976). 'The binding of juvenile hormone to larval epidermis: influence of a carrier protein from the hemolymph of *Plodia interpunctella*', in *The Juvenile Hormones* (Ed. L. I. Gilbert), pp. 342–353, Plenum Press, New York.

Ferrell, D., and Verloop, A. (1975). 'Current status of research on Dimilin (TH-6040)', *ACS Abstracts Chicago Meeting*, PEST 35.

Firstenberg, D. E., and Silhacek, D. L. (1973). 'Juvenile hormone regulation of oxidative metabolism in isolated insect mitochondria', *Experientia* **29**, 1420–1422.

Fristrom, J. W., Chihara, C. J., Kelly, L., and Nishiura, J. T. (1976). 'The effects of juvenile hormone on imaginal discs of *Drosophila in vitro*: the role of the inhibition of protein synthesis', in *The Juvenile Hormones* (Ed. L. I. Gilbert), pp. 432–448, Plenum Press, New York.

Gavin, J. A., and Williamson, J. H. (1978). 'Effects of alpha amanitin and juvenile hormone analogue (ZR-515) on labelling of RNA and protein in adult *Drosophila*', *J. Insect Physiol.* **24**, 413–416.

Gijswijt, M. T., Deul, D. H., and DeJong, B. J. (1979). 'Inhibition of chitin synthesis by benzoyl-phenylurea insecticides, III. Similarity in action in *Pieris brassicae* (L.) with Polyoxin D', *Pestic. Biochem. Physiol.* **12**, 87–94.

Gilbert, L. I. (1972). 'Insect hormones: transport, binding proteins and action', *Int. Cong. Ser. No.* **273**, Endocrinology, Proc. 4th Int. Cong. of Endocrinol., pp. 306–310.

Gilbert, L. I. (1974). 'Endocrine action during insect growth', *Recent Prog. in Hormone Res.* **30**, 347–390.

Gilbert, L. I. (Ed.) (1976). *The Juvenile Hormones*, 572 pp., Plenum Press, New York.

Gilbert, L. I., Goodman, W., and Bollenbacher, W. E. (1977). 'Biochemistry of regulatory lipids and sterols in insects', in *Biochemistry of Lipids II* (Ed. T. W. Goodwin), Int. Review of Biochem. **14**, pp. 1–50, Univ. Park Press, Baltimore.

Gilbert, L. I., Goodman, W., and Granger, N. (1978). 'Regulation of juvenile hormone titre in the Lepidoptera', in *Comparative Endocrinology* (Eds. P. J. Gaillard and H. H. Boer), pp. 471–486, Elsevier North Holland Biomedical Press, Amsterdam.

Gilbert, L. I., Goodman, W., and Nowock, J. (1976). 'The possible roles of binding proteins in juvenile hormone metabolism and action', in *Actualités sur les Hormones D'Invertebrés*, pp. 413–434, Colloq. Int. CNRS No. 251, Paris.

Gilbert, L. I., and King, D. S. (1973). 'Physiology of growth and development: endocrine aspects', in *The Physiology of Insecta*, **I** (Ed. M. Rockstein), pp. 249–370, Academic Press, New York.

Gill, S. S., and Hammock, B. D. (1979). 'Hydration of *cis*- and *trans*-epoxymethyl stearates by the cytosolic epoxide hydrase of mouse liver', *Biochem. Biophys. Res. Commun.* **89**, 965–971.

Gill, S. S., and Hammock, B. D. (1980). 'Distribution and properties of a mammalian soluble epoxide hydrase', *Biochem. Pharmacol.* **29**, 389–395.

Gill, S. S., Hammock, B. D., and Casida, J. E. (1974). 'Mammalian metabolism and environmental degradation of the juvenoid 1-(4'-ethylphenoxy)-3,7-dimethyl-6,7-epoxy-*trans*-2-octene and related compounds', *J. Agr. Food Chem.* **22**, 386–395.

Gill, S. S., Hammock, B. D., Yamamoto, I., and Casida, J. E. (1972). 'Preliminary chromatographic studies on the metabolites and photodecomposition products of the juvenoid 1-(4'-ethylphenoxy)-6,7-epoxy-3,7-dimethyl octene', in *Insect Juvenile Hormones: Chemistry and Action* (Eds. J. J. Menn and M. Beroza), pp. 177–189, Academic Press, New York.

Goodman, W., Bollenbacher, W. E., Zvenko, H., and Gilbert, L. I. (1976). 'A competitive protein binding assay for juvenile hormone', in *The Juvenile Hormones* (Ed. L. I. Gilbert), pp. 75–95, Plenum Press, New York.

Goodman, W., and Gilbert, L. I. (1974). 'Haemolymph protein binding of juvenile hormone in *Manduca sexta*', *Amer. Zool.* **14**, 1289.

Goodman, W., and Gilbert, L. I. 1978). 'The hemolymph titer of juvenile hormone binding protein and binding sites during the fourth larval instar of *Manduca sexta*', *Gen. Comp. Endocrinol.* **35**, 27–34.

Goodman, W., O'Hern, P. A., Zaugg, R. H., and Gilbert, L. I. (1978a). 'Purification and characterization of a juvenile hormone binding protein from the hemolymph of the fourth instar tobacco hornworm, *Manduca sexta*', *Mol. Cell. Endocrinol.* **11**, 225–242.

Goodman, W., Schooley, D. A., and Gilbert, L. I. (1978b). 'Specificity of the juvenile hormone binding protein: the geometrical isomers of juvenile hormone I', *Proc. Natl. Acad. Sci. USA* **75**, 185–189.

Grosscurt, A. C. (1977). 'Mode of action of diflubenzuron as an ovicide and some factors influencing its potency', in *1977 British Crop Protection Conf.-Pests and Diseases* **1**, p. 141, British Crop Protection Council, London.

Grosscurt, A. C. (1978). 'Effects of diflubenzuron on mechanical penetrability, chitin formation, and structure of the elytra of *Leptinotarsa decemlineata*', *J. Insect Physiol.* **24**, 827–831.

Hafferl, W., Zurflüh, R., and Dunham, L. (1971). 'Radiochemical synthesis part II. The preparation of ^{14}C-labeled juvenile hormone', *J. Label. Comp.* **7**, 331–339.

Hajjar, N. P. (1978). *Mechanism of the Insecticidal Action of Diflubenzuron*, Ph.D. Thesis, Univ. of California, Berkeley, University Microfilms International, Ann Arbor, Michigan.

Hajjar, N. P. (1979). 'Diflubenzuron inhibits chitin synthesis in *Culex pipiens* L. larvae', *Mosq. News* **39**, 381–384.

Hajjar, N. P., and Casida, J. E. (1978). 'Insecticidal benzoylphenyl ureas: structure-activity relationships as chitin synthesis inhibitors', *Science* **200**, 1499–1500.

Hajjar, N. P., and Casida, J. E. (1979). 'Structure-activity relationships of benzoylphenyl ureas as toxicants and chitin synthesis inhibitors in *Oncopeltus fasciatus*', *Pestic. Biochem. Physiol.* **11**, 33–45.

Hammock, B. D. (1973). 'Chemical and biological studies on aryl geranyl epoxide ether juvenoids', Ph.D. Dissertation, Univ. California, Berkeley. 269 pp.

Hammock, B. D. (1975). 'NADPH dependent epoxidation of methyl farnesoate to juvenile hormone in the cockroach *Blaberus giganteus* L.', *Life Sci.* **17**, 323–328.

Hammock, B. D., El-Tantawy, M., Gill, S. S., Hasegawa, L., Mullin, C. A., and Ota, K. (1980a). 'Extramicrosomal epoxide hydration', in *Microsomes, Drug Oxidations, and Chemical Carcinogenesis* (Eds. M. J. Coon *et al.*), Vol. II, pp. 655–6665, Academic Press, New York.

Hammock, B. D., Gill, S. S., and Casida, J. E. (1974a). 'Insect metabolism of a phenyl epoxygeranyl ether juvenoid and related compounds', *Pestic. Biochem. Physiol.* **4**, 393–406.

Hammock, B. D., Gill, S. S., and Casida, J. E. (1974b). 'Synthesis and morphogenetic activity of derivatives and analogs of aryl geranyl ether juvenoids', *J. Agr. Food Chem.* **22**, 379–385.

Hammock, B. D., Gill, S. S., Hammock, L., and Casida, J. E. (1975a). 'Metabolic O-dealkylation of 1-(4'-ethylphenoxy)-3,7-dimethyl-7-methoxy or ethoxy-*trans*-2-octene, potent juvenoids', *Pestic. Biochem. Physiol.* **5**, 12–18.

Hammock, B. D., Gill, S. S., Mumby, S. M., and Ota, K. (1980b). 'Current status and possible implication of research on a soluble mammalian epoxide hydrase', in *Molecular Basis of Environmental Toxicity* (Ed. R. S. Bhatnagar), pp. 229–272, Ann Arbor Science Publishers, Ann Arbor, Mich.

Hammock, B. D., Gill, S. S., Stamoudis, V., and Gilbert, L. I. (1976). 'Soluble mammalian epoxide hydratase: action on juvenile hormone and other terpenoid epoxides', *Comp. Biochem. Physiol.* **53B**, 263–265.

Hammock, B. D., and Mumby, S. M. (1978). 'Inhibition of epoxidation of methyl farnesoate to juvenile hormone III by cockroach corpora allata homogenates', *Pestic. Biochem. Physiol.* **9**, 39–47.

Hammock, B. D., Mumsby, S. M., and Lee, P. W. (1977a). 'Mechanisms of resistance to the juvenoid methoprene in the house fly, *Musca domestica* L.', *Pestic. Biochem. Physiol.* **7**, 261–272.

Hammock, B., Nowock, J., Goodman, W., Stamoudis, V., and Gilbert, L. I. (1975b). 'The influence of hemolymph-binding protein on juvenile hormone stability and distribution in *Manduca sexta* fat body and imaginal discs in vitro', *Mol. Cell. Endocrinol.* **3**, 167–184.

Hammock, B. D., and Quistad, G. B. (1976). 'The degradative metabolism of juvenoids by insects', in *The Juvenile Hormones* (Ed. L. I. Gilbert), pp. 374–393, Plenum Press, New York.

Hammock, B. D., and Sparks, T. C. (1977). 'A rapid assay for insect juvenile hormone esterase activity', *Anal. Biochem.* **82**, 573–579.

Hammock, B. D., Sparks, T. C., and Mumby, S. M. (1977b). 'Selective inhibition of JH esterases from cockroach hemolymph', *Pestic. Biochem. Physiol.* **7**, 517–530.

Hammock, L. G., Hammock, B. D., and Casida, J. E. (1974c). 'Detection and analysis of epoxides with 4-(*p*-nitrobenzyl)-pyridine', *Bull. Environ. Contam. Toxicol.* **12**, 759–764.

Handler, A. M., and Postlethwait, J. H. (1978). 'Regulation of vitellogenin synthesis in *Drosophila* by ecdysterone and juvenile hormone', *J. Exp. Zool.* **206**, 247–254.

Hangartner, W. W., Suchý, M., Wipf, H.-K., and Zurflueh, R. C. (1976). 'Synthesis and laboratory and field evaluation of a new, highly active and stable insect growth regulator', *J. Agr. Food Chem.* **24**, 169–175.

Hartmann, R. (1978). 'The juvenile hormone-carrier in the haemolymph of the acridine grasshopper, *Gomphocerus rufus* L.: blocking of the juvenile hormone's action by means of repeated injections of an antibody to the carrier', *Roux's Arch. Develop. Biol.* **184**, 310–324.

Hatzios, K. K., and Penner, D. (1978). 'The effect of diflubenzuron [1-(4-chlorophenyl)-3-(2,6-difluorobenzoyl)urea] on soybean [*Glycine max* (L.) Merr.] photosynthesis, respiration, and leaf ultrastructure', *Pestic. Biochem.* **9**, 65–69.

Hawkins, D. R., Weston, K. T., Chasseaud, L. F., and Franklin, E. R. (1977). 'Fate of methoprene (isopropyl (2*E*, 4*E*)-11-methoxy-3,7,11-trimethyl-2,4-dodecadienoate) in rats', *J. Agr. Food Chem.* **25**, 398–403.

Henrick, C. A., Staal, G. B., and Siddall, J. B. (1973). 'Alkyl 3,7,11-trimethyl-2,4-dodecadienoates, a new class of potent insect growth regulators with juvenile hormone activity', *J. Agr. Food Chem.*. **21**, 354–359.

Henrick, C. A., Willy, W. E., McKean, D. R., Baggiolini, E., and Siddall, J. B. (1975). 'Approaches to the synthesis of the insect juvenile hormone analog ethyl 3,7,11-trimethyl-2,4-dodecadienoate and its photochemistry', *J. Org. Chem.* **40**, 8–14.

Himeno, M., Takahashi, J., and Komano, T. (1979). 'Effect of juvenile hormone on macromolecular synthesis of an insect cell line', *Agr. Biol. Chem.* **43**, 1285–1292.

Hiruma, K., Shimada, H., and Yagi, S. (1978a). 'Activation of the prothoracic gland by juvenile hormone and prothoracicotropic hormone in *Mamestra brassicae*', *J. Insect Physiol.* **24**, 215–220.

Hiruma, K., Yagi, S., and Agui, N. (1978b). 'Action of juvenile hormone on the cerebral neurosecretory cells of *Mamestra brassicae in vivo* and *in vitro*', *App. Entomol. Zool.* **13**, 149–157.

Hoffman, L. J., Ross, J. H., and Menn, J. J. (1973). 'Metabolism of 1-(4'-ethylphenoxy)-6,7-epoxy-3,7-dimethyl-2-octene (R-20458) in the rat', *J. Agr. Food Chem.* **21**, 156–163.

Hooper, G. H. S. (1976). 'Esterase mediated hydrolysis of naphthyl esters, malathion, methoprene, and cecropia juvenile hormone in *Culex pipiens pipiens*', *Insect Biochem.* **6**, 255–266.

Hunter, E., and Vincent, J. F. V. (1974). 'The effects of a novel insecticide on insect cuticle', *Experientia* **30**, 1432–1433.

Hwang-Hsu, K., Reddy, G., Kumaran, A. K., Bollenbacher, W. E., and Gilbert, L. I. (1979). 'Correlations between juvenile hormone esterase activity ecdysone titre and cellular reprogramming in *Galleria mellonella*', *J. Insect Physiol.* **25**, 105–111.

Ishaaya, I., and Casida, J. E. (1974). 'Dietary TH 6040 alters composition and enzyme activity of house fly larval cuticle', *Pestic. Biochem. Physiol.* **4**, 484–490.

Ivie, G. W. (1976). 'Epoxide to olefin: a novel of biotransformation in the rumen', *Science* **191**, 959–961.

Ivie, G. W. (1977). 'Metabolism of insect growth regulators in animals', in *Fate of Pesticides in the Large Animal* (Eds. G. W. Ivie and H. W. Dorough), pp. 111–125, Academic Press, New York.

Ivie, G. W. (1978). 'Fate of diflubenzuron in cattle and sheep', *J. Agr. Food Chem.* **26**, 81–89.

Ivie, G. W., Bull, D. L., and Veech, J. A. (1979). 'Metabolism of diflubenzuron by mammals, insects, and soil fungi, and its fate in water', *ACS Abstracts Honolulu Meeting*, Spring 1979, PEST 112.

Ivie, G. W., Bull, D. L., and Veech, J. A. (1980a). 'Fate of diflubenzuron in water', *J. Agr. Food Chem.* **28**, 330–337.

Ivie, G. W., and Casida, J. E. (1971). 'Photosensitizers for the accelerated degradation of chlorinated cyclodienes and other insecticide chemicals exposed to sunlight on bean leaves', *J. Agr. Food Chem.* **19**, 410–416.

Ivie, G. W., MacGregor, J. T., and Hammock, B. D. (1980b). 'Mutagenicity of psoralen epoxides', *Mutation Res.* **79**, 73–77.

Ivie, G. W., and Wright, J. E. (1978). 'Fate of diflubenzuron in the stable fly and house fly', *J. Agr. Food Chem.* **26**, 90–94.

Ivie, G. W., Wright, J. E., and Smalley, H. E. (1976). 'Fate of the juvenile hormone mimic 1-(4'-ethylphenoxy)-3,7-dimethyl-6-7-epoxy-*trans*-2-octene (Stauffer R-20458) following oral and dermal exposure to steers', *J. Agr. Food Chem.* **24**, 222–227.

Jennings, R. C., and Ottridge, A. P. (1979). 'The synthesis of precocene I epoxide (2,2-dimethyl-3,4-epoxy-7-methoxy-2H-1-benzopyran', *J. Chem. Soc. Chem. Commun.* 920–921.

Jerina, D. M., Daly, J. W., Witkop, B., Zaltzman-Nirenberg, P., and Udenfriend, S. (1968). 'The role of arene oxide–oxepin systems in the metabolism of aromatic substrates. III. Formation of 1,2-naphthalene oxide from naphthalene by liver microsomes', *J. Amer. Chem. Soc.* **90**, 6525–6527.

Jones, G., Wing, K. D., Jones, D., and Hammock, B. D. (1980). 'The source and action of head factors regulating juvenile hormone esterase in larvae of the cabbage looper, *Trichoplusia ni*', *J. Insect Physiol.* accepted.

Judy, K. J., Schooley, D. A., Dunham, L. L., Hall, M. S., Bergot, B. J., and Siddall, J. B. (1973). 'Isolation, structure, and absolute configuration of a new natural insect juvenile hormone from *Manduca sexta*', *Proc. Natl. Acad. Sci. USA* **70**, 1509–1513.

Julin, A. M., and Sanders, H. O. (1978). 'Toxicity of the IGR, diflubenzuron, to freshwater invertebrates and fishes', *Mosq. News* **38**, 256–259.

Kalbfeld, J., Hoffman, L. J., Chan, J. H., and Hermann, D. A. (1973). 'Synthesis of 1-(4'-ethylphenoxy)-^{14}C(U)-6,7-epoxy-3,7-dimethyl-2-octene. A juvenile hormone analog', *J. Label. Comp.* **9**, 615–618.

Kamimura, H., Hammock, B. D., Yamamoto, I., and Casida, J. E. (1972). 'A potent juvenile hormone mimic, 1-(4'-ethylphenoxy)-6,7-epoxy3,7-dimethyl-2-octene, labeled with tritium in either the ethylphenyl- or geranyl-derived moiety', *J. Agr. Food Chem.* **20**, 439–442.

Kelly, T. J., and Fuchs, M. S. (1978). 'Precocene is not a specific antigonadotropic agent in adult female *Aedes aegypti*', *Physiol. Entomol.* **3**, 297–301.

Kensler, T. W., and Mueller, G. C. (1978). 'Inhibition of mitogenesis in bovine lymphocytes by juvenile hormones', *Life Sci.* **22**, 505–510.

Ker, R. F. (1977). 'Investigation of locust cuticle using the insecticide diflubenzuron', *J. Insect Physiol.* **23**, 39–48.

Ker, R. F. (1978). 'The effect of diflubenzuron on the growth of insect cuticle', *Pestic. Sci.* **9**, 259–265.

Kiguchi, K., and Riddiford, L. M. (1978). 'A role of juvenile hormone in pupal development of the tobacco hornworm, *Manduca sexta*', *J. Insect Physiol.* **24**, 673–680.

Klages, G., and Emmerich, H. (1979a). 'Juvenile hormone metabolism and juvenile hormone esterase titer in hemolymph and peripheral tissues of *Drosophila hydei*', *J. Comp. Physiol.* **132**, 319–325.

Klages, G., and Emmerich, H. (1979b). 'Juvenile hormone binding proteins in the haemolymph of third instar larvae of *Drosophila hydei*', *Insect Biochem.* **9**, 23–30.

deKort, C. A. D. (1981). 'Regulation of the juvenile hormone titre', *Ann. Rev. Entomol.* submitted.

deKort, C. A. D., Kramer, S. J., and Wieten, M. (1978). 'Regulation of juvenile hormone titres in the adult Colorado beetle: interaction with carboxylesterases and carrier proteins', in *Comparative Endocrinology* (Eds. P. J. Gaillard and H. H. Boer), pp. 507–510, Elsevier/North Holland Biomedical Press, Amsterdam.

deKort, C. A. D., Wieten, M., Kramer, S. J., and Goewie, E. (1977). 'Juvenile hormone degradation and carrier proteins in honey bee larvae', *Proc. Koninklijke Nederlandse Akademie van Wetenschappen, Amsterdam* **80C**, 297–301.

Kramer, K. J., and Childs, C. N. (1977). 'Interaction of juvenile hormone with carrier proteins and hydrolases from insect haemolymph', *Insect Biochem.* **7**, 397–403.

Kramer, K. J., Dunn, P. E., Peterson, R. C., and Law, J. H. (1976a). 'Interaction of juvenile hormone with binding proteins in insect hemolymph', in *The Juvenile Hormones* (Ed. L. I. Gilbert), pp. 327–341, Plenum Press, New York.

Kramer, K. J., Dunn, P. E., Peterson, R. C., Seballos, H. L., Sanburg, L. L., and Law, J. H. (1976b). 'Purification and characterization of the carrier protein for juvenile hormone from hemolymph of the tobacco hornworm, *Manduca sexta* Johannson (Lepidoptera: Sphingidae)', *J. Biol. Chem.* **251**, 4979–4985.

Kramer, K. J., Sanburg, L. L., Kézdy, F. J., and Law, J. H. (1974). 'The juvenile hormone binding protein in the hemolymph of *Manduca sexta* Johannson (Lepidoptera: Sphingidae)', *Proc. Natl. Acad. Sci. USA* **71**, 493–497.

Kramer, S. J. (1978). 'Regulation of the activity of JH-specific esterases in the Colorado potato beetle, *Leptinotarsa decemlineata*', *J. Insect Physiol.* **24**, 743–747.

Kramer, S. J., and deKort, C. A. D. (1976a). 'Age-dependent changes in juvenile hormone esterase and general carboxyesterase activity in the hemolymph of the Colorado potato beetle, *Leptinotarsa decemlineata*', *Mol. Cell. Endocrinol.* **4**, 43–53.

Kramer, S. J., and deKort, C. A. D. (1976b). 'Some properties of hemolymph esterases from *Leptinotarsa decemlineata* Say', *Life Sci.* **19**, 211–218.

Kramer, S. J. and deKort, C. A. D. (1978). 'Juvenile hormone carrier lipoproteins in the haemolymph of the Colorado potato beeetle *Leptinotarsa decemlineata*', *Insect Biochem.* **8**, 87–92.

Kramer, S. J., and Law, J. H. (1980). 'Synthesis and transport of juvenile hormones in insects', *Acc. Chem. Res.* **13**, 297–303.

Kramer, S. J., Wieten, M., and deKort, C. A. D. (1977). 'Metabolism of juvenile hormone in the Colorado potato beetle, *Leptinotarsa decemlineata*', *Insect Biochem.* **7**, 231–236.

Krishnakumaran, A., and Schneiderman, H. A. (1965). 'Prothoracicitropic activity of compounds that mimic juvenile hormone', *J. Insect Physiol.* **11**, 1517–1532.

Kryspin-Sorensen, I., Gelbic, I., and Slama, K. (1977). 'Juvenoid action on the total body metabolism in larvae of a noctuid moth', *J. Insect Physiol.* **23**, 531–53.

Landers, M. H., and Happ, G. M. (1979). 'The effects of the precocenes on vitellogenesis and other juvenile hormone related processes in *Drosophila melanogaster*', *Amer. Zool.* **19**, 917.

Laskowska-Bożek, H., and Zielińska, Z. M. (1978). 'Interference of a synthetic C_{18} juvenile hormone with mammalian cells *in vitro*. II. Effects on cell cycle', *Folia Histochem. Cytochem.* **16**, 225–232.

Law, J. H. (1978). 'Interaction of juvenile hormone with the hemolymph carrier protein', *Comp. Endocrinol.* (Eds. P. J. Gaillard and H. H. Boer), pp. 511–514, Elsevier/North Holland Biomedical Press, Amsterdam.

Lewis, R. J., and Tatken, R. L. (Eds.) (1979). *Registry of Toxic Effects of Chemical Substances,* 1978 Edition, p. 1279, U.S. Dept. Health, Education, and Welfare, NIOSH, Cincinnati.

Liechty, L., and Sedlak, B. J. (1978). 'Ultrastructure of precocene-induced effects on corpora allata of adult milkweed bug, *Oncopeltus fasciatus*', *Gen. Comp. Endocrinol.* **36**, 433–436.

Loher, W. (1960). 'The chemical acceleration of the maturation process and its hormonal control in the male of the desert locust', *Proc. Roy. Soc. London* **153**, 380–397.

Long, F. A., and Pritchard, J. G. (1956). 'Hydrolysis of substituted ethylene oxides in H_2O^{18} solutions', *J. Amer. Chem. Soc.* **78**, 2663–2667.

Lu, A. Y. H., and Miwa, G. T. (1980). 'Molecular properties and biological functions of microsomal epoxide hydrase', *Ann. Rev. Pharmacol. Toxicol.* **20**, 513–531.

Mane, S. D., and Rembold, H. (1977). 'Developmental kinetics of juvenile hormone inactivation in queen and worker castes of the honey bee, *Apis mellifera*', *Insect Biochem.* **7**, 463–467.

Mansagar, E. R., Still, G. G., and Frear, D. S. (1979). 'Fate of [^{14}C]diflubenzuron on cotton and in soil', *Pestic. Biochem. Physiol.* **12**, 172–182.

Marks, E. P., and Sowa, B. A. (1974). 'An *in vitro* model system for the production of insect cuticle', in *Mechanism of Pesticide Action* (Ed. G. K. Kohn), pp. 144–155, Amer. Chem. Soc., Washington, D.C.

Marx, J. L. (1977). 'Chitin synthesis inhibitors: new class of insecticides', *Science,* **197**, 1170–1172.

Masner, P., Bowers, W. S., Kälin, M., and Mühle, T. (1979). 'Effect of precocene II on the endocrine regulation of development and reproduction in the bug, *Oncopeltus fasciatus*', *Gen. Comp. Endocrinol.* **37**, 155–166.

Mayer, R. T., and Burke, M. D. (1976). 'Albumin and cytochrome P_{450} binding characteristics of juvenile hormone and its analogs', *Pestic. Biochem. Physiol.* **6**, 377–385.
Mayer, R. T., Meola, S. M., Coppage, D. L., and DeLoach, J. R. (1980). 'Utilization of imaginal tissues from pupae of the stable fly for the study of chitin synthesis and screening of chitin synthesis inhibitors', *J. Econ. Entomol.* **73**, 76–80.
McCaleb, D. C., and Kumaran, A. K. (1978). 'Effect of factors that influence metamorphosis on JH esterase activity in *Galleria mellonella*', *Amer. Zool.* **18**, 626.
McCaleb, D. C., Reddy, G., and Kumaran, A. K. (1980). 'Some properties of the hemolymph juvenile hormone esterases in *Galleria* larvae and *Tenebrio* pupae', *Insect Biochem.* **10**, 273–277.
McKague, A. B., and Pridmore, R. B. (1978). 'Toxicity of Altosid and Dimilin to juvenile rainbow trout and coho salmon', *Bull. Environ. Contam. Toxicol.* **20**, 167–169.
Menn, J. J., and Beroza, M. (Eds.) (1972). *Insect Juvenile Hormones Chemistry and Action*, 341 pp., Academic Press, New York.
Metcalf, R. L., Lu, P.-Y., and Bowlus, S. (1975). 'Degradation and environmental fate of 1-(2,6-difluorobenzoyl)-3-(4-chlorophenyl)urea', *J. Agr. Food Chem.* **23**, 359–364.
Metcalf, R. L., and Sanborn, J. R. (1975). 'Pesticides and environmental quality in Illinois', *Ill. Nat. Hist. Survey Bull.* **31**, 393.
Meyer, A. S., Hanzmann, E., Schneiderman, H. A., Gilbert, L. I., and Boyette, M. (1970). 'The isolation and identification of the two juvenile hormones from the cecropia silk moth', *Arch. Biochem. Biophys.* **137**, 190–213.
Miller, J. A., and Miller, E. C. (1977). 'Ultimate chemical carcinogens as reactive mutagenic electrophiles', in *Origins of Human Cancer, Book B* (Eds. H. H. Hiatt, J. D. Watson, and J. A. Winsten), pp. 605–627, Cold Spring Harbor Laboratory, Cold Spring Harbor, Mass.
Miller, R. W., Cecil, H. C., Carey, A. M., Corley, C., and Kiddy, C. A. (1979). 'Effects of feeding diflubenzuron to young male Holstein cattle', *Bull. Environ. Contam. Toxicol.* **23**, 482–486.
Miller, S., and Collins, J. M. (1975). 'The nature of the changes in the pattern of RNA synthesis by the juvenile hormone analogue Altosid', *J. Insect Physiol.* **21**, 1295–1303.
Mitlin, N., Wiygul, G., and Haynes, J. W. (1977). 'Inhibition of DNA synthesis in boll weevils (*Anthonomus grandis* Boheman) sterilized by Dimilin', *Pestic. Biochem. Physiol.* **7**, 559–563.
Mitsui, T., Riddiford, L. M., and Bellamy, G. (1979). 'Metabolism of juvenile hormone by the epidermis of the tobacco hornworm *Manduca sexta*', *Insect Biochem.* **9**, 637–643.
Moore, R. F., Leopold, R. A., and Taft, H. M. (1978). 'Boll weevils: mechanism of transfer of diflubenzuron from male to female', *J. Econ. Entomol.* **71**, 587–590.
Morello, A., and Agosin, M. (1979). 'Metabolism of juvenile hormone with isolated rat hepatocytes', *Biochem. Pharmacol.* **28**, 1533–1539.
Mulder, R., and Gijswijt, M. J. (1973). 'The laboratory evaluation of two promising new insecticides which interfere with cuticle deposition', *Pestic. Sci.* **4**, 737–745.
Müller, P. J., Masner, P., Kälin, M., and Bowers, W. S. (1979). '*In vitro* inactivation of corpora allata of the bug *Oncopeltus fasciatus* by precocene II', *Experientia* **35**, 704–705.
Mullin, C. A., and Hammock, B. D. (1980). 'A rapid radiometric assay for mammalian cytosolic epoxide hydrolase', *Anal. Biochem.* **106**, 476–485.
Mullin, C. A., and Wilkinson, C. F. (1980a). 'Purification of an epoxide hydratase from the midgut of the southern armyworm (*Spodoptera eridania*)', *Insect Biochem.* accepted.

Mullin, C. A., and Wilkinson, C. F. (1980b). Insect epoxide hydrolase: properties of a purified enzyme from the southern armyworm (*Spodoptera eridania*)', *Pestic. Biochem. Physiol.* submitted.

Mumby, S. M., and Hammock, B. D. (1979a). 'A partition assay for epoxide hydrases acting on insect juvenile hormone and an epoxide containing juvenoid', *Anal. Biochem.* **92**, 16–21.

Mumby, S. M., and Hammock, B. D. (1979b). 'Stability of epoxide containing juvenoids to dilute aqueous acid', *J. Agr. Food Chem.* **27**, 1223–1228.

Mumby, S. M., and Hammock, B. D. (1979c). 'Substrate selectivity and stereochemistry of enzymatic epoxide hydration in the soluble fraction of mouse liver', *Pestic. Biochem. Physiol.* **11**, 275–284.

Mumby, S. M., Hammock, B. D., Sparks, T. C., and Ota, K. (1979). 'Synthesis and bioassay of carbamate inhibitors of the juvenile hormone hydrolyzing esterases from the housefly, *Musca domestica*', *J. Agr. Food Chem.* **27** 763–765.

National Cancer Institute (1979). *Bioassay of p-Chloroaniline for Possible Carcinogenicity*, pp. vii–viii, NCI-CG-TR-189, U.S. Dept. Health, Education, and Welfare, NCI, NIH, Bethesda, Maryland.

Nelson, J. O., and Matsumura, F. (1973). 'Dieldrin (HEOD) metabolism in cockroaches and houseflies', *Arch. Environ. Contam. Toxicol.* **1**, 224–244.

Nijhout, H. F. (1975). 'Dynamics of juvenile hormone action in larvae of the tobacco hornworm, *Manduca sexta* (L)', *Biol. Bull.* **149**, 568–579.

Nishioka, T., Fujita, T., and Nakajima, M. (1979. 'Effect of chitin synthesis inhibitors on cuticle formation of the cultured integument of *Chilo suppressalis*', *J. Pestic. Sci.* **4**, 367–374.

Norris, J. M., Humiston, C. G., Schwetz, B. A., Kociba, R. J., Jersey, C. G., and Wade, C. E. (1974). 'The toxicological properties of 4-((4,8-dimethyldecyl)oxyl) 1,2-(methylenedioxy)-benzene an insect juvenile hormone mimic', *Toxicol. Appl. Pharmacol.* **29**, 129.

Nowock, J., and Gilbert, L. I. (1976). '*In vitro* analysis of factors regulating the juvenile hormone titer of insects', in *Invertebrate Tissue Culture* (Eds. E. Kurstak and K. Maramorosch), pp. 203–212, Academic Press, New York.

Nowock, J., Goodman, W., Bollenbacher, W. E., and Gilbert, L. I. (1975). 'Synthesis of juvenile hormone binding proteins by the fat body of *Manduca sexta*', *Gen. Comp. Endocrinol.* **27**, 230–239.

Nowock, J., Hammock, B. D., and Gilbert, L. I. (1976). 'The binding protein as a modulator of juvenile hormone stability and uptake', in *The Juvenile Hormones* (Ed. L. I. Gilbert), pp. 354–373, Plenum Press, New York.

Oberlander, H., and Leach, C. E. (1974). 'Inhibition of chitin synthesis in *Plodia interpunctella*', in *Proc. 1st Int. Working Conf. on Stored Products Entomol.*, p. 651, Savannah, Georgia, Oct. 7–11, 1974.

Oesch, F. (1973). 'Mammalian epoxide hydrases: inducible enzymes catalyzing the inactivation of carcinogenic and cytotoxic metabolites derived from aromatic and olefinic compounds', *Xenobiotica* **3**, 305–340.

Ohta, T., Kuhr, R. J., and Bowers, W. S. (1977). 'Radiosynthesis and metabolism of the insect antijuvenile hormone, precocene II', *J. Agr. Food Chem.* **25**, 478–481.

O'Neill, M. P., Holman, G. M., and Wright, J. E. (1977). 'β-Ecdysone levels in pharate pupae of the stable fly, *Stomoxys calcitrans* and interaction with the chitin inhibitor diflubenzuron', *J. Insect Physiol.* **23**, 1243–1244.

Ota, K., and Hammock, B. D. (1980). 'Differential properties of cytosolic and microsomal epoxide hydrolases in mammalian liver', *Science,* **207**, 1479–1481.

Pallos, F. M., Menn, J. J., Letchworth, P. E., and Miaullis, J. B. (1971). 'Synthetic mimics of insect juvenile hormone', *Nature (London)* **232**, 486.

Pawson, B. A., Scheidl, F., and Vane, F. (1972). 'Environmental stability of juvenile hormone mimicking agents', in *Insect Juvenile Hormones Chemistry and Action* (Eds. J. J. Menn and M. Beroza), pp. 191–214, Academic Press, New York.

Peter, M. G., Gunawan, S., and Emmerich, H. (1979a). 'Preparation of optically pure juvenile hormone I labeled in the ester methyl group with tritium at very high specific activity', *Experientia* **35**, 1141–1142.

Peter, M. G., Gunawan, S., Gellissen, G., and Emmerich, H. (1979b). 'Differences in hydrolysis and binding of homologous juvenile hormones in *Locusta migratoria* hemolymph', *Z. Naturf.* **34C**, 588–598.

Peterson, R. C., Reich, M. F., Dunn, P. E., Law, J. H., and Katzenellenbogen, J. A. (1977). 'Binding specificity of the juvenile hormone carrier protein from the hemolymph of the tobacco hornworm, *Manduca sexta* Johannson (Lepidoptera: Sphingidae)', *Biochemistry* **16**, 2305–2311.

Pimprikar, G. D., and Georghiou, G. P. (1979). 'Mechanisms of resistance to diflubenzuron in the house fly, *Musca domestica* (L.)', *Pestic. Biochem. Physiol.* **12**, 10–22.

Post, L. C., deJong, B. J., and Vincent, W. R. (1974). '1-(2,6-disubstituted benzoyl)-3-phenylurea insecticides: inhibitors of chitin synthesis', *Pestic. Biochem. Physiol.* **4**, 473–483.

Post, L. C., and Mulder, R. (1974). 'Insecticidal properties and mode of action of 1-(2,6-dihalogenbenzoyl)-3-phenylureas', in *Mechanism of Pesticide Action* (Ed. G. K. Kohn), pp. 136–143, Amer. Chem. Soc., Washington, D.C.

Post, L. C., and Vincent, W. R. (1973). 'A new insecticide inhibits chitin synthesis', *Naturwissenschaften* **60**, 431–432.

Postlethwait, J. H., and Gray, P. (1975). 'Regulation of acid phosphatase activity in the ovary of *Drosophila melanogaster*', *Develop. Biol.* **47**, 196–205.

Postlethwait, J. H., Handler, A. M., and Gray, P. W. (1976). 'A genetic approach to the study of juvenile hormone control vitellogenesis in *Drosophila melanogaster*', in *The Juvenile Hormones* (Ed. L. I. Gilbert), pp. 449–469, Plenum Press, New York.

Pratt, G. E. (1975). 'Inhibition of juvenile hormone carboxyesterase of locust haemolymph by organophosphates *in vitro*', *Insect Biochem.* **5**, 595–607.

Pritchard, J. G., and Long, F. A. (1956). 'Kinetics and mechanism of the acid-catalyzed hydrolysis of substituted ethylene oxides', *J. Amer. Chem. Soc.* **78**, 2667–2670.

Quistad, G. B., Schooley, D. A., Staiger, L. E., Bergot, B. J., Sleight, B. H., and Macek, K. J. (1976a). 'Environmental degradation of the insect growth regulator methoprene. IX. Metabolism by bluegill fish', *Pestic. Biochem. Physiol.* **6**, 523–529.

Quistad, G. B., Staiger, L. E., and Schooley, D. A. (1974a). 'Environmental degradation of the insect growth regulator methoprene (Isopropyl ($2E$, $4E$)-11-methoxy-3,7,11-trimethyl-2,4-dodecadienoate). I. Metabolism by alfalfa and rice', *J. Agr. Food Chem.* **22**, 582–589.

Quistad, G. B., Staiger, L. E., and Schooley, D. A. (1974b). 'Cholesterol and bile acids via acetate from the insect juvenile hormone analog methoprene', *Life Sci.* **15**, 1797–1804.

Quistad, G. B., Staiger, L. E., and Schooley, D. A. (1975a). 'Environmental degradation of the insect growth regulator methoprene (isopropyl ($2E,4E$)-11-methoxy-3,7,11-trimethyl-2,4-dodecadienoate). III. Photodecomposition', *J. Agr. Food Chem.* **23**, 299–303.

Quistad, G. B., Staiger, L. E., Bergot, B. J., and Schooley, D. A. (1975b). 'Environmental degradation of the insect growth regulator methoprene. VII. Bovine metabolism to cholesterol and related natural products', *J. Agr. Food Chem.* **23**, 743–749.

Quistad, G. B., Staiger, L. E., and Schooley, D. A. (1975c). 'Environmental degradation of the insect growth regulator methoprene. VIII. Bovine metabolism to natural products in milk and blood', *J. Agr. Food Chem.* **23**, 750–753.

Quistad, G. B., Staiger, L. E., and Schooley, D. A. (1975d). 'Environmental degradation of the insect growth regulator methoprene. V. Metabolism by houseflies and mosquitoes', *Pestic. Biochem. Physiol.* **5**, 233–241.

Quistad, G. B., Staiger, L. E., and Schooley, D. A. (1975e). 'Comparative metabolism of the insect juvenile hormone analog methoprene', *ACS Abstracts Chicago Meeting,* Fall, PEST 59.

Quistad, G. B., Staiger, L. E., and Schooley, D. A. (1976b). 'Environmental degradation of the insect growth regulator methoprene. X. Chicken metabolism', *J. Agr. Food Chem.* **24**, 644–648.

Reddy, G., Hwang-Hsu, K., and Kumaran, A. K. (1979). 'Factors influencing juvenile hormone esterase activity in the wax moth, *Galleria mellonella*', *J. Insect Physiol.* **25**, 65–71.

Reddy, G., and Krishnakumaran, A. (1974). 'Oxidase activity in waxmoth larvae during metamorphosis: effect of juvenile hormone and injury', *Insect Biochem.* **4**, 355–362.

Reddy, G., and Kumaran, A. K. (1980). 'Changes in juvenile hormone esterase activity during postembryonic development in *Tenebrio molitor*', *Physiol. Zool.* submitted.

Retnakaran, A., and Joly, P. (1976). 'Neurosecretory control of juvenile hormone inactivation in *Locusta migratoria* (L)', in *Actualitiés sur les Hormones D'Invertebrés*, pp. 317–323, Colloq. Int. CRNS No. 251, Paris.

Riddiford, L. M., and Truman, J. W. (1978). 'Biochemistry of insect hormones and insect growth regulators', in *Biochemistry in Insects* (Ed. M. Rockstein), pp. 307–357, Academic Press, New York.

Röller, H., Dahm, K. H., Trost, B. M., and Sweeley, C. C. (1967). 'Structure of juvenile hormone', *Agnew. Chem. Int. Ed., Engl.* **6**, 179–180.

Rowlands, D. G. (1976). 'Uptake and metabolism by stored wheat grains of an insect juvenile hormone and two insect hormone mimics', *J. Stored Prod. Res.* **12**, 35–41.

Rudnicka, M., Sehnal, F., Jarolim, V., and Kochman, M. (1979). 'Hydrolysis and binding of the juvenile hormone in the haemolymph of *Galleria mellonella*', *Insect Biochem.* **9**, 569–575.

Ruzo, L. O., Zabik, M. J., and Schuetz, R. D. (1974). 'Photochemistry of bioactive compounds. 1-(4-chlorophenyl)-3-(2,6-dihalobenzoyl)ureas', *J. Agr. Food Chem.* **22**, 1106–1108.

Salama, H. S., Motagally, Z. A., and Skatulla, U. (1976). 'On the mode of action of Dimilin as a moulting inhibitor in some lepidopterous insects', *Z. Ang. Entomol.* **80**, 396–407.

Sams, G. R., Cocchiaro, G. F., and Bell, W. J. (1978). 'Metabolism of juvenile hormone in cultures of ovaries and fat body in the cockroach *Periplaneta americana*', *In vitro* **14**, 956–960.

Sanburg, L. L., Kramer, K. J., Kézdy, F. J., and Law, J. H. (1975a). 'Juvenile hormone-specific esterases in the haemolymph of the tobacco hornworm, *Manduca sexta*', *J. Insect. Physiol.* **21**, 873–887.

Sanburg, L. L., Kramer, K. J., Kézdy, F. J., Law, J. H., and Oberlander, H. (1975b). 'Role of juvenile hormone esterases and carrier proteins in insect development', *Nature (London)* **253**, 266–267.

Schaefer, C. H., and Dupras, E. F. (1973). 'Insect developmental inhibitors. 4. Persistence of ZR-515 in water', *J. Econ. Entomol.* **66**, 923–925.

Schaefer, C. H., and Dupras, E. F., Jr. (1976). 'Factors affecting the stability of Dimilin in water and the persistence of Dimilin in field waters', *J. Agr. Food Chem.* **24**, 733–739.

Schaefer, C. H., and Dupras, E. F., Jr. (1977). 'Residues of diflubenzuron [1-(4-chlorophenyl)-3-(2,6-difluorobenzoyl)urea] in pasture soil, vegetation, and water following aerial applications', *J. Agr. Food Chem.* **25**, 1026–1030.

Schaefer, C. H., and Dupras, E. F., Jr. (1979). 'Factors affecting the stability of SIR-8514 (2-chloro(N-[[[4-(trifluoromethoxy)phenyl]amino]-carbonyl]benzamide) under laboratory and field conditions', *J. Agr. Food Chem.* **27**, 1031–1034.

Scheller, K., Karlson, P., and Bodenstein, D. (1978). 'Effects of ecdysterone and the juvenile hormone analogue methoprene on protein, RNA, and DNA synthesis in wing discs of *Calliphora vicina*', *Z. Naturf.* **33C**, 253–260.

Schmialek, P., Borowski, M., Geyer, A., Miosga, V., Nündel, M., Rosenberg, E., and Zapf, B. (1973). 'Epidermis of the pupae of *Tenebrio molitor* L. as target organ for juvenile hormone analogue 10,11-epoxy-6,7-*trans*-2,3-*trans*-farnesylpropenylether', *Z. Naturf.* **28C**, 173–177.

Schmialek, P., Geyer, A., Miosga, V., Nündel, M., and Zapf, B. (1975). 'Juvenilhormonbindende Substanzen mit allosterischen Eigenschaften in den Ovarien von *Tenebrio molitor* L.', *Z. Naturf.* **30C**, 730–733.

Schmialek, P., Geyer, A., Miosga, V., Nündel, M., and Zapf, B. (1976). 'Synthesis of 10,11-epoxy-6,7-*trans*-2,3-*trans*-farnesyl-[2,3-^{3}H]propenylether, a juvenile hormone efficient compound with very high specific radioactivity', *Insect Biochem.* **6**, 19–20.

Schneider, F., and Aubert, J. (Eds.) (1971). 'Swiss symposium on the juvenile hormones', *Mitt. Schweiz. Ent. Ges.* **44**, 1–208.

Schooley, D. A. (1977). 'Analysis of the naturally occurring juvenile hormones—their isolation, identification, and titer determination at physiological levels', in *Analytical Biochemistry of Insects* (Ed. R. B. Turner), pp. 241–287, Elsevier Science Publications, New York.

Schooley, D. A., and Bergot, B. J. (1979). 'Biochemical studies on juvenile hormone antagonists', Paper 70, Pesticide Chemistry Division, 178th National American Society National Meeting, Sept. 9–14, Washington, D.C.

Schooley, D. A., Bergot, B. J., Dunham, L. L., and Siddall, J. B. (1975a). 'Environmental degradation of the insect growth regulator methoprene (isopropyl (2E,4E)-11-methoxy-3,7,11-trimethyl-2,4-dodecadienoate). II. Metabolism by aquatic microorganisms', *J. Agr. Food Chem.* **23**, 293–298.

Schooley, D. A., Bergot, B. J., Goodman, W., and Gilbert, L. I. (1978). 'Synthesis of both optical isomers of insect juvenile hormone III and their affinity for the juvenile hormone-specific binding protein of *Manduca sexta*', *Biochem. Biophys. Res. Commun.* **81**, 743–749.

Schooley, D. A., Creswell, K. M., Staiger, L. E., and Quistad, G. B. (1975b). 'Environmental degradation of the insect growth regulator isopropyl (2E,4E)-11-methoxy-3,7,11-trimethyl-2,4-dodecadienoate (methoprene). IV. Soil metabolism', *J. Agr. Food Chem.* **23**, 369–373.

Schooley, D. A., and Quistad, G. B. (1979). 'Metabolism of insect growth regulators in aquatic organisms', in *Pesticide and Xenobiotic Metabolism in Aquatic Organisms* (Eds. M. A. Q. Khan, J. J. Lech, and J. J. Menn), pp. 161–176, Amer. Chem. Soc., Washington, D.C.

Schooneveld, H. (1979a). 'Precocene-induced collapse and resorption of *corpora allata* in nymphs of *Locusta migratoria*', *Experientia* **35**, 363–364.

Schooneveld, H. (1979b). 'Precocene-induced necrosis and haemocyte-mediated breakdown of corpora allata in nymphs of the locust (*Locusta migratoria*)', *Cell Tissue Res.* **203**, 25–33.

Schooneveld, H., Kramer, S. J., Privee, H., and van Huis, A. (1979). 'Evidence of controlled corpus allatum activity in the adult Colorado potato beetle', *J. Insect Physiol.* **25**, 449–453.

Schwarz, M., Miller, R. W., Wright, J. E., Chamberlain, W. F., and Hopkins, D. E. (1974). 'Compounds related to juvenile hormone. Exceptional activity of arylterpenoid compounds in four species of flies', *J. Econ. Entomol.* **67**, 598–601.

Sehnal, F. (1976). 'Action of juvenoids on different groups of insects', in *The Juvenile Hormones* (Ed. L. I. Gilbert), pp. 301–321, Plenum Press, New York.

Seubert, W., and Fass, E. (1964). 'Untersuchungen über den bakteriellen abbau von isoprenoiden V. Der mechanismus des isoprenoidabbaues', *Biochem. Z.* **341**, 35–44.

Seuferer, S. L., Braymer, H. D., and Dunn, J. J. (1979). 'Metabolism of diflubenzuron by soil microorganisms and mutagenicity of the metabolites', *Pestic. Biochem. Physiol.* **10**, 174–180.

Shirk, P. D., Dahm, K. H., and Röller, H. (1976). 'The accessory sex glands as the repository for juvenile hormone in male Cecropia moths', *Z. Naturf.* **31C**, 199–200.

Siddall, J. B. (1976). 'Insect growth regulators and insect control: a critical appraisal', *Environ. Hlth. Perspec.* **14**, 119–126.

Siddall, J. B., and Slade, M. (1971). 'Absence of acute oral toxicity of *Hyalophora cecropia* juvenile hormone in mice', *Nature, New Biol.* **229**, 158.

Singh, S. (1973). 'Metabolism and environmental degradation of the juvenoid 1-(4'-ethylphenoxy)-3,7-dimethyl-6,7-epoxy-2-octene', Ph.D. Dissertation, University of California, Berkeley, 227 pp.

Slade, M., Brooks, G. T., Hetnarski, H. K., and Wilkinson, C. F. (1975). 'Inhibition of the enzymatic hydration of the epoxide HEOM in insects', *Pestic. Biochem. Physiol.* **5**, 35–46.

Slade, M., Hetnarski, H. K., and Wilkinson, C. F. (1976). 'Epoxide hydrase activity and its relationship to development in the southern armyworm, *Prodenia eridania*', *J. Insect Physiol.* **22**, 619–622.

Slade, M., and Wilkinson, C. F. (1973). 'Juvenile hormone analogs: a possible case of mistaken identity?', *Science* **181**, 672–674.

Slade, M., and Wilkinson, C. F. (1974). 'Degradation and conjugation of Cecropic juvenile hormone by the southern armyworm (*Prodenia eridania*)', *Comp. Biochem. Physiol.* **49B**, 99–103.

Slade, M., and Zibitt, C. H. (1971). 'Metabolism of cecropia juvenile hormone in lepidopterans', in *Chemical Releasers in Insects* 3 (Ed. A. S. Tahori), pp. 45–58, Proc. 2nd Int. IUPAC Cong. Pestic. Chem., Gordon & Breach Science Publishers, New York.

Slade, M., and Zibitt, C. H. (1972). 'Metabolism of cecropia juvenile hormone in insects and mammals', in *Insect Juvenile Hormone: Chemistry and Action* (Eds. J. J. Menn and M. Beroza), pp. 155–176, Academic Press, New York.

Sláma, K., and Hodkova, M. (1975). 'Insect hormones and bioanalogues: their effect on respiratory metabolism in *Dermestes vulpinus* F. (*Coleoptera*)', *Biol. Bull, Woods Hole*, **148**, 320–332.

Sláma, K., and Jarolím, V. (1980). 'Fluorimetric method for the determination of juvenoid esterase activity in insects', *Insect Biochem.* **10**, 73–80.

Sláma, K., Kahovcová, J., and Romaňuk, M. (1978). 'Action of some aromatic juvenogen esters on insects', *Pestic Biochem. Physiol.* **9**, 313–321.

Sláma, K., Romaňuk, M., and Sorm, F. (1974). *Insect Hormones and Bioanalogues*, 477 pp., Springer-Verlag, New York.

Smalley, H. E., Wright, J. E., Crookshank, H. R., and Younger, R. L. (1974). 'Toxicity studies of an insect juvenile hormone analogue in domestic animals', *Toxicol. Appl. Pharmacol.* **29**, 129.

Soderlund, D. M., Messeguer, A., and Bowers, W. S. (1980). 'Precocene II metabolism in insects: synthesis of potential metabolites and identification of initial *in vitro* biotransformation products', *J. Agr. Food Chem.*, **28**, 724–731.

Solomon, K. R., and Metcalf, R. L. (1974). 'The effect of piperonyl butoxide and triorthocresyl phosphate on the activity and metabolism of Altosid (isopropyl 11-methoxy-3,7,11-trimethyldodeca-2,4-dienoate) in *Tenebrio molitor* L. and *Oncopeltus fasciatus* (Dallas)', *Pestic. Biochem. Physiol.* **4**, 127–134.

Solomon, K. R., and Walker, W. F. (1974). 'Juvenile hormone synergists: a possible case of hasty conclusion?', *Science* **185**, 461–462.

Sowa, B. A., and Marks, E. P. (1975). 'An *in vitro* system for the quantitative measurement of chitin synthesis in the cockroach: inhibition by TH 6040 and polyoxin D', *Insect Biochem.* **5**, 855–859.

Sparks, T. C., and Hammock, B. D. (1979a). 'A comparison of the induced and naturally occurring juvenile hormone esterases from last instar larvae of *Trichoplusia ni*', *Insect Biochem.* **9**, 411–421.

Sparks, T. C., and Hammock, B. D. (1979b). 'Induction and regulation of juvenile hormone esterases during the last larval instar of the cabbage looper, *Trichoplusia ni*', *J. Insect Physiol.* **25**, 551–560.

Sparks, T. C., and Hammock, B. D. (1980a). 'Insect growth regulators: resistance and the future', in *Pest Resistance to Pesticides: Challenges and Prospects* (Eds. G. P. Georghiou and T. Saito), Plenum Press, New York, in press.

Sparks, T. C., and Hammock, B. D. (1980b). 'Comparative inhibition of the juvenile hormone esterases from *Trichoplusia ni, Musca domestica,* and *Tenebrio molitor*', *Pestic. Biochem. Physiol.* submitted.

Sparks, T. C., Willis, W. S., Shorey, H. H., and Hammock, B. D. (1979a). 'Hemolymph juvenile hormone esterase activity in synchronous last instar larvae of the cabbage looper, *Trichoplusia ni*', *J. Insect Physiol.* **25**, 125–132.

Sparks, T. C., Wing, K. D., and Hammock, B. D. (1979b). 'Effects of the antihormone-hormone mimic ETB on the induction of insect juvenile hormone esterase in *Trichoplusia ni*', *Life Sci.* **25**, 445–450.

Staal, G. B. (1975). 'Insect growth regulators with juvenile hormone activity', *Ann. Rev. Entomol.* **20**, 417–460.

Staal, G. B. (1977). 'Insect control with insect growth regulators based on insect hormones', *Pontificicae Academiae Scientiarum Scripta Varia, Rome* **41**, 353–383.

Staiger, L. E., Quistad, G. B., and Schooley, D. A. (1980). Unpublished data.

Still, G. G., and Leopold, R. A. (1975). 'The elimination of 1-(4-chlorophenyl)-3-(2,6-difluorobenzoyl)-urea by the cotton boll weevil', *ACS Abstracts Chicago Meeting* Fall, PEST 5.

Still, G. G., and Leopold, R. A. (1978). 'Elimination of *N*-(4-chlorophenyl)amino-carbonyl-2,6-difluorobenzamide by the boll weevil', *Pestic. Biochem. Physiol.* **9**, 304–312.

Terriere, L. C. (1980). 'Enzyme induction gene amplification and insect resistance to insecticides', in *Pest Resistance to Pesticides: Challenges and Prospects* (Eds. G. P. Georghiou and T. Saito), Plenum Press, New York, in press.

Terriere, L. C., and Yu, S. J. (1973). 'Insect juvenile hormones: induction of detoxifying enzymes in the housefly and detoxication by housefly enzymes', *Pestic. Biochem. Physiol.* **3**, 96–107.

Terriere, L. C., and Yu, S. J. (1974). 'The induction of detoxifying enzymes in insects', *J. Agr. Food Chem.* **22**, 366–373.

Terriere, L. C., and Yu, S. J. (1977). 'Juvenile hormone analogs: *in vitro* metabolism in relation to biological activity in blow flies and flesh flies', *Pestic. Biochem. Physiol.* **7**, 161–168.

Tobe, S. S., and Stay, B. (1979). 'Modulation of juvenile hormone synthesis by an analogue in the cockroach', *Nature (London)* **281**, 481–482.

Tokiwa, T., Uda, F., Uemura, J., and Nakazawa, M. (1975). 'Study on isopropyl 11-methoxy-3,7,11-trimethyldodeca-2,4-dienoate (IGR). I. The absorption and excretion of [5-^{14}C]-isopropyl 11-methoxy-3,7,11-trimethyldodeca-2,4-dienoate (^{14}C-IGR) in rats', *Oyo Yakuri* **10**, 471–474.

Trautmann, K. H. (1972). '*In vitro* study of the carrier proteins of ^{3}H labelled juvenile hormone active compounds in the haemolymph of *Tenebrio molitor* L. larvae', *Z. Naturf.* **27B**, 263–273.

Trautmann, K. H., Schuler, A., Suchý, M., and Wipf, H.-K. (1974). 'A method for the qualitative and quantitative determination of three natural insect juvenile hormones', *Z. Naturf.* **29C**, 161–168.

Tungikar, V. B., Sharma, R. N., and Das, K. G. (1978). 'Metabolism of hydroprene by red cotton bug *Dysdercus koenigii*', *Indian J. Exp. Biol.* **16**, 1264–1266.

Unnithan, G. C., and Nair, K. K. (1979). 'The influence of corpus allatum activity on the susceptibility of *Oncopeltus fasciatus* to precocene', *Ann. Entomol. Soc. Amer.* **72**, 38–40.

Unsworth, B., Hennen, S., and Krishnakumaran, A. (1974). 'Teratogenic evaluation of terpenoid derivatives', *Life Sci.* **15**, 1649–1655.

van Eck, W. H. (1979). 'Mode of action of two benzoylphenyl ureas as inhibitors of chitin synthesis in insects', *Insect Biochem.* **9**, 295–300.

Verloop, A., and Ferrell, C. D. (1977). 'Benzoylphenylureas—a new group of larvicides interfering with chitin deposition', in *Pesticide Chemistry in the 20th Century* (Ed. J. R. Plimmer), pp. 237–275, *ACS Symp. Ser.* **37**, Washington, D.C.

Vijverberg, A. J., and Ginsel, L. A. (1976). 'Juvenile hormone and DNA synthesis in imaginal disks of *Calliphora erythrocephala:* results of a new incubation technique', *J. Insect Physiol.* **22**, 181–186.

Vince, R. K., and Gilbert, L. I. (1977). 'Juvenile hormone esterase activity in precisely timed last instar larvae and pharate pupae of *Manduca sexta*', *Insect Biochem.* **7**, 115–120.

Walker, C. H., and El Zorgani, G. (1974). 'The comparative metabolism and excretion of HCE, a biodegradable analogue of dieldrin, by vertebrate species', *Arch. Environ. Contam. Toxicol.* **2**, 97–116.

Weirich, G. F., and Culver, M. G. (1979). 'S-adenosylmethionine: juvenile hormone acid methyltransferase in male accessory reproductive glands of *Hyalophora cecropia* (L)', *Arch. Biochem. Biophys.* **198**, 175–181.

Weirich, G., and Wren, J. (1973). 'The substrate specificity of juvenile hormone esterase from *Manduca sexta* haemolymph', *Life Sci.* **13**, 213–226.

Weirich, G., and Wren, J. (1976). 'Juvenile-hormone esterase in insect development: a comparative study', *Physiol. Zool.* **49**, 341–350.

Weirich, G., Wren, J., and Siddall, J. B. (1973). 'Developmental changes of the juvenile hormone esterase activity in haemolymph of the tobacco hornworm, *Manduca sexta*', *Insect Biochem.* **3**, 397–407.

Wellinga, K., Mulder, R., and van Daalen, J. J. (1973). 'Synthesis and laboratory evaluation of 1-(2,6-disubstituted benzoyl)-3-phenylureas, a new class of insecticides. I. 1-(2,6-dichlorobenzoyl)-3-phenyl-ureas', *J. Agr. Food Chem.* **21**, 348–354.

White, A. F. (1972). 'Metabolism of the juvenile hormone analogue methyl farnesoate-10,11-epoxide in two insect species', *Life Sci.* **11**, 201–210.

Whitmore, D., Jr., Gilbert, L. I., and Ittucheriah, P. I. (1974). 'The origin of hemolymph carboxylesterases "induced" by the insect juvenile hormone', *Mol. Cell. Endocrinol.* **1**, 37–54.

Whitmore, D., Jr., Whitmore, E., and Gilbert, L. I. (1972). 'Juvenile hormone induction of esterases: a mechanism for the regulation of juvenile hormone titer', *Proc. Natl. Acad. Sci. USA* **69**, 1592–1595.

Whitmore, E., and Gilbert, L. I. (1972). 'Haemolymph lipoprotein transport of juvenile hormone', *J. Insect Physiol.* **18**, 1153–1167.

Whitmore, E., and Gilbert, L. I. (1974). 'Haemolymph proteins and lipoproteins in Lepidoptera: a comparative electrophoretic study', *Comp. Biochem. Physiol.* **47B**, 63–78.

Wigglesworth, V. B. (1936). 'The function of the corpus allatum in the growth and reproduction of *Rhodnius prolixus*', *Quart. J. Microscop. Sci.* **79**, 91–121.

Wigglesworth, V. B. (1970). *The Physiology of Insect Metamorphosis*, 163 pp., Cambridge Univ. Press, New York.

Wilkinson, C. F. (1980). 'Role of mixed-function oxidases in insecticide resistance', in *Pest Resistance to Pesticides: Challenges and Prospects* (Eds. G. P. Georghiou and T. Saito), Plenum Press, New York, in press.

Williams, C. M. (1956). 'The juvenile hormone of insects', *Nature (London)* **178**, 212–213.

Williams, C. M. (1967). 'Third-generation pesticides', *Sci. Am.* **217**, 13–17.

Wilson, T. G., and Gilbert, L. I. (1978). 'Metabolism of juvenile hormone I in *Drosophila melanogaster*', *Comp. Biochem. Physiol.* **60A**, 85–89.

Wing, K. D., Sparks, T. C., Lovell, V. M., and Hammock, B. D. (1980). 'The compartmentalization and interrelationship of proteins influencing juvenile hormone metabolism in *Trichoplusia ni*', *J. Insect Biochem.* submitted.

Wright, J. E. (1976). 'Environmental and toxicological aspects of insect growth regulators', *Environ. Health. Perspect.* **14**, 127–132.

Wright, J. E., and Smalley, H. E. (1977). 'Biological activity of insect juvenile hormone analogues against the stable fly and toxicity studies in domestic animals', *Arch. Environ. Contam. Toxicol.* **5**, 191–197.

Wright, J. E., and Spates, G. E. (1975). 'Penetration and persistence of an insect growth regulator in the pupa of the stable fly, *Stomoxys calcitrans*', *J. Insect Physiol.* **21**, 801–805.

Yamamoto, H. Y., and Higashi, R. M. (1978). 'Violaxanthin de-epoxidase, lipid composition and substrate specificity', *Arch. Biochem. Biophys.* **190**, 514–522.

Yawetz, A., and Agosin, M. (1979). 'Epoxide hydrase in *Trypanosoma cruzi* epimastigotes', *Biochim. Biophys. Acta* **585**, 210–219.

Yu, S. J., and Terriere, L. C. (1975a). 'Microsomal metabolism of juvenile hormone analogs in the house fly, *Musca domestica* L', *Pestic. Biochem. Physiol.* **5**, 418–430.

Yu, S. J., and Terriere, L. C. (1975b). 'Activities of hormone metabolizing enzymes in house flies treated with some substituted urea growth regulators', *Life Sci.* **17**, 619–626.

Yu, S. J., and Terriere, L. C. (1977a). 'Metabolism of [^{14}C]hydroprene (ethyl 3,7,11-trimethyl-2,4-dodecadienoate) by microsomal oxidases and esterases from three species of Diptera', *J. Agr. Food Chem.* **25**, 1076–1080.

Yu, S. J., and Terriere, L. C. (1977b). 'Esterase and oxidase activity of house fly microsomes against juvenile hormone analogues containing branched chain ester groups and its induction by phenobarbital', *J. Agr. Food Chem.* **25**, 1333–1336.

Yu, S. J., and Terriere, L. C. (1977c). 'Ecdysone metabolism by soluble enzymes from three species of Diptera and its inhibition by the insect growth regulator TH6040', *Pestic. Biochem. Physiol.* **7**, 48–55.

Yu, S. J., and Terriere, L. C. (1978a). 'Metabolism of juvenile hormone I by microsomal oxidase, esterase, and epoxide hydrase of *Musca domestica* and some comparisons with *Phormia regina* and *Sarcophaga bullata*', *Pestic. Biochem. Physiol.* **9**, 237–246.

Yu, S. J., and Terriere, L. C. (1978b). 'Juvenile hormone epoxide hydrase in house flies, flesh flies and blow flies', *Insect Biochem.* **8**, 349–352.

Zielińska, Z. M., Laskowska-Bożek, H., and Jasterboff, P. (1978). 'Interference of a synthetic C_{18} juvenile hormone with mammalian cells *in vitro*. I. Effects on growth and morphology', *Folia Histochem. Cytochem.* **16**, 205–224.

Zurflueh, R. C. (1976). 'Phenylethers as insect growth regulators: laboratory and field experiments', in *The Juvenile Hormones* (Ed. L. I. Gilbert), pp. 61–74, Plenum Press, New York.

CHAPTER 2

Experimental approaches to studying the fate of pesticides in soil

Johann A. Guth

INTRODUCTION	85
TRANSFORMATION STUDIES IN SOIL	86
Laboratory Model Systems	86
Field Experiments	98
CHEMICAL VERSUS MICROBIAL TRANSFORMATION	101
PHOTOCHEMICAL TRANSFORMATION STUDIES	103
MICROBIAL STUDIES	105
PRESENT PROBLEM AREAS	106
Microbiological Activity of Test Soils	106
The Standard Soils Concept	107
Duration of *in vitro* Experiments	108
Influence of Other Chemicals on Pesticide Transformation	109
CONCLUSIONS	109
REFERENCES	110

INTRODUCTION

Large amounts of pesticides are either applied directly to the soil or reach it indirectly after application to a target organism. Therefore, knowledge of the fate of pesticides in the soil, an important part of our environment, is necessary both for agricultural practice and environmental safety. The transformation of pesticides in soil can be caused by microbiological, chemical, and photochemical processes.

The microbial breakdown of pesticides must be considered the most important cause of degradation in soil. Pesticides can be microbially degraded either by being used as a substrate for growth and energy or by co-metabolism without supplying energy for the microorganisms.

Chemical transformations in soil, such as hydrolysis and oxidation, are widespread phenomena, but reduction or isomerization have also been observed. The chemical reactions may be catalysed by clay surfaces, metal oxides, metal ions, and organic surfaces. Also extracellular enzymes and free radicals in soils may play a role in the degradation of many chemicals.

Although considerable data on the photodecomposition of pesticides in water and air are available, information on photochemical degradation in soils is rather limited. There has always been discussion whether photolysis of pesticides in soil has much practical importance. However, published information exists on the significant influence of light on the degradation of agrochemicals on soil surfaces.

Numerous experimental approaches have been described in the literature to study the fate of pesticides in soil. The methods available for measuring photochemical breakdown and transformation in soil and for distinguishing between chemical and microbial degradation will be reviewed and their advantages and limitations will be discussed. It is not the intention of this chapter to deal with the analytical methodology used for pesticide transformation experiments. The final part of this review will consider a few points of special importance for transformation studies, such as biological activity of soils, the use of standard soils, duration of experiments, and the influence of other chemicals on pesticide transformation.

TRANSFORMATION STUDIES IN SOIL

Laboratory Model Systems

A wide variety of experimental approaches have been used to study pesticide transformations in laboratory soil. They range from very simple systems, such as polyethylene bags, closed plastic bottles, glass jars closed with plastic screw caps, bottles sealed with paraffin, open soil pots, etc., to models in which the transformation processes can be followed under controlled and standardized conditions. The latter model systems can be classified in four categories:

(1) soil perfusion systems;
(2) soil biometers;
(3) gas flow-through systems;
(4) integrated systems.

The first three approaches can be used for aerobic and anaerobic soil studies and have been reviewed by Kaufman (1977a). Other reviews on this subject were published by Hill and Arnold (1978) and Newby *et al.* (1979).

Soil perfusion systems were originally developed to study soil microbiological processes, such as nitrification, denitrification, and microbial utilization of water-insoluble substrates such as sulphur or coal (Audus, 1946; Chase, 1946; Lees and Quastel, 1946; Lees, 1947; Temple, 1951; Sperber and Sykes, 1964). The system shown in Figure 2.1 was developed by Kaufman (1966) and successfully applied to examine microbial degradation of pesticides. Other examples of perfusion apparatus described by Kimura and Yamaguchi (1978) and by Fung and Uren (1977) are shown in Figures 2.2 and 2.3. In such investigations, soils containing the pesticides were perfused with distilled water or untreated soils

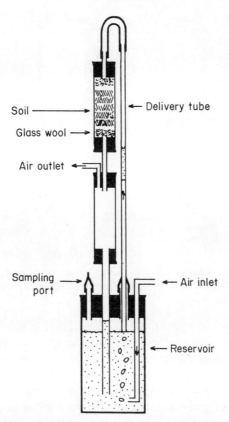

Figure 2.1 Diagram of a soil perfusion unit (Kaufman, 1966, reproduced by permission of the Weed Science Society of America)

were washed with aqueous solutions of pesticides. Degradation can be measured by either determining the rate of disappearance of the pesticide or by the formation of breakdown products.

Soil perfusion units can be useful in assessing the relative biodegradability of pesticides (Kaufman, 1977b), the factors affecting microbial degradation, the microbial degradation of pesticide combinations (Kaufman *et al.*, 1970; Kaufman, 1977b), and soil enrichment with subsequent isolation of effective microorganisms.

Perfusion systems have a number of advantages but also certain limitations which are both summarized in Table 2.1. The predominance of bacteria and the unnatural conditions in the system are important disadvantages since the results obtained are not representative of what actually happens in the soil environment.

Biometer-type systems have been frequently used to study degradation and metabolism of ^{14}C-labelled pesticides in soils. The first 'soil biometer' flask was developed by Bartha and Pramer (1965) and is shown in Figure 2.4. Essentially,

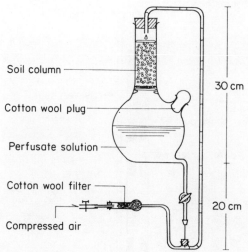

Figure 2.2 Perfusion apparatus according to Kimura and Yamaguchi (1978). (Reproduced by permission of Pergamon Press)

the system contains the treated soil, a CO_2 trapping solution, a method for adding additional trapping solution, and a technique to prevent contamination by ambient CO_2. Table 2.2 summarizes the advantages and disadvantages of the system.

The major limitation certainly is the trend to anaerobicity. As the microbial population of the soil consumes oxygen and produces carbon dioxide which is absorbed by the trapping solution, without air exchange the nitrogen concentration in the gas phase of the system will steadily increase and finally might reach anaerobic conditions. This problem can be avoided by simply leaving the syringe needle open so that a negative pressure builds up in the flask which automatically will draw fresh air into the system, but this procedure will still lead to an increasing nitrogen concentration. The latter point can be overcome by attaching the system to an oxygen source.

These shortcomings of the Bartha–Pramer flask have resulted in the development of useful modifications which do not have the above limitations. The system of Anderson (1975) shown in Figure 2.5 consists of an Erlenmeyer flask and a trapping tower which contains glass wool treated with paraffin oil to absorb volatile compounds and an upper layer of soda lime to absorb CO_2. It allows exchange of air with the outside atmosphere and it also allows the trapping of volatile products. However, the question could be raised whether diffusion is fast enough to guarantee completely free air circulation.

These problems, the latter in particular, have been carefully studied by Marvel et al. (1976, 1978) during several years' work with their modified system shown in Figure 2.6. The authors stated that their simple apparatus requires minimal

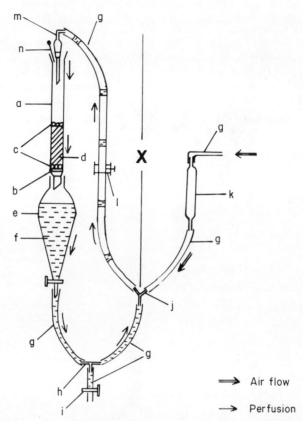

Figure 2.3 Perfusion apparatus according to Fung and Uren (1977). a, chromatography column with integral sinter; b, sinter (porosity 0); c, glasswool; d, soil; e, 250 cm^3 separating funnel; f, perfusing solution; g, polyvinyl chloride tubing; h, T-piece Kartell connector; i, tap; j, Y-piece Kartell connector; k, trap; l, Hoffman screw clip; m, leak-tube; n, air-leak. (Reprinted with permission from Fung, K. K. H., and Uren, N. C. (1977). *J. Agr. Food Chem.* **25**, 966–969. Copyright 1977 American Chemical Society)

space, needs little personal attention, consists of simple equipment and is applicable to studies in natural waters and sediments as well as flooded soils. In addition, the modifications allow the separate analysis of evolved $^{14}CO_2$ and atmospheric carbon dioxide. The necessary movement of air by diffusion through the trapping tower has been established by studies with *Escherichia coli*, a strict aerobe. Comparisons between flasks with and without trapping towers demonstrate that the growth of this aerobic microbe was not significantly different in both flasks and therefore it was concluded that the system readily maintained the desired aerobic conditions.

The system shown in Figure 2.7 was developed by Loos *et al.* (1980). It involves a free-drainage mixture of soil, sand, and organic substrate over a layer of pea

Table 2.1 Advantages and disadvantages of soil perfusion systems

Advantages	Disadvantages
Adaptability for use with either positive or negative pressure systems	Bacteria appear to be the predominant organisms in soil perfusion systems
Ease of maintaining sterility if required	Conditions in the system are far away from practice
Perfusate can easily be sampled	Data obtained are not representative for the actual soil environment
Rate of perfusion can easily be controlled	
Adaptable to use gases of known composition	
Infrequent adjustment requirements	
Self-supporting and compact design	
Relatively inexpensive parts	

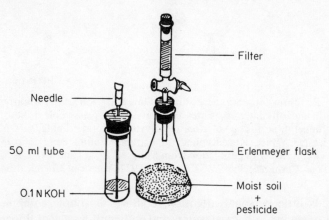

Figure 2.4 Incubation flask for the measurement of the persistence of pesticides in soil (Reproduced from Bartha and Pramer (1965). *Soil Sci.* **100**, 68–70 by permission of the Williams & Wilkins Co., Baltimore)

gravel which permits watering to water-holding capacity and recirculation of drainage water for easy maintenance of constant moisture. The flask is continually open to air and is fitted with ascarite and KOH traps to remove carbon dioxide from incoming and outgoing air, respectively. Evolved $^{14}CO_2$ is absorbed in the potassium hydroxide trap which can be periodically removed for analysis by liquid scintillation counting. The system needs considerable time,

Table 2.2 The advantages and limitations of the Bartha–Pramer biometer flask

Advantages	Limitations
Both $^{14}CO_2$ and total CO_2 production can be monitored	Volatile products may not be absorbed by KOH solution or, when trapped, have to be separated from $^{14}CO_2$ by additional analytical effort
All measurements can be made on the same soil sample	
No significant exposure to atmospheric CO_2	The atmosphere of the flask will become richer in N_2 and thus tend to anaerobic conditions
Difference in CO_2 amounts produced by treated and untreated soil can be used to measure oxidation of an unlabelled pesticide*	Pesticide degradation to $^{14}CO_2$ is less extensive than in flow-through systems

* The results will be misleading if pesticides stimulate microbial activity in addition to serving as substrate. The same is valid when organic solvents or formulation additives are added together with the pesticides.

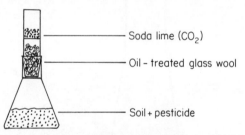

Figure 2.5 Culture-flask system for studying the metabolism of ^{14}C-labelled pesticides in soil (Anderson, 1975)

effort, and attention in setting it up and also lacks any mechanism for readily distinguishing between volatilized products and evolved radiolabelled carbon dioxide.

Flow-through systems have also been used for pesticide transformation investigations in soil. They have varied from a few flasks connected in series to the complex manifold developed by Parr and Smith (1969) which is shown in Figure 2.8. The advantages listed in Table 2.3 make these flow-through systems superior to other incubation approaches. Particularly, the control of the aeration flow rate is very important for organic matter decomposition since CO_2 production was found to increase markedly with increasing air flow rate (Parr and Reuszer, 1962). Kaufman (1977a) has also observed a more rapid and extensive degradation of ^{14}C-labelled pesticides in flow-through systems than in

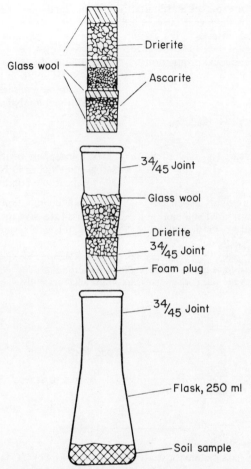

Figure 2.6 Aerobic soil metabolism incubation system according to Marvel *et al.* (1978). (Reprinted with permission from Marvel *et al.* (1978). *J. Agr. Food Chem.* **26**, 1116–1120. Copyright 1978 American Chemical Society)

biometer flasks used in comparable experiments. For absorption of volatile compounds, ethylene glycol, ethanolamine, or solids, such as silica gel or polyurethane plugs (Kearney and Kontson, 1976), have been used.

The complexity of and the great costs for equipment are considered to be disadvantages for flow-through approaches. However, this is not necessarily so, since the equipment used for controlled and constant aeration, once it has been installed, need not be removed and can therefore be used at all times for further experiments without additional investment. The parts which have to be changed, as in the biometer system, are the absorption devices and the incubation flask at the end of a sampling interval.

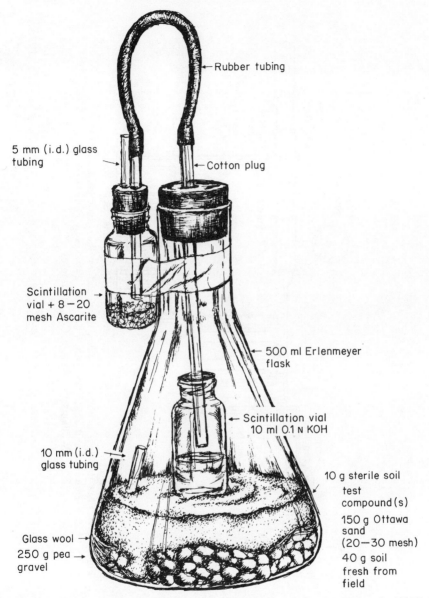

Figure 2.7 Soil and sand incubation system developed by Loos *et al.* (1980). (Reproduced by permission of Pergamon Press)

Another flow-through system was developed by Goswami and Koch (1976). The basic design of the manifold is illustrated in Figure 2.9. A single apparatus consists of 12 reaction flasks connected in parallel through a series of gang valves

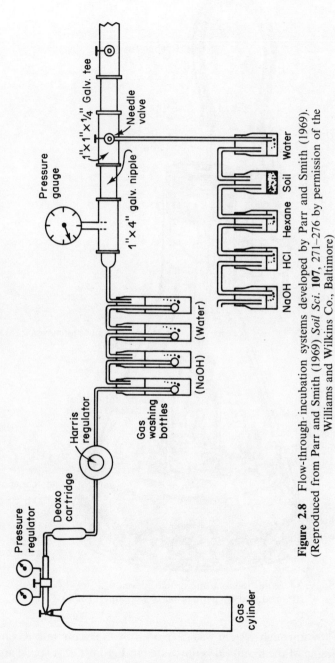

Figure 2.8 Flow-through-incubation systems developed by Parr and Smith (1969). (Reproduced from Parr and Smith (1969) *Soil Sci.* **107**, 271–276 by permission of the Williams and Wilkins Co., Baltimore)

Table 2.3 Advantages and disadvantages of flow-through systems

Advantages	Disadvantages
Soil respiration experiments can be conducted under more precisely defined conditions	They are complex in their design
Different gaseous environments can be used	Require more expensive equipment
Can be operated on either positive or negative pressure	
More rapid and extensive degradation of pesticides than in biometer flasks	
Permit characterization of both $^{14}CO_2$ and volatile products	
Exposure to atmospheric CO_2 can be prevented	
Total CO_2 production can be monitored and used as indication of soil microbiological activity	

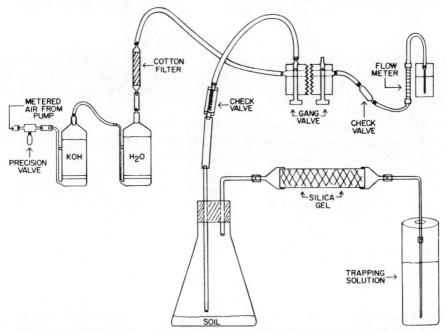

Figure 2.9 Apparatus for trapping $^{14}CO_2$ from degradation of ^{14}C-labelled pesticides (Goswami and Koch, 1976). (Reproduced by permission of Pergamon Press)

with a single source of CO_2-free moist air. According to the authors, the apparatus is inexpensive to construct, requires no special glassware or material, can handle numerous samples at one time, and requires limited attention.

The simple apparatus shown in Figure 2.10 was used by Kearney and Kontson (1976) to examine vapour and metabolic losses of two dinitroaniline herbicides from soil. The polyurethane foam plug effectively trapped volatiles from the soil surface, while allowing $^{14}CO_2$ to pass through the plugs and become absorbed in the alkali solution.

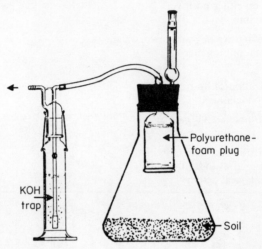

Figure 2.10 Flow-through system developed by Kearney and Kontson (1976). (Reprinted with permission from Kearney and Kontson (1976). *J. Agr. Food Chem.* **24**, 424–426. Copyright 1976 American Chemical Society)

The soil metabolism equipment used in CIBA-GEIGY's ecological chemistry laboratories is illustrated in Figure 2.11 (Guth, 1980). It consists of a gas washing bottle filled with distilled water to saturate the incoming air in aerobic experiments, a needle valve for adjustment of gas flow rate (air or nitrogen), a one litre glass bottle (Steilbrust) containing the treated soil, several gas absorption bottles filled with ethylene glycol, 0.1N sulphuric acid and 2N sodium hydroxide to trap volatiles and radioactive carbon dioxide, and finally a flow meter measuring the gas flow rate. This equipment is attached to a gas manifold system which allows it to be connected up to 60 soil reaction flasks. The purge gas is delivered from a gas pressure cylinder. This system can be used for aerobic, sterile as well as anaerobic experiments and all of its parts are commercially available.

Integrated systems with a greater degree of complexity have been developed in other laboratories. One example is that of Best and Weber (1974) shown in Figure 2.12 which allows simultaneous measurement of degradation, respiration, volatilization, plant uptake and leaching processes. A similar approach

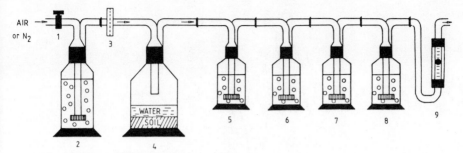

Figure 2.11 Soil metabolism apparatus used in the author's laboratory. 1, needle valve; 2, gas washing bottle containing water; 3, ultramembrane (sterile conditions only), pore size 0.2 µm; 4, soil metabolism flask; 5, ethylcellosolve trap; 6, sulphuric acid trap; 7, 8, sodium hydroxide trap; 9, flow meter. (Reproduced from Guth (1980) by permission of Academic Press)

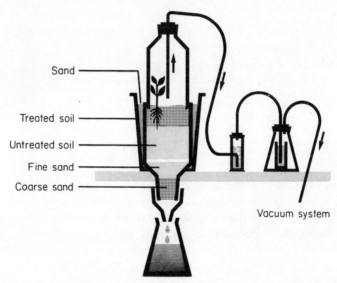

Figure 2.12 The soil–plant incubation apparatus used by Best and Weber (1974) for studies of the degradation and movement of s-triazines in soil. (Reproduced by permission of the Weed Science Society of America)

was used by Lichtenstein *et al.* (1974) when studying translocation and metabolism of phorate (Figure 2.13). Harvey and Reiser (1973) incubated plants in soil treated with radiolabelled pesticide and, using an enclosed system with air flow, recovered volatile material from the air stream. Beall *et al.* (1976) have described the use of an enclosed 'agroecosystem' in which plants can be grown in soil and pesticide degradation examined under varying regimes of rainfall, wind velocity, and light.

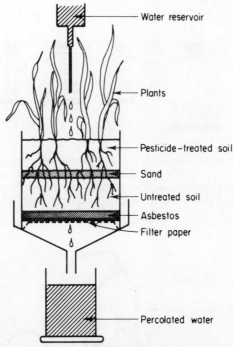

Figure 2.13 Model system for studies of the degradation of pesticides in soil in the presence of plants. (Reprinted with permission from Lichtenstein *et al.* (1974). *J. Agr. Food Chem.* **22**, 991–996. Copyright 1974 American Chemical Society)

Although these integrated systems have tried to introduce the complexity of the environment, they are really unable to simulate actual environmental conditions. In addition, complex systems of such types produce a large number of variable results which, due to this complexity, may be difficult to interpret. Noted examples were some complex terrestrial–aquatic ecosystems used in the past which led to some confusion in the scientific community. It also seems unreasonable to monitor various processes in a complex system routinely for each compound when certain processes are not of significance. Rather, the importance of each process for the environmental fate of a pesticide should be estimated in simple models which are especially designed to study a certain process.

Field Experiments

Field data are less easy to obtain than laboratory data and 'balance studies' are difficult to establish under natural field conditions. In spite of the difficulties in carrying out field soil metabolism experiments a few systems have been successfully applied. The approach used by Roberts (1976) under outdoor conditions is shown in Figure 2.14. Plants are grown in boxes which are placed in a pit in such a way that the soil surface in the boxes is at the same level as the soil in the surrounding field. The floor of the pit slopes towards a central well containing a pipe linked to a drain. This pipe can be closed and the water can be periodically monitored for radioactivity. The plywood boxes (60 × 60 × 60 cm) are fitted with slats at the bottom to allow drainage. Flaps are fitted on one side so that soil samples may be taken horizontally at different depths. The boxes are filled with top soil and allowed to settle for 1–2 months. Plants can be grown in the boxes which can be treated with the formulated pesticide at any time required.

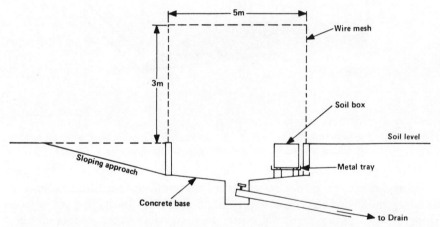

Figure 2.14 Diagram of an outdoor enclosure for radiochemical experiments (Roberts, 1976). (Reproduced by permission of the British Crop Protection Council)

Standardized lysimeters as shown in Figure 2.15 have been used for pesticide balance and metabolism studies (Führ *et al.*, 1976). The lysimeters are made of polyvinyl chloride walls which are 6 mm thick. They are packed with 40 cm of surface soil. A 5 cm layer of washed quartz gravel at the base of the lysimeter prevents impoundment of the percolate. In the centre of the lysimeter, a separated control vessel is installed containing untreated soil and plants. Percolates from the treated portion and the control vessel can be collected separately.

However, the above systems do not simulate actual environmental conditions in every respect. The fact that artificial soil layers are used in these lysimeters instead of naturally grown, intact soil profiles allows no clear statement on the leaching behaviour of a pesticide under field conditions to be made. There is only one way of simulating actual environmental conditions in every respect, that is to use monolith lysimeter systems like the one recently described by Jarczyk (1978).

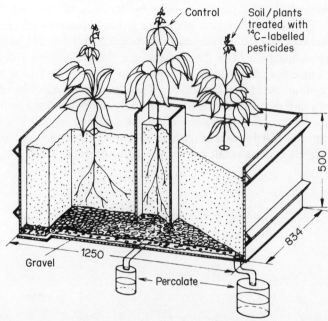

Figure 2.15 The experimental lysimeter (measurements in mm) according to Führ et al. (1976). (Reproduced by permission of J. D. Sauerländer's Verlag)

Degradation of pesticides in soil under field conditions can also be studied by applying radiolabelled compounds to small field plots and taking soil samples from various depths at regular intervals. This procedure does not generally allow one to measure the contribution of volatilization to the overall disappearance of the compound or to collect volatile products. However, with a method described by Hill and Arnold (1978), volatile products can be sampled also in field metabolism experiments. The treated soil surface is covered by 60 cm high bell jars mounted on stainless steel rings embedded 5 cm into the soil. The air is sucked out of the bell jars and through trapping solutions for collection of volatile radiolabelled products. To reduce the heating effects inside the jar, volatiles are sampled only during a 24 h period from mid-day to mid-day every 48 h. However, it cannot be recommended that such an elaborate approach be routinely used for every compound. Using a less expensive laboratory model it should first be determined whether volatilization of the compound is of sufficient importance that it must be monitored in the field.

The use of radiolabelled pesticides, enabling a more detailed investigation of transformations *in situ* under various field conditions is not always possible because of restrictions in their usage at field sites and the difficulty of maintenance and monitoring at locations apart from the laboratory. Particularly in a time of a steadily growing public concern about radioactivity, the situation will certainly become more difficult in the future. The limitations placed on field studies by the necessity for controls on radiochemicals could be removed by the use of ^{13}C-enriched pesticides as suggested recently by Roberts (1976).

In this context it might be interesting to mention that in a recent review (Guth, 1980) the necessity of field studies for the establishment of pathways of transformation was discussed. The data presented indicate that transformation pathways of pesticides in soil can be predicted from studies with soils incubated under controlled laboratory conditions. Thus, there seems to be no need for field experiments with radiolabelled pesticides when only information on pathways of transformation is required.

CHEMICAL VERSUS MICROBIAL TRANSFORMATION

A considerable amount of work has been done to distinguish between chemical and microbiological transformation of pesticides in soil. The task is difficult since the transition from one mechanism to the other is neither abrupt nor well defined. The experimental approach commonly used is to carry out studies in sterile and non-sterile soil and compare the results obtained. Soils can be sterilized by chemicals, by irradiation, and by autoclaving. Each of these sterilization methods has certain shortcomings which are summarized in Table 2.4. More research is needed either to improve the existing sterilization procedures or to develop better new methods. However, the question as to whether a given pesticide is degraded by microbial or chemical mechanisms can be considered of more academic than of practical significance and certainly is not important from an environmental safety standpoint. Despite the above shortcomings, experiments with autoclaved and/or γ-irradiated soils give an indication of the mechanisms involved. The sterilization aspects have also been treated by Pramer and Bartha (1972), Cawse (1975), Powlson and Jenkinson (1976), Hill and Arnold (1978), and Torstensson (1980).

Another possibility of differentiating between chemical and microbial transformation has been suggested by Meikle *et al.* (1973). They suggested the use of widely different activation energies for microbiological (4–6 kcal mol^{-1}) and chemical degradation processes (18–25 kcal mol^{-1}) as the basis for determining which process is responsible. Although this approach has been successfully applied, for example, by Rahn and Zimdahl (1973) and Gingerich and Zimdahl (1976) it should be considered with caution. A fundamental supposition for the applicability of this proposal is that the reactions follow first order kinetics. For other reaction orders the apparent activation energies will depend on the concentrations at which the comparisons are made. Also, confusing results will be obtained when biological and abiotic decomposition occur simultaneously. The latter was confirmed by Hamaker (1972) who found out that activation energies do not suggest two distinct categories but fall within a range of 1–32 kcal mol^{-1}. In addition, Sizer (1943) reported activation energies for

Table 2.4 Shortcomings of soil sterilization methods

Sterilization method	Shortcomings of method	Reference
Steam sterilization (autoclaving)	High temperatures and pressures alter soil surface adsorption characteristics and cause organic matter decomposition	Furmidge and Osgerby (1967) Warcup (1957)
	Increased release into the soil solution and extractability of nitrogen, phosphorus, sulphur, manganese	Eno and Popenoe (1964)
	Destroys the free radical generating system in soil	Kaufman (1977a)
γ-Irradiation	Increased release into the soil solution and increased extractability of nitrogen, phosphorus, sulphur, manganese	Bowen and Cawse (1964)
	Enzymic activity derived from microbes and/or plants being retained by the sterilized soil	McLaren et al. (1962) Vela and Wyss (1962) Getzin and Rosenfield (1968) Hill and Arnold (1978)
Chemical sterilization*	Due to adsorption, considerable amounts need to be added to soil which affects the chemical reactivity and adsorptive properties of soil	Hill and Arnold (1978)
	Chemicals may not eliminate the whole population of soil microorganisms	
	Antibiotics and organic biocides such as phenol are degraded and of limited use as soil sterilants	Torstensson (1980)
	Some chemical sterilizing agents can be slightly phytotoxic	Skipper and Westermann (1973)

* Chemical sterilants used were, for example: sodium or potassium azide, ethylene or propylene oxide, mercuric salts, phenolics, antibiotics, organic solvents, soil fumigants.

enzyme systems between 2.7 and 65 kcal mol^{-1} and for the only microbial system quoted, bacterial dehydrogenase, values of 15–25 kcal. Hence it can be concluded that the suggestion of Meikle et al. (1973) is not valid.

Recently, it has been proposed as another alternative that a combination of photolysis on soil surfaces and aqueous hydrolysis could be used to evaluate the contribution of abiotic mechanisms to the overall breakdown of pesticides in soil (Burkhard and Guth, 1979).

PHOTOCHEMICAL TRANSFORMATION STUDIES

Considerable information is available on photochemical procedures for solutions, thin films, and vapour phase, but published knowledge on experimental approaches for measuring photolysis in soil or on soil surfaces is rather limited. The author is only aware of the few laboratory systems listed in Table 2.5. Some other studies were carried out under natural sunlight and thus fully depended on the meteorological variabilities of the environment (Nilles and Zabik, 1974, 1975; Liang and Lichtenstein, 1976; Parochetti and Dec, 1978; Plimmer, 1978; Smith et al., 1978). Others used germicidal lamps which are rich in radiation between 240 and 260 nm (Funderburk et al., 1966; Liang and Lichtenstein, 1976). The latter experiments are of no practical value for estimating the photochemical stability of pesticides on soil surfaces since ultraviolet light below 290 nm does not reach the earth.

Although all systems given in Table 2.5 can be used for soil photolysis studies in the laboratory, the Hanau Suntest apparatus shown in Figure 2.16 is especially recommended since it has several advantages. The system is commercially

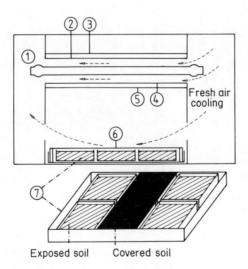

Figure 2.16 The Hanau Suntest apparatus (Burkhard and Guth, 1979). 1, xenon lamp; 2, u.v. reflecting mirror (infrared transmissive); 3, light reflecting mirror (infrared transmissive); 4, quartz glass with infrared reflecting coating; 5, u.v. filter with a radiation cut-off at 290 nm; 6, quartz glass; 7, metal boxes containing soil samples

Table 2.5 Laboratory approaches for studying photochemical transformations of pesticides on soil surfaces

Apparatus	Light source	Energy	Filters	Soil samples Exposed as	Distance from light source	Reference
Enclosed chamber	Two FS 40 T 12 Westinghouse lamps	40 W 82% of energy at 310 nm	None	1.5 cm soil layer in petri dishes covered with polished quartz	33 cm	Parochetti and Hein (1973)
Rayonet Srinivasan–Griffin photoreactor with merry-go-round apparatus	8 RPR 3000 Å and 8 RPR 3500 Å lamps	$2-4 \times 10^4$ erg·cm^{-2}·s^{-1}	None	Thin layers of soil, 500 μm thick	Not given	Nilles and Zabik (1974, 1975)
No special	Medium pressure mercury arc 450 W	450 W	Double-mantled Pyrex tube, water-cooled	Soil thin layer (30 μm) in glass tray with water cooling and covered with Pyrex glass	14 cm	Hautala (1976, 1978)
Environmental control chamber with turntable	Two FS 20 (20 W) and six F30T 8 BLB (30 W) Westinghouse fluorescent lamps	2000 μW·cm^{-2}	None	Thin layers of soil (0.5 mm) on glass plates	Not given	Kennedy and Talbert (1977)
Hanau Suntest apparatus (Figure 2.1)	Xenon lamp	950 J·m^{-2}·s^{-1}	See Figure 2.1	5–10 mm soil layers in shallow metal boxes covered with quartz glass	20 cm	Burkhard and Guth (1979)

available for a reasonable price. In addition, the xenon burner combined with appropriate mirrors and filters emits light with a spectral energy distribution which very closely approaches that of sunlight (Figure 2.17). The light intensity emitted is equivalent to approximately double the energy of natural sunlight so that information on the photochemical reactivity of pesticides on soil surfaces can be obtained faster than outside the laboratory. Furthermore, as with all laboratory model systems, experiments can be done at any time of the year independent of the meteorological variations of a particular environment.

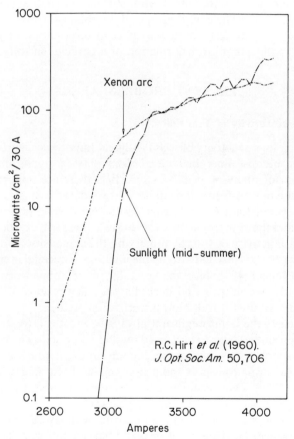

Figure 2.17 Spectral energy distribution of sunlight and xenon arc

MICROBIAL STUDIES

A considerable amount of work has been directed to the metabolism of pesticides in microbial cultures. Laboratory studies of microbial metabolism have utilized organisms either obtained from culture collections or isolated from soil, aquatic systems, sewage, etc. When radiotracer methodology was still in its infancy microbial studies were quite useful to get information on the biodegradability of certain pesticides. However, the question may be raised as to the relevance of these studies for the persistence and fate of pesticides under field conditions. Pure culture studies carried out with unrealistically high concentrations very often do not reveal the true fate of pesticides in natural soils and can never reproduce conditions *in vivo* (Cripps and Roberts, 1978). It should be kept in mind by scientists as well as by legislators that microbial studies although giving an interesting insight into biotransformation mechanisms have only academic importance. The comparison of data obtained from sterile and non-sterile soil experiments certainly gives sufficient information of the microbial contribution to the overall transformation of a pesticide in soil.

PRESENT PROBLEM AREAS

Microbiological Activity of Test Soils

The microbiological activity of soils is of great importance, particularly for *in vitro* systems, because microbial metabolism plays a significant role in the transformation of many pesticides in soil. It is therefore important that soil sampled for use in transformation studies is treated as a living tissue and hence handled with care (Pramer and Bartha, 1972). Air-drying, prolonged storage, and freezing and thawing should be avoided since these will drastically influence the biochemical activity of soils by modifying the composition and population density of microorganisms and also by inactivating extracellular enzymes in the soil. Lay and Ilnicki (1975) have made similar observations when studying the effect of soil storage on propanil degradation. They proposed that fresh soil samples should be used for all biotransformation studies.

In special cases the latter might not always be practical, for example, when transformation studies have to be carried out in foreign soils which require some time for transport. A possible way to circumvent eventual losses of biological activity of soils during transport and processing has been suggested by Chisaka and Kearney (1970). They reactivated five Japanese rice soils by cultivating rice plants before they started the degradation experiments.

This principle was also used in our laboratories with several plant species. The activating influence of maize and cotton on the degradation rate of chlorotoluron in soil is shown in Table 2.6 (Laanio, 1974). The results demonstrate that the $^{14}CO_2$ evolution was markedly increased, particularly after use of cotton, when compared to soil which was not pretreated by plants. In additional experiments, the half-life of the urea herbicide thiazafluron in a German standard soil was determined to be 45 weeks, whereas in another stored soil reactivated with rape for 15 weeks a half-life of only 7 weeks was observed (Keller, 1976, 1977). Hill

Table 2.6 Reactivation of biological activity of soil by plant roots measured by the rate of mineralization of ^{14}C-labelled chlorotoluron (Laanio, 1974)

Pretreatment of soil* (4 weeks growth)	Amount (%) of $^{14}CO_2$ formed within 12 weeks of aerobic incubation
None	<1
Maize	8
Cotton	25

* Clay loam, pH = 6.2, 5.4% organic matter, cation exchange capacity = 25 meq/100 g, water holding capacity = 55.3%.

(1978) also measured enhanced rates of degradation of ethirimol and of a thiadiazolidin–oxime carbamate in soil in the presence of barley and maize plants.

It is clear from some of the above data that the extent of reactivation of soil is greatly influenced by the plants used. It is further known that the microbial population of soil can be altered both qualitatively and quantitatively by the presence of plant roots (Rovira and McDougall, 1967). Thus, the activating properties of plants and the species differences observed are probably due to the build-up of microorganisms. On the other hand, when microbial functions (basic respiration, nitrification) of plant-reactivated soils were measured, depression effects were observed in comparison to soils which were only rewetted with water (Ellgehausen, 1978). It seems that, although this activation technique has been used in several laboratories, knowledge of the fundamental processes involved is still lacking.

There is certainly more research needed in this area, particularly on the activating influence of plants. At the moment, it is preferable to use fresh soil samples from the field for any transformation studies of pesticides in soil. In the future, appropriate methods should be developed which can satisfactorily measure the biodegradative capacity of soil.

The Standard Soils Concept

The use of standard soils for the various ecological studies carried out with pesticides is in principle a fascinating idea. Standard soils were originally selected in Germany to measure the leaching of pesticides in different laboratories under standardized conditions in order to obtain comparable and reproducible results. They were later used also for soil degradation studies but soon it became obvious that in some cases relatively low rates of transformation were observed in the standard soils in comparison to data obtained with soils of similar properties which were collected freshly from agricultural areas. The herbicide diallate, for example, degraded 2.5–5 times faster in the fresh soils than in standard soil 2.2 as shown in Figure 2.18 (Guth, 1978). Experiments currently being carried out with the phosphorus insecticide parathion show the same tendency. The differences found were mainly caused by different microbiological activities. The microbial biomass of the Speyer standard soil 2.2 was only approximately one-fourth to one-fifth of that of the two fresh soils (Domsch, 1978). Therefore the concept of using standard soils should be considered with caution. Standard soils are certainly quite useful if non-biological processes, such as adsorption, leaching, volatilization, etc., have to be studied but they are of questionable value for biotransformation experiments. Another critical point which should not be overlooked is a guarantee of the same quality of soil over long time periods. It is evident that even large soil sources will sooner or later be exhausted so that from time to time new standards have to be established.

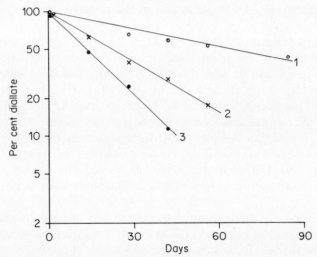

Figure 2.18 Rate of degradation of diallate in 3 soils under aerobic conditions at $22 \pm 2\,°C$ (Guth, 1978). 1. Loamy sand, pH 5.6; organic carbon, 2.2%; clay, 8.8%; silt, 6.4%; sand, 84.8% (Speyer Standard Soil 2.2). 2. Loamy sand, pH 8.0; organic carbon, 1.5%; clay, 3.0%; silt, 14.6%; sand, 82.4%. 3. Sandy loam, pH 7.8; organic carbon, 1.8%; clay, 6.3%; silt, 24.4%; sand, 69.3%. Reprinted with permission of EWRC, Oxford and Wageningen, The Netherlands

Duration of *in vitro* Experiments

The microbiological activity of soils is not only critical in the beginning of transformation experiments, it is equally important for the duration of a soil metabolism study. Some registration authorities require that studies be carried out for a period of one year but it seems questionable that soils retain their biological activity under *in vitro* conditions for such a long time. It is known from soil respiration experiments that CO_2 production and O_2 consumption of soils, one measure of microbial activity, steadily decrease under controlled conditions

(Davies and Marsh, 1977; Marsh et al., 1977; Marsh and Davies, 1978). In our laboratories observations were also made that $^{14}CO_2$ formation from radio-labelled pesticides significantly decreased between 3 and 6 months after the experiments were started, but was stimulated by simply stirring up the soil sample. Such effects will certainly appear earlier and more drastically if soils with lower initial biological activity are used for laboratory experiments. This is another strong argument for the use of fresh soils with full biological potential for transformation studies. It is recommended that the bioactivity of soils under *in vitro* conditions should be monitored . This could be done, for example, by measuring at least one parameter (e.g. CO_2 production) from basic soil respiration. In this way, more information on the duration of biological activity of soils in laboratory systems could be obtained.

Although the data reported in these paragraphs are mainly derived from kinetic experiments it has significant importance for pathways of transformation. Low biological activity of soils can cause a reduced rate of pesticide degradation and metabolite formation; this could create problems in identifying major metabolites and terminal residues.

Influence of Other Chemicals on Pesticide Transformation

Pesticides often have a relatively low water solubility and consequently they are added to soil dissolved in limited amounts of organic solvents. Another possibility is to apply a pesticide as a formulated product suspended or emulsified in water. The addition of these other chemicals, however, should be carefully controlled. Published data exist on possible interactions between various chemicals (Kaufman, 1972, 1977b; Priest and Stephens, 1975; Wallnöfer et al., 1977) and although only the rates of degradation were affected, this point could also have consequences for transformations in soil. Consideration should therefore be given to the amounts and types of solvents or formulation ingredients used to avoid possible toxicity to the microbial population. Chlorinated organic solvents, such as chloroform, etc, should not be used since they show considerable fumigation effects (Lynch and Panting, 1979).

CONCLUSIONS

A great variety of experimental approaches has been used to study the fate of pesticides in the soil environment. Experiments have been carried out in laboratory model systems and also under field conditions.

The laboratory methods for measuring transformations in soil include flow-through systems, biometer flasks, soil perfusion apparatus, and so-called integrated systems. Flow-through apparatus were found to be superior to the other three systems mentioned and obviously have no great disadvantages. Experiments in biometer flasks will also yield representative results if attention is paid to certain critical points. Soil perfusion systems are useful for microbial studies but are far away from natural conditions. Also the complex, integrated systems show some limitations and, in the end, cannot simulate actual environmental situations.

Transformation studies with radiolabelled pesticides have also been conducted under field conditions but, for the future, difficulties have to be expected due to the growing public opposition against the use of radioisotopes. This point might not be critical since data presented in a recent review (Guth, 1980) indicate that transformations in field soils can be predicted from laboratory results.

Several laboratory systems have been described to measure the photochemical contribution the overall transformation of pesticides in the soil environment. The commercially available Hanau Suntest apparatus has several advantages compared to other systems and can be recommended for further use.

Much work has been done to distinguish between chemical and microbial transformation in soil. All of the sterilization methods have certain shortcomings but can at least give an indication whether microbial or chemical processes are involved in the degradation of a particular pesticide. However, it should be kept in mind that this question is of rather academic importance.

Finally, problems were discussed which at present are of special importance for transformation studies. Air-drying, prolonged storage, freezing, and thawing will significantly alter biological activity of soils and it is therefore advisable to use fresh soils from the field for biotransformation experiments in the laboratory. Soils can also be reactivated to some extent by growing various plant species before pesticides are applied but knowledge of the mechanisms involved is lacking.

Although the use of standard soils for environmental studies is a fascinating idea, the question of the biological activity has certainly been overlooked in this concept. According to the experience collected so far, standard soils cannot be recommended for use in transformation experiments of pesticides.

Duration of experiments and also the use of other chemicals, such as organic solvents and formulation ingredients, can also adversely affect microbiological activity of soils. It is therefore necessary to examine carefully the potential influence also of the latter two factors. In particular, the biological activity during long term studies should be regularly monitored.

REFERENCES

Anderson, J. P. E. (1975). 'Einfluss von Temperatur und Feuchte auf Verdampfung, Abbau und Festlegung von Diallat im Boden', *Z. Pflkrankh Pflschutz, Sonderheft VII*, 141–146.

Audus, L. J. (1946). 'A new soil perfusion apparatus', *Nature* **158**, 419.

Bartha, R., and Pramer, D. (1965). 'Features of a flask and method for measuring the persistence and biological effects of pesticides in soil', *Soil Sci.* **100**, 68–70.

Beall, M. L., Nash, R. G., and Kearney, P. C. (1976). 'Agroecosystem. A laboratory model ecosystem to simulate agricultural field conditions for monitoring pesticides',

Proc. EPA Conference on Modelling and Simulation, Environ. Protect. Agency 790–793.
Best, J. A., and Weber, J. B. (1974). 'Disappearance of s-triazines as affected by soil pH using a balance-sheet approach', *Weed Sci.* **22**, 364–373.
Bowen, H. J. M., and Cawse, P. A. (1964). 'Some effects of gamma radiation on the composition of the soil solution and soil organic matter', *Soil Sci.* **98**, 358–361
Burkhard, N., and Guth, J. A. (1979). 'Photolysis of organophosphorus insecticides on soil surfaces', *Pestic. Sci.* **10**, 313–319.
Cawse, P. A. (1975). 'Microbiology and biochemistry of irradiated soils', in *Soil Biochemistry* (Eds. E. A. Paul and A. D. McLaren), Vol. III, 213–267, Marcel Dekker, New York.
Chase, F. E. (1946). 'A preliminary report on the use of the Lees and Quastel soil perfusion technique in determining the nitrifying capacity of field soils', *Sci. Agr.* **28**, 315–320.
Chisaka, H., and Kearney, P. C. (1970). 'Metabolism of propanil in soils', *J. Agr. Food Chem.* **18**, 854–858.
Cripps, R. E., and Roberts, T. R. (1978). 'Microbial degradation of herbicides', in *Pesticide Microbiology. Microbiological Aspects of Pesticide Behaviour in the Environment* (Eds. I. R. Hill and S. J. L. Wright), pp. 669–730, Academic Press, London.
Davies, H. A., and Marsh, J. A. P. (1977). 'The effect of herbicides on respiration and transformation of nitrogen in two soils. II. Dalapon, pyrazone and trifluralin', *Weed Res.* **17**, 373–378.
Domsch, K. H. (1978). Personal communication.
Ellgehausen, H. (1978). Unpublished data. CIBA-GEIGY Ltd., Agricultural Division.
Eno, C. F., and Popenoe, H. (1964). 'Gamma radiation compared with steam and methyl bromide as a soil sterilising agent', *Proc. Soil Sci. Soc. Amer.* **28**, 533–535.
Führ, F., Cheng, H. H., and Mittelstaedt, W. (1976). 'Pesticide balance and metabolism studies with standardised lysimeters', *Landwirtschaftl. Forsch., Sonderheft* **32**, 272–278.
Funderburk, H. H., Negi, N. S., and Lawrence, J. M. (1966). 'Photochemical decomposition of diquat and paraquat', *Weeds* **14**, 240–243.
Fung, K. K. H., and Uren, N. C. (1977). 'Microbial transformation of S-methyl N-[(methylcarbamoyl)oxy]thioacetimidate (methomyl) in soils', *J. Agr. Food Chem.* **25**, 966–969.
Furmidge, C. G. L., and Osgerby, J. M. (1967). 'Persistence of herbicides in soil', *J. Sci. Food Agr.* **18**, 269–273.
Getzin, L. W., and Rosenfield, I. (1968). 'Organophosphorus insecticide degradation by heat-labile substances in soil', *J. Agr. Food Chem.* **16**, 598–601.
Gingerich, L. L., and Zimdahl, R. L. (1976). 'Soil persistence of isopropalin and oryzalin', *Weed Sci.* **24**, 431–434.
Goswami, K. P., and Koch, B. L. (1976). 'A simple apparatus for measuring degradation of ^{14}C-labelled pesticides in soil', *Soil Biol. Biochem.* **8**, 527–528.
Guth, J. A. (1978). Unpublished data. CIBA-GEIGY Ltd., Agricultural Division.
Guth, J. A. (1980). In Hance, R. J. (Ed.) '*Interactions between Herbicides and the Soil*', In press. Academic Press, London.
Hamaker, J. W. (1972). 'Decomposition: Quantitative aspects', in *Organic Chemicals in the Soil Environment* (Eds. C. A. I. Goring and J. W. Hamaker), Vol. I, pp. 253–340, Marcel Dekker, Inc., New York.
Harvey, J., and Reiser, R. W. (1973). 'Metabolism of methomyl in tobacco, corn and cabbage', *J. Agr. Food Chem.* **21**, 775–783.

Hautala, R. R. (1976). 'Photolysis of pesticides on soil surfaces', *Symposium on Nonbiological Transport and Transformation of Pollutants on Land and Water: Processes and Critical Data Required for Predictive Description.* Gaithersburg, Maryland (Abstract Volume).
Hautala, R. R. (1978). 'Surfactant effects on pesticide photochemistry in water and soil', *EPA-Report 600/3-78-060*, U.S. EPA, Athens, Georgia.
Hill, I. R. (1978). 'Microbial transformation of pesticides', in *Pesticide Microbiology. Microbiological Aspects of Pesticide Behaviour in the Environment* (Eds. I. R. Hill and S. J. L. Wright), pp. 137–202, Academic Press, London.
Hill, I. R., and Arnold, D. J. (1978). 'Transformations of pesticides in the environment—the experimental approach', in *Pesticide Microbiology. Microbiological Aspects of Pesticide Behaviour in the Environment* (Eds. I. R. Hill and S. J. L. Wright), pp. 203–245, Academic Press, London.
Jarczyk, H. J. (1978). 'Behaviour of pesticides in soil as determined by undisturbed soil cores and lysimeter systems', *4th International Congress of Pesticide Chemistry (IUPAC)*, Zürich, July 24–28, Abstract Volume V–35.
Kaufman, D. D. (1966). 'An inexpensive, positive pressure soil perfusion system', *Weeds* **14**, 90–91.
Kaufman, D. D. (1972). 'Degradation of pesticide combinations', *Pestic. Chem.* **6**, 175–205.
Kaufman, D. D. (1977a). 'Approaches to investigating soil degradation and dissipation of pesticides', *Symposium on Terrestrial Microcosms and Environmental Chemistry*, Corvallis, Oregon, June 13–17.
Kaufman, D. D. (1977b). 'Biodegradation and persistence of several acetamide, acylanilide, azide, carbamate and organophosphate pesticide combinations', *Soil Biol. Biochem.* **9**, 49–57.
Kaufman, D. D., Kearney, P. C., Von Endt, D. W., and Miller, D. E. (1970). 'Methylcarbamate inhibition of phenylcarbamate metabolism in soil', *J. Agr. Food Chem.* **19**, 513–519.
Kearney, P. C., and Kontson, A. (1976). 'A simple system to simultaneously measure volatilization and metabolism of pesticides from soils', *J. Agr. Food Chem.* **24**, 424–426.
Keller, A. (1976, 1977). Unpublished data. CIBA-GEIGY Ltd., Agricultural Division.
Kennedy, J. M., and Talbert, R. E. (1977). 'Comparative persistence of dinitroaniline type herbicides on the soil surface', *Weed Sci.* **25**, 373–381.
Kimura, R., and Yamaguchi, M. (1978). 'Microbial decomposition of 2-oxo-4-methyl-6-ureido-hexahydropyrimidine in soil', *Soil Biol. Biochem.* **10**, 503–508.
Laanio, T. L. (1974). Unpublished data. CIBA-GEIGY Ltd., Agricultural Division.
Lay, M. M., and Ilnicki, R. D. (1975). 'Effect of soil storage on propanil degradation', *Weed Res.* **15**, 63–66.
Lees, H. (1947). 'A simple automatic percolator', *J. Agr. Sci.* **37**, 27–28.
Lees, H., and Quastel, J. H. (1946). 'Biochemistry of nitrification in soil. Addendum by H. Lees. A soil perfusion apparatus', *Biochem. J.* **40**, 812–815.
Liang, T. T., and Lichtenstein, E. P. (1976). 'Effects of soils and leaf surfaces on the photodecomposition of [^{14}C]azinphosmethyl', *J. Agr. Food Chem.* **24**, 1205–1210.
Lichtenstein, E. P., Fuhremann, T. W., and Schulz, K. R. (1974). 'Translocation and metabolism of [^{14}C]-phorate as affected by percolating water in a model soil-plant ecosystem', *J. Agr. Food Chem.* **22**, 991–996.
Loos, M. A., Kontson, A., and Kearney, P. C. (1980). 'Inexpensive soil flask for [^{14}C]-pesticide degradation studies', *Soil Biol. Biochem.* accepted for publication.
Lynch, J. M., and Panting, L. M. (1979). 'Cultivation and soil biomass', *Soil Biol. Biochem.* **12**, 29–33.

Marsh, J. A. P., and Davies, H. A. (1978). 'The effect of herbicides on respiration and transformation of nitrogen in two soils. III. Lenacil, terbacil, chlorthiamid and 2,4,5-T', *Weed Res.* **18**, 57–62.

Marsh, J. A. P., Davies, H. A., and Grossbard, E. (1977). 'The effect of herbicides on respiration and transformation of nitrogen in two soils. I. Metribuzin and glyphosate', *Weed Res.* **17**, 77–82.

Marvel, J. T., Brightwell, B. B., Malik, J. M., Sutherland, M. L., and Rueppel, M. L. (1978). 'A simple apparatus and quantitative method for determining the persistence of pesticides in soil', *J. Agr. Food Chem.* **26**, 1116–1120.

Marvel, J. T., Sutherland, M. L., Brightwell, B. B., Malik, J. M., and Rueppel, M. L. (1976). 'A simple apparatus and quantitative method for determining the persistence of pesticides in soil and soil/water mixtures', *172nd National Meeting of the American Chemical Society, San Francisco, August 1976*, PEST 02.

McLaren, A. D., Luse, R. A., and Skujins, J. J. (1962). 'Sterilization of soil by irradiation and some further observations on soil enzyme activity', *Proc. Soil Sci. Soc. Amer.* **26**, 371–377.

Meikle, R. W., Youngson, C. R., Hedhund, R. T., Goring, C. A. I., Hamaker, J. W., and Addington, W. W. (1973). 'Measurement and prediction of picloram disappearance rates from soil', *Weed Sci.* **21**, 549–555.

Newby, L. C., Ballantine, L. G., Ellgehausen, H., and Guth, J. A. (1979). 'Herbicide soil metabolism. Laboratory methods', *Weed Sci. Soc. Amer. Meeting, San Francisco*, Abs. No. 247.

Nilles, G. P., and Zabik, M. J. (1974). 'Photochemistry of bioactive compounds. Multiphase photodegradation of basalin', *J. Agr. Food Chem.* **22**, 684–688.

Nilles, G. P., and Zabik, M. J. (1975). 'Photochemistry of bioactive compounds. Multiphase photodegradation and mass spectral analysis of basagran', *J. Agr. Food Chem.* **23**, 410–415.

Parochetti, J. V., and Dec, G. W. Jr. (1978). 'Photodecomposition of eleven dinitroaniline herbicides', *Weed Sci.* **26**, 153–156.

Parochetti, J. V., and Hein, E. R. (1973). 'Volatility and photodecomposition of trifluralin, benefin and nitralin', *Weed Sci.* **21**, 469–473.

Parr, J. F., and Reuszer, H. W. (1962). 'Organic matter decomposition as influenced by oxygen level and flow rate of gases in the constant aeration method', *Proc. Soil Sci. Soc. Amer.* **26**, 552–556.

Parr, J. F., and Smith, S. (1969). 'A multipurpose manifold assembly: use in evaluating microbiological effects of pesticides', *Soil Sci.* **107**, 271–276.

Plimmer, J. R. (1978). 'Photolysis of TCDD and trifluralin on silica and soil', *Bull. Environ. Contam. Toxicol.* **20**, 87–92.

Powlson, D. S., and Jenkinson, D. S. (1976). 'The effects of biocidal treatments on metabolism in soil. II. Gamma irradiation, autoclaving, air-drying and fumigation', *Soil Biol. Biochem.* **8**, 179–188.

Pramer, D., and Bartha, R. (1972). 'Preparation and processing of soil samples for biodegradation studies', *Environ. Lett.* **2**, 217–224.

Priest, B., and Stephens, R. J. (1975). 'Studies on the breakdown of p-chlorophenyl-methyl carbamate I. In soil. *Pestic. Sci.* **6**, 53–59.

Rahn, P. R., and Zimdahl, R. L. (1973). 'Soil degradation of two phenyl pyridazinone herbicides', *Weed Sci.* **21**, 314–317.

Roberts, T. R. (1976). 'Experimental models for studying the fate of pesticides in plants', *Proc. BCPC Symposium: Persistence of Insecticides and Herbicides*, 159–168.

Rovira, A. D., and McDougall, B. M. (1967). 'Microbiological and biochemical aspects of the rhizosphere', in *Soil Biochemistry* (Eds. A. D. McLaren and G. H. Peterson), Vol. I, 417–463.

Sizer, I. W. (1943). 'Effects of temperature on enzyme kinetics', *Adv. Enzymol.* **35**, 35–62.
Skipper, H. D., and Westermann, D. T. (1973). 'Comparative effects of propylene oxide, sodium azide and autoclaving on selected soil properties', *Soil Biol. Biochem.* **5**, 409–414.
Smith, C. A., Iwata, Y., and Gunther, F. A. (1978). 'Conversion and disappearance of methidathion on thin layers of dry soil', *J. Agr. Food Chem.* **26**, 959–962.
Sperber, J. I., and Sykes, B. J. (1964). 'A perfusion apparatus with variable aeration', *Plant and Soil* **20**, 127–130.
Temple, K. L. (1951). 'A modified design of the Lees soil percolation apparatus', *Soil Sci.* **71**, 209–210.
Torstensson, L. (1980). In Hance, R. J. (Ed.) *Interactions between Herbicides and the Soil*, Academic Press, London, In press.
Vela, G. R., and Wyss, O. (1962). 'The effect of gamma radiation on nitrogen transformations in soil', *Bact. Proc.* **24**, A17.
Wallnöfer, P., Poschenrieder, G., and Engelhardt, G. (1977). 'Verhalten von Pestizidkombinationen im Boden', *Z. Pflkrankh. Pflschutz.*, Sonderheft VIII, 199–207.
Warcup, J. H. (1957). 'Chemical and biological aspects of soil sterilisation', *Soil Fertil.* **20**, 1–5.

Progress in Pesticide Biochemistry, Volume 1
Edited by D. H. Hutson and T. R. Roberts
© 1981 John Wiley & Sons, Ltd.

CHAPTER 3

The metabolism of the synthetic pyrethroids in plants and soils

Terry R. Roberts

INTRODUCTION.	115
Plant and Soil Metabolism Studies.	117
METABOLISM IN PLANTS	119
Permethrin.	119
Decamethrin	123
Cypermethrin.	125
Fenvalerate.	128
Phenothrin.	130
DEGRADATION IN SOIL	132
Permethrin.	132
Cypermethrin.	137
Fenpropathrin.	139
Fenvalerate.	140
Phenothrin.	142
CONCLUSIONS.	145
REFERENCES.	145

INTRODUCTION

During the last decade a new class of agricultural insecticides, the synthetic pyrethroids, has emerged as a complement to the organochlorines, organophosphates, and carbamates. The history of the emergence of the synthetic pyrethroids from research programmes on natural pyrethrins is now well documented (Elliott and Janes, 1973). The major constituents of natural pyrethrum (pyrethrin I and pyrethrin II) combined high insecticidal activity with low acute mammalian toxicity. However, the problem of isolation on a large scale from *Chrysanthemum cinerariaefolium* limited their availability. More important, their photochemical instability rendered them totally unsuitable for consideration as agricultural insecticides.

The earlier synthetic analogues such as allethrin (**1**), resmethrin (**2**), and phenothrin (**3**) also exhibited high biological activity but showed little improvement in photochemical stability compared with natural products.

The first photostable pyrethroid, permethrin (NRDC 143) (**4**), was reported in 1973 by Elliott and Janes (1973). Permethrin is structurally closely related to

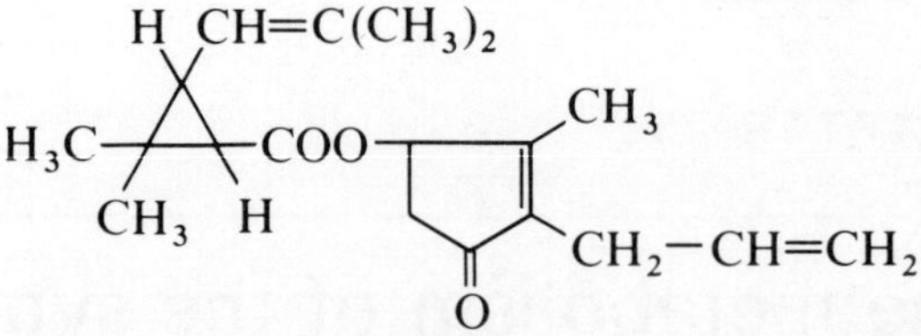

[Structure (1)]

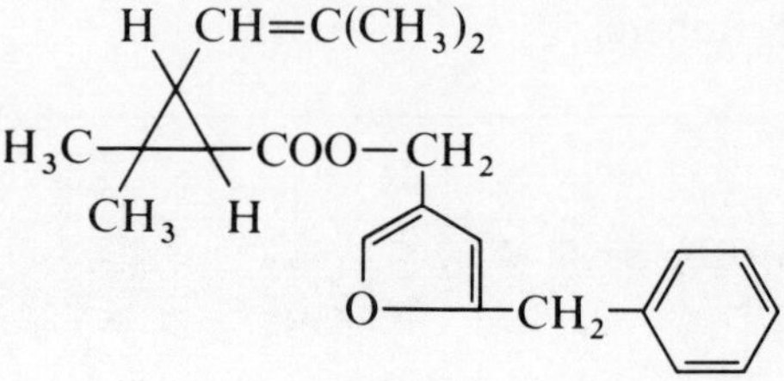

[Structure (2)]

[Structure (3)]

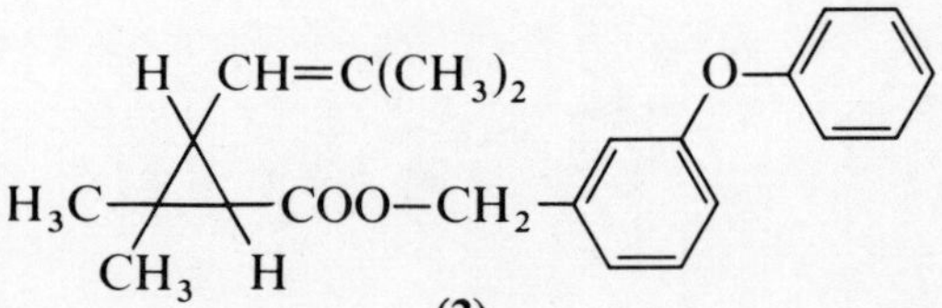

phenothrin but the dichlorovinyl side chain results in enhanced insecticidal activity and much greater photochemical stability. The introduction of an α-cyano group into the 3-phenoxybenzyl moiety, notably in the compounds cypermethrin (5), fenpropathrin (6), and decamethrin (7), served to further enhance biological activity. In fact, decamethrin, which is the single, most active isomer ([S],1R,cis) of structure (7), is currently one of the most potent insecticides known and is 1000 times more active against house flies than pyrethrin I (Elliott, 1977).

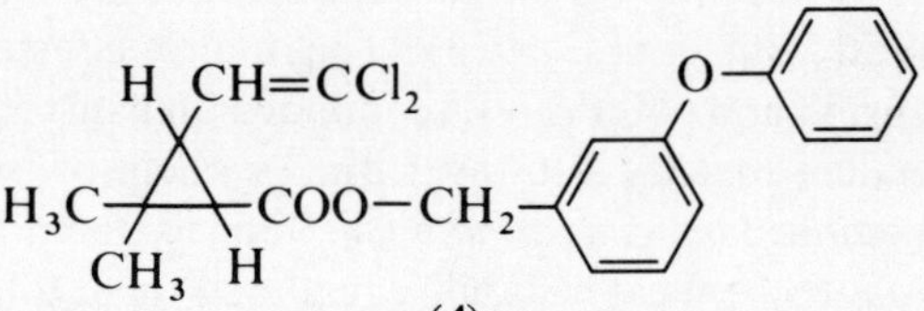

[Structure (4)]

Another compound, synthesized by Ohno et al. (1974, 1976), using 3-phenoxybenzaldehyde cyanohydrin and 2-(4-chlorophenyl)-3-methylbutyric acid is fenvalerate (8). This structure bears little resemblance to that of pyrethrin I and the fact that it too is a highly insecticidal compound emphasizes the considerable scope within this field of chemistry for novel structures.

(5) $R^1 = H$, $R^2 = CH=CCl_2$
(6) $R^1 = CH_3$, $R^2 = CH_3$
(7) $R^1 = H$, $R^2 = CH=CBr_2$

(8)

Of the synthetic compounds discussed, those listed in Table 3.1 are currently used as agricultural insecticides. Consequently, with these compounds, studies of their metabolic fate in plants and degradation in soil have been required and the information currently available in the published literature is included in this chapter. Since the use of pyrethroids in agriculture is relatively recent there is considerably less information available (on fewer compounds) than there is on mammalian metabolism (Hutson, 1979) of pyrethrins and synthetic analogues.

Plant and Soil Metabolism Studies

In mammalian metabolism studies there is considerable interest in species variations and the metabolism of a pesticide in several animal species is usually compared. It is fair to say that valid comparisons between the metabolism of a series of compounds in a single species at different laboratories can also usually be made.

This is not the case with plant and soil studies since there is considerable variation in the experimental conditions used in different laboratories. For example, the methods of application of chemicals to plants include treatment of individual leaves, stem injection, root uptake, or an overall spray and either the formulated or the unformulated compound could be used. Plant experiments can be carried out in the laboratory, glasshouse, growth chamber, or outdoors and this can lead to marked variations in the rate of metabolism and in the distribution of products remaining in the plant. Photochemical conversions (where possible) are likely to be observed only when treated plants are grown outdoors and the extent of such conversions will depend on the climatic conditions.

Table 3.1 Names and structures of synthetic pyrethroids

Trivial name	Structure	Systematic name
Cypermethrin	H_3C, CH_3 on cyclopropane with $CH=CCl_2$ and $COO-CH(CN)-$(3-phenoxyphenyl)	[S,R]-α-cyano-3-phenoxybenzyl (1R,1S,*cis,trans*)-2,2-dimethyl-3-(2,2-dichlorovinyl)cyclopropane carboxylate
Decamethrin	H_3C, CH_3 on cyclopropane with $CH=CBr_2$ and $COO-CH(CN)-$(3-phenoxyphenyl)	[S]-α-cyano-3-phenoxybenzyl (1R,*cis*)-2,2-dimethyl-3-(2,2-dibromovinyl)cyclopropane carboxylate
Fenpropathrin	H_3C, CH_3, H_3C, CH_3 tetramethylcyclopropane with $COO-CH(CN)-$(3-phenoxyphenyl)	[S,R]-α-cyano-3-phenoxybenzyl 2,2,3,3-tetramethylcyclopropane carboxylate
Fenvalerate	(4-Cl-C₆H₄)-CH(CH(CH₃)₂)-COO-CH(CN)-(3-phenoxyphenyl)	[S,R]-α-cyano-3-phenoxybenzyl [S,R]-2-(4-chlorophenyl)-3-methylbutyrate
Permethrin	H_3C, CH_3 on cyclopropane with $CH=CCl_2$ and $COO-CH_2-$(3-phenoxyphenyl)	3-Phenoxybenzyl (1R,1S,*cis,trans*)-2,2-dimethyl-3-(2,2-dichlorovinyl)cyclopropane carboxylate
Phenothrin	H_3C, CH_3 on cyclopropane with $CH=C(CH_3)_2$ and $COO-CH_2-$(3-phenoxyphenyl)	3-Phenoxybenzyl (1R,1S,*cis,trans*)-2,2-dimethyl-3-(2,2-dimethylvinyl)cyclopropane carboxylate

In soil studies, not only can the nature of the soil vary considerably from site to site (even within the same geographical location), but the condition of the soil used in different experiments can vary. For laboratory studies fresh field soils should be used where possible and information on their sampling regime and prior history of pesticide usage could be important factors affecting the ability of the soil to degrade a particular pesticide. Recommendations for conducting soil degradation studies have recently been made (Hill and Arnold, 1978).

In view of the foregoing comments it is often difficult to make valid detailed comparisons of the fate of a series of related compounds such as the pyrethroids in plants or soils. However, the overall routes of metabolism can be usefully compared, especially when studies with several compounds have been conducted under similar conditions in the same laboratory.

METABOLISM IN PLANTS

Permethrin

Permethrin (**4**) is a mixture of two pairs of diastereoisomers referred to as the *cis*- and *trans*-isomers. As permethrin was one of the first synthetic pyrethroids to be suitable for use as an agricultural chemical it also gave rise to the first reports of its metabolism in plants.

Ohkawa *et al.* (1977) used separate samples of *cis*-[^{14}C]permethrin and *trans*-[^{14}C]permethrin labelled separately in the dichlorovinyl group and in the methylene carbon atom (**9**). Using these four labelled preparations, the short term metabolism on permethrin in snap bean seedlings (*Phaseolus vulgaris* L.) was studied under glasshouse conditions. Each radiolabelled chemical was applied to a single leaf of 14 day old seedlings at a concentration of 15µg (in 100 µl methanol) per leaf. At time intervals up to 14 days after treatment, five leaves were removed and rinsed with methanol prior to extraction by homogenizing with methanol. The leaf washings and extracts were examined separately.

$$H_3C \underset{CH_3}{\overset{H\ CH=\overset{14}{C}Cl_2}{\bigtriangleup}} COO-\overset{14}{C}H_2-\text{(aryl)}$$
(**9**)

Autoradiography of whole treated plants prior to extraction showed that very little translocation (<1%) of radiolabelled *cis*- or *trans*-permethrin or its metabolites had occurred at 14 days after treatment. After 14 days the amounts of radioactivity present were 13–17% (of the amount applied) in the surface wash, 46–58% in the methanol extract, and 8–14% unextracted by methanol in the plant residue.

Some interconversion of the *cis*- and *trans*-isomers occurred and the *cis*-isomer was slightly more persistent than the *trans*-isomer. The initial half-lives on bean plants were 9 and 7 days respectively for the *cis*- and *trans*-isomers under the conditions used.

A large number of metabolites were present in the plant extracts, the major ones from the alcohol moiety being 3-phenoxybenzyl alcohol (PBAlc), 3-phenoxybenzoic acid (PBAcid), 2'-HO-per (**10**), 4'-HO-per (**11**), 2'-HO-PBAlc (**12**), and 4'-HO-PBAlc (**13**). At least eight minor compounds remained unidentified. The polar compounds present underwent hydrolysis with β-glucosidase to yield PBAlc, 2'-HO-PBAlc, and 4'-HO-PBAlc as aglycones.

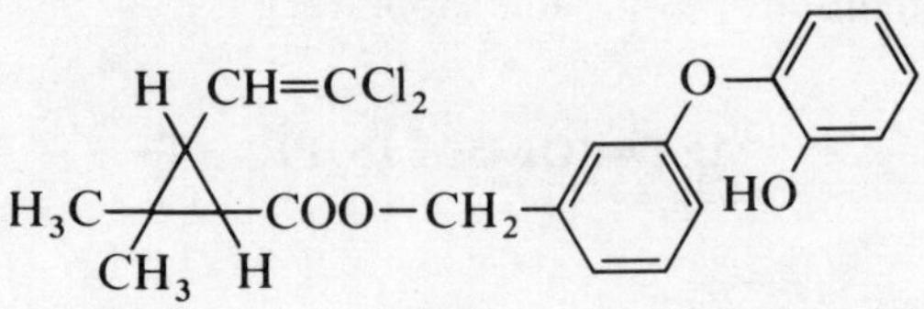

(**10**)

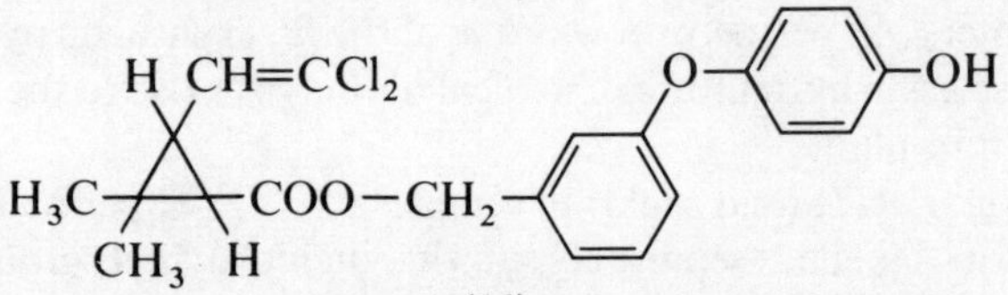

(**11**)

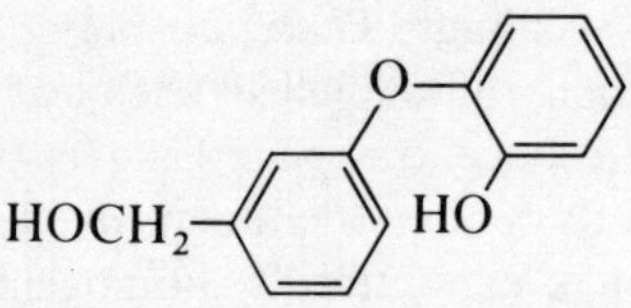

(**12**) 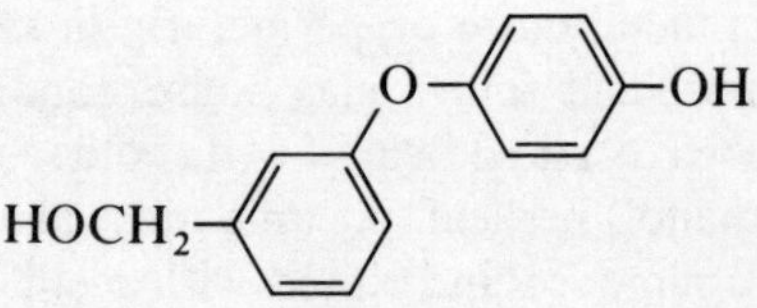   (**13**)

The cyclopropane carboxylic acids *cis*-Cl$_2$CA and *trans*-Cl$_2$CA (**14**) were the major metabolites formed from the acid moiety by hydrolysis of permethrin. Both acids were present mainly in conjugated form. As the conjugates underwent hydrolysis with β-glucosidase it was inferred that these conjugates were mainly glucosides.

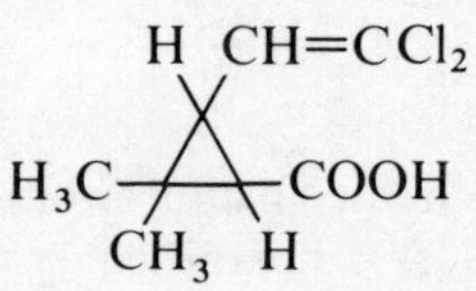

(**14**)

In separate experiments, Casida and co-workers have also studied the metabolism of permethrin in snap beans in the glasshouse. In addition the metabolism studies were extended to cotton growing in the glasshouse and outdoors (Gaughan *et al.*, 1977; Gaughan and Casida, 1978). Four radiolabelled compounds were used, namely *cis*- and *trans*-permethrin labelled separately in the carboxy or methylene carbon atom.

In the glasshouse, individual leaves on snap beans (*cv.* Contender) and on cotton plants (*cv.* Stoneville 7A) were treated separately with each labelled isomer preparation. Similar treatments were made to cotton plants growing under field conditions near Davis, California. In some experiments beans were treated by stem injection. The experiments on beans were less detailed than those on cotton. However, it was noted that good recoveries of radioactivity were obtained up to 3 weeks after topical applications to bean leaves. The metabolites detected were the same as those reported by Ohkawa *et al.*, although rates of degradation were slower in Gaughan and Casida's experiments.

In the glasshouse and outdoor experiments with cotton, samples of leaves were taken 3 and 6 weeks after treatment. Only minor differences in the nature of products formed under glasshouse and outdoor conditions were observed. About 30% of the radiolabel was lost from the plants under outdoor conditions within 1 week and the more rapid loss of the *trans*-isomer from plants was confirmed. The major degradation pathway was again hydrolysis. The major hydrolysis products, *cis*- and *trans*-Cl_2CA, PBAlc, and PBAcid, were mainly present as conjugates. Some conjugates could readily be cleaved with β-glucosidase whereas other were not. Little information was obtained on the nature of the latter compounds. However, they are unlikely to be amino acid conjugates since they did not undergo methylation.

The hydroxy esters, 4'-HO-per and 2'-HO-per, were mainly present as free compounds. In addition, permethrin hydroxylated in one of the cyclopropyl methyl groups was formed and this had not been detected in earlier studies with beans. The HO-Cl_2CA (**15**) was also present mainly as a mixture of conjugates.

$$\text{H} \quad \text{CH}=\text{CCl}_2$$
$$\text{HOCH}_2 - \overset{}{\underset{\text{CH}_3 \quad \text{H}}{\bigtriangleup}} - \text{COOH}$$

(**15**)

The authors of this work concluded that the types of products formed from permethrin in plants were similar to those formed in animals except for the nature of the conjugates.

The information on the metabolic pathways of permethrin following application to plants is summarized in Figure 3.1.

Although there is considerable information on the metabolism of permethrin in cotton and bean foliage, this could usefully be complemented by longer term experiments in plants grown to maturity. This would give additional information on the nature of any products present in the cotton or bean seed. Although residues of pesticide metabolites usually occur in plants following direct application, they can also arise by uptake of degradation products from the soil. If this occurs, the concentrations are usually much lower than those arising from direct

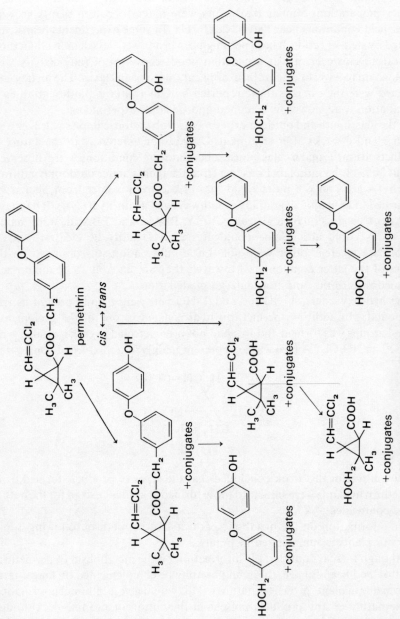

Figure 3.1 The metabolic pathway of permethrin in plants

application. However, it is possible that a soil degradation product which is not a plant metabolite could be taken up by the plant.

Experiments designed to examine this possibility for permethrin have been carried out by Leahey and Carpenter (1980). Sugar beet, wheat, lettuce, and cotton were grown in soil treated with [^{14}C]permethrin labelled in the cyclopropane ring or phenyl ring. In a series of experiments, individual pots (23 cm diameter) were filled with sandy loam soil and separate batches were treated with [^{14}C-*cyclopropyl*]permethrin and [^{14}C-*phenyl*]permethrin at doses equivalent to a spray application of 2 kg ha^{-1}. The applied compound was thoroughly mixed with the top 8 cm of soil and crops were sown 30, 60, and 120 days after treatment. The crops were grown to maturity but plants were thinned at intervals during the growing period. In mature plants from seed sown 30 days after soil treatment small radioactive residues (up to 0.86 mg kg^{-1}) were detected and in general higher residues occurred in crops grown in soil treated with [^{14}C-*cyclopropyl*]permethrin. The 30 day sugar beet sample was chosen for a detailed study of the compounds present and both Cl$_2$CA and a cyclopropane dicarboxylic acid (16) were shown to be present. The latter compound was identified by g.c.–mass spectrometry after methylation, and an authentic standard was subsequently synthesized (Huff, 1980). It is likely that both carboxylic acids were formed in soil and were subsequently taken up by the plants, since considerably lower radioactive residues were present in the parallel experiment with [^{14}C-*phenyl*]permethrin.

$$\text{HOOC} \underset{\underset{\text{CH}_3 \ \ \text{H}}{}}{\overset{\overset{\text{H} \ \ \text{CH}=\text{CCl}_2}{}}{\triangle}} \text{COOH}$$

(16)

Decamethrin

Decamethrin (7) is a single isomer (the [S],1*R*,*cis*-isomer) and a recent study of its metabolism in cotton has been reported by Ruzo and Casida (1979).

For this work samples of [^{14}C]decamethrin labelled in the dibromovinyl, benzylic, and cyano carbons were used. Cotton (*cv*. Stoneville 7A) was treated in the glasshouse and outdoors in Davis, California, in the same way as with the earlier experiments with permethrin from the same laboratory. Samples were taken for analysis 2 and 6 weeks after treatment in the glasshouse and outdoors.

Under glasshouse conditions the initial half-life of decamethrin was approximately 1 week. The overall rate of loss and amounts of extractable and unextracted radioactivity were similar for each radiolabelling position. Conversion of decamethrin into the *trans*-isomer occurred and after 6 weeks the *trans*:*cis* ratio was 0.44. Decamethrin degraded more rapidly under outdoor conditions and in addition to the more rapid loss of parent compound there was a higher rate of *cis*- to *trans*-conversion. Greater quantities of unextracted products were also present in outdoor samples.

Trace amounts (<1% of the applied radioactivity) of hydroxylated derivatives of decamethrin (4'-HO-dec (**17**), *t*-HO-dec (**18**), 4'-HO(*t*-HO)dec (**19**)) were detected with all three radiolabelled samples. However, the major metabolites were free and conjugated dibromovinylcyclopropane carboxylic acid (Br_2CA), PBAlc, and PBAcid together with smaller quantities of *t*-HO-Br_2CA and 4'-HO-PBAcid. The above-mentioned compounds are analogous to those formed from permethrin in cotton (Gaughan and Casida, 1978).

However, a major difference between the structures of the two pyrethroids is the presence of the α-cyano group in decamethrin. Consequently the first product of hydrolysis is the cyanohydrin PBcy (**20**) rather than PBAlc (from permethrin). 3-Phenoxybenzaldehyde (PBAld) was a decamethrin metabolite, and this was presumably formed from the unstable cyanohydrin.

Several types of conjugated metabolite (based on their behaviour on hydrolysis) were isolated but they were not fully characterized. One type cleaved readily with β-glucosidase or hydrochloric acid to yield Br_2CA and PBAcid from the appropriate labelled material. Two other types were resistant to β-glucosidase but were cleaved with hydrochloric acid to yield Br_2CA from the dibromovinyl label), PBAlc, PBAcid, and PBcy (from the benzyl label), and

$$\text{HO-CH}\begin{array}{c}\\|\\\text{CN}\end{array}\!\!\!\left\langle\!\!\begin{array}{c}\text{phenyl-O-phenyl}\end{array}\!\!\right\rangle$$

(20)

PBcy (from the cyano label). The authors concluded that PBcy must therefore have been present as a conjugate but the possibility that decamethrin itself could have been bound or conjugated to a tissue component was not ruled out. Such binding would have given rise to the same hydrolysis products (Br_2CA, PBAlc, PBAcid, PBcy) upon hydrolysis of the bound decamethrin.

The proposed conjugates of PBcy are of interest since they could be similar to the naturally occurring cyanogenic glucosides. The information on cyanogenic glucosides has been reviewed by Seigler (1977).

Decamethrin metabolism has also been studied in cotton and bean leaf discs. Fresh leaves taken from plants were placed under water and 10 mm diameter discs were pressed out with a cork borer. [^{14}C]Decamethrin in ethanol was added to flasks containing 25 discs in 2 ml distilled water and the mixture was incubated for 5 h at 30 °C. The supernatant was decanted and upon analysis was shown to contain only decamethrin. The discs were washed and extracted with methanol:chloroform (2:1 v/v).

Limited conversion (~6%) of decamethrin into glycoside conjugates (Br_2CA–glyc or PBAlc–glyc depending on the radiolabel position) occurred in the discs. Both [^{14}C]Br_2CA and [^{14}C]PBcy were also used as substrates for leaf disc metabolism experiments and much more extensive metabolism and/or conjugation took place. Three types of glycoside, separable on t.l.c., were formed from Br_2CA. For PBAlc and PBAcid, conjugates were formed when PBcy was used as substrate but very little PBcy conjugate was formed. The PBcy also degraded to PBAld which could be oxidized to PBAcid or reduced to the alcohol. Differences were also observed between cotton and bean.

The information from these experiments on decamethrin is summarized in Figure 3.2. Further work is necessary before the presence of PBcy conjugates can be confirmed, preferably with identification of the intact conjugates. Moreover, experiments with leaf discs and with whole plants have revealed the complex mixture of conjugated metabolites which can be formed and future work will no doubt throw more light on their nature and identity.

Cypermethrin

Cypermethrin (5) is the α-cyano analogue of permethrin and is a mixture of eight isomers in total. The *cis*- and *trans*-isomers each comprise two diastereoisomeric pairs.

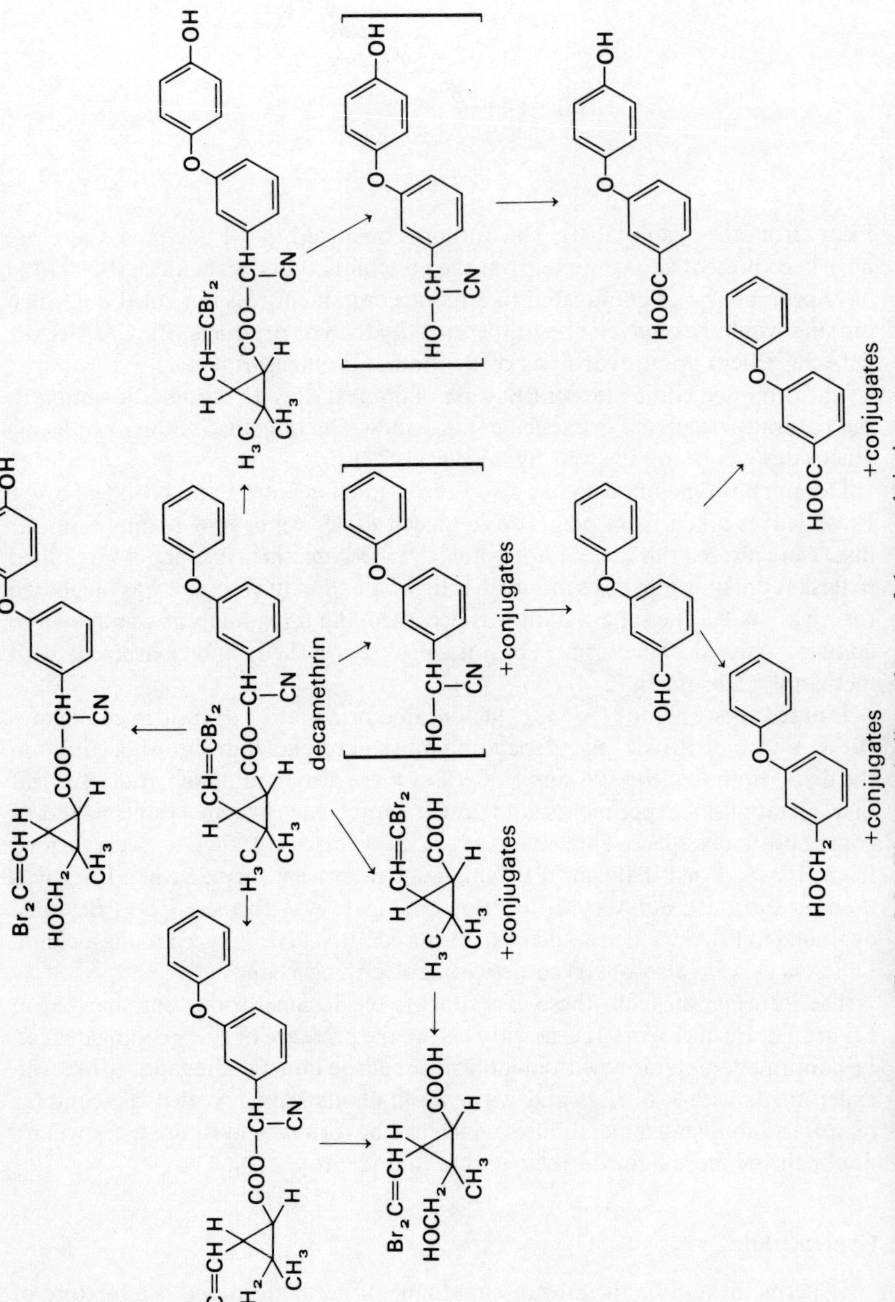

Figure 3.2 The metabolic pathway of decamethrin in cotton

[^{14}C]Cypermethrin (a *cis* : *trans* mixture) labelled in the cyclopropane ring has been applied to lettuce growing in soil under outdoor conditions (Wright *et al.*, 1980). The plants were sprayed twice at 0.3 kg ha^{-1} with the formulated insecticide and were harvested for analysis 21 days after the last treatment. The plants contained a total radioactive residue of 0.83 mg kg^{-1} (equivalent to cypermethrin) which comprised cypermethrin (33% of the total radiolabel present) and polar materials (54%) which were shown to be mainly conjugates of Cl$_2$CA. Evidence for this included transesterification of the conjugate fraction, using methanol/hydrogen chloride, to the methyl ester of Cl$_2$CA and acetylation of the fraction to a mixture of products the major one having the same chromatographic properties as an authentic sample of the tetra-acetyl glucose ester of Cl$_2$CA.

Since Cl$_2$CA was the major plant metabolite of the acid moiety from cypermethrin and it appeared to be present in more than one conjugated form, separate studies were then initiated to study its fate in plants. Abscised cotton leaves were supplied with solutions of *cis*-[^{14}C]Cl$_2$CA. Two major products were formed, the less polar one being identified as the glucose ester of Cl$_2$CA. The other compound could be hydrolysed by acid to Cl$_2$CA together with the sugars glucose, arabinose, and xylose in the approximate ratio 2 : 1 : 1. Acetylation followed by mass spectrometry gave evidence for the presence of disaccharide conjugates of Cl$_2$CA, these being tentatively the glucosylxylose and glucosylarabinose esters. These short term experiments emphasize the complex possibilities for conjugation of metabolites that can occur in plants.

In separate experiments, the metabolism of *cis*- and *trans*-isomers of cypermethrin in apple fruit and foliage was studied (Roberts and Dutton, 1980) Samples of the two isomers labelled separately in the cyclopropyl and benzyl rings were applied to individual leaves and fruits on apple trees growing in an orchard in Kent, each tree being surrounded by a net cage to prevent access by birds. The leaves were treated three times and the apples twice and the leaves and apples were harvested 4 and 3 weeks respectively after the final treatment.

Between 36 and 42% of the radioactivity recovered from the leaves was present as the unchanged insecticide and up to 30% conversion of the *cis*- into *trans*-isomer had occurred. There appeared to be no conversion of the *trans*-isomer into the *cis*-isomer. Metabolites detected in the leaves included PBAcid, PBAld, *cis*-Cl$_2$CA, *trans*-Cl$_2$CA, the amide analogue of cypermethrin (**21**), *cis*- and

(**21**)

trans-4'-HO-cyper (**22**). These free metabolites accounted for between 7 and 15% of the total radioactivity present in leaves. In addition a mixture of polar compounds was isolated from leaves and these underwent hydrolysis to PBAcid, PBAlc, PBAld, and 4'-HO-PBAcid. These compounds were presumably present as conjugates although the reason for formation of PBAld as a hydrolysis product is not clear. It is possible that the polar fractions contained small amounts of bound cypermethrin.

$$\text{H}_3\text{C} \underset{\text{CH}_3}{\overset{\text{H} \quad \text{CH}=\text{CCl}_2}{\diagdown \diagup}} \text{COO}-\underset{\text{CN}}{\overset{}{\text{CH}}}- \text{C}_6\text{H}_4-\text{O}-\text{C}_6\text{H}_4-\text{OH}$$

(**22**)

The conjugation of [^{14}C]-3-phenoxybenzoic acid (PBAcid) itself in cotton, vine, and other plant species was studied using abscissed leaves in order to obtain more information on the nature of the conjugates produced (More *et al.*, 1978). The PBAcid was rapidly converted into the glucose ester and disaccharide conjugates, particularly the glucosylarabinose and glycosylxylose esters.

Comparison of h.p.l.c. of some of the polar conjugates, isolated from apple leaf treated with [^{14}C]cypermethrin (after acetylation) with authentic standards of acetylated sugar conjugates of PBAcid and Cl$_2$CA indicated that the glucose ester of Cl$_2$CA and mono- and disaccharide conjugates of PBAcid were probably present. However, rigorous identification was not possible in view of the small amounts of purified metabolites available.

Less extensive metabolism occurred on the apple fruit. More than 98% of the total radiolabel recovered was associated with the peel and up to 77% of this was parent insecticide. Small amounts of the same free compounds detected in leaf extracts were also present in the peel together with polar materials. This information on the metabolism of cypermethrin in plants is summarized in Figure 3.3.

Fenvalerate

Fenvalerate (**8**) is another synthetic pyrethroid in commercial use for the control of cotton pests in several countries and for use on a wide variety of other crops. Ohkawa *et al.* (1980) have studied the metabolism of fenvalerate in bean plants under laboratory conditions.

Comparative experiments were carried out using [^{14}C]fenvalerate labelled in the cyano group and the [*S*]-acid ester isomer labelled separately in the cyano, carbonyl and benzylic carbon atoms. Separate leaves on kidney bean plants (*Phaseolus vulgaris* L.) were treated with these labelled preparations and samples

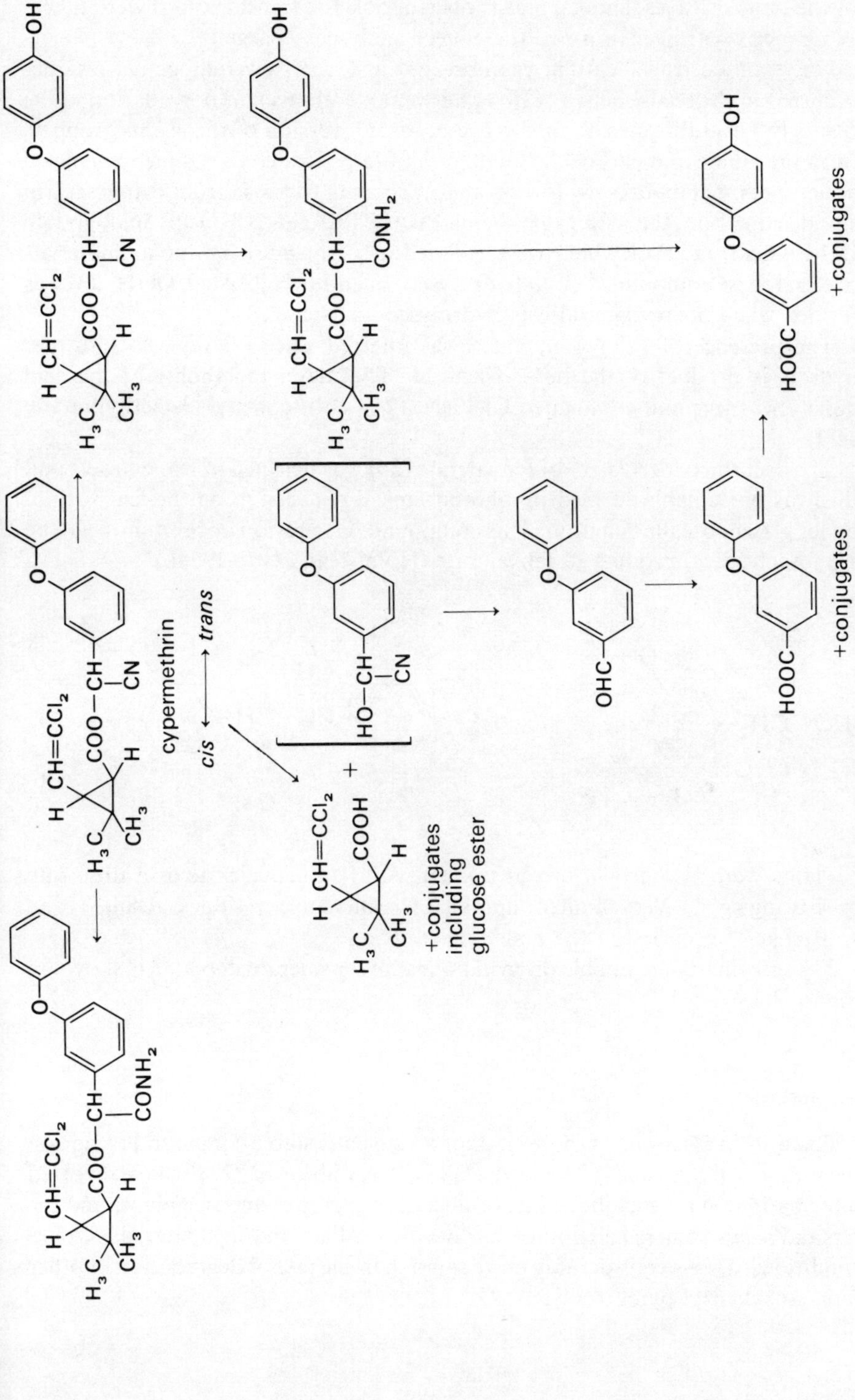

Figure 3.3 The metabolic pathway of cypermethrin in plants

of the treated leaves, shoots, and roots (to look for translocation) were taken. When pods with seeds formed these were analysed separately.

Only limited translocation was observed and very low radioactive residues occurred in the seeds. Fenvalerate underwent metabolism or degradation in the plants by several routes. A minor route was conversion of the cyano group to form the amide and carboxylic analogues of fenvalerate. The 3-phenoxybenzyl moiety was metabolized to form similar products to those from cypermethrin and decamethrin, the major ones being PBAlc, PBAcid, 2'-HO-PBAcid, and 4'-HO-PBAcid, most of which were isolated as conjugates (i.e. polar materials released these compounds upon hydrolysis). In addition PBAlc-COOH (23) was formed when polar material was hydrolysed.

The presence of PBcy conjugates was inferred since PBAld was a further hydrolysis product of the polar fractions. The major metabolite of the acid moiety was free and conjugated Cl-Vacid (2-(4-chlorophenyl)-3-methylbutyric acid).

The decarboxy derivative of fenvalerate (24) was detected in leaf extracts and this was presumably formed by photochemical conversion on the leaf surface under glasshouse illumination. This compound has previously been shown to be a photochemical product of fenvalerate (Holmstead *et al.*, 1978b).

Plants were also grown in soil treated with [^{14}C]fenvalerate at a dose rate of 1.0 mg kg^{-1}. Very limited uptake of radioactive residues resulted and <10 μg kg^{-1} occurred in the pods and seeds.

The information available on metabolism of fenvalerate in plants is shown in Figure 3.4.

Phenothrin

Phenothrin (3) is one of the least photochemically stable compounds amongst the synthetic pyrethroids (Barlow *et al.*, 1977). Nambu *et al.* (1980) have reported information on the metabolic fate of phenothrin in bean and rice plants and the *cis*- and *trans*-isomers had initial half-lives of less than one day under glasshouse conditions. This is considerably more rapid than the rate of degradation of other 3-phenoxybenzyl pyrethroids.

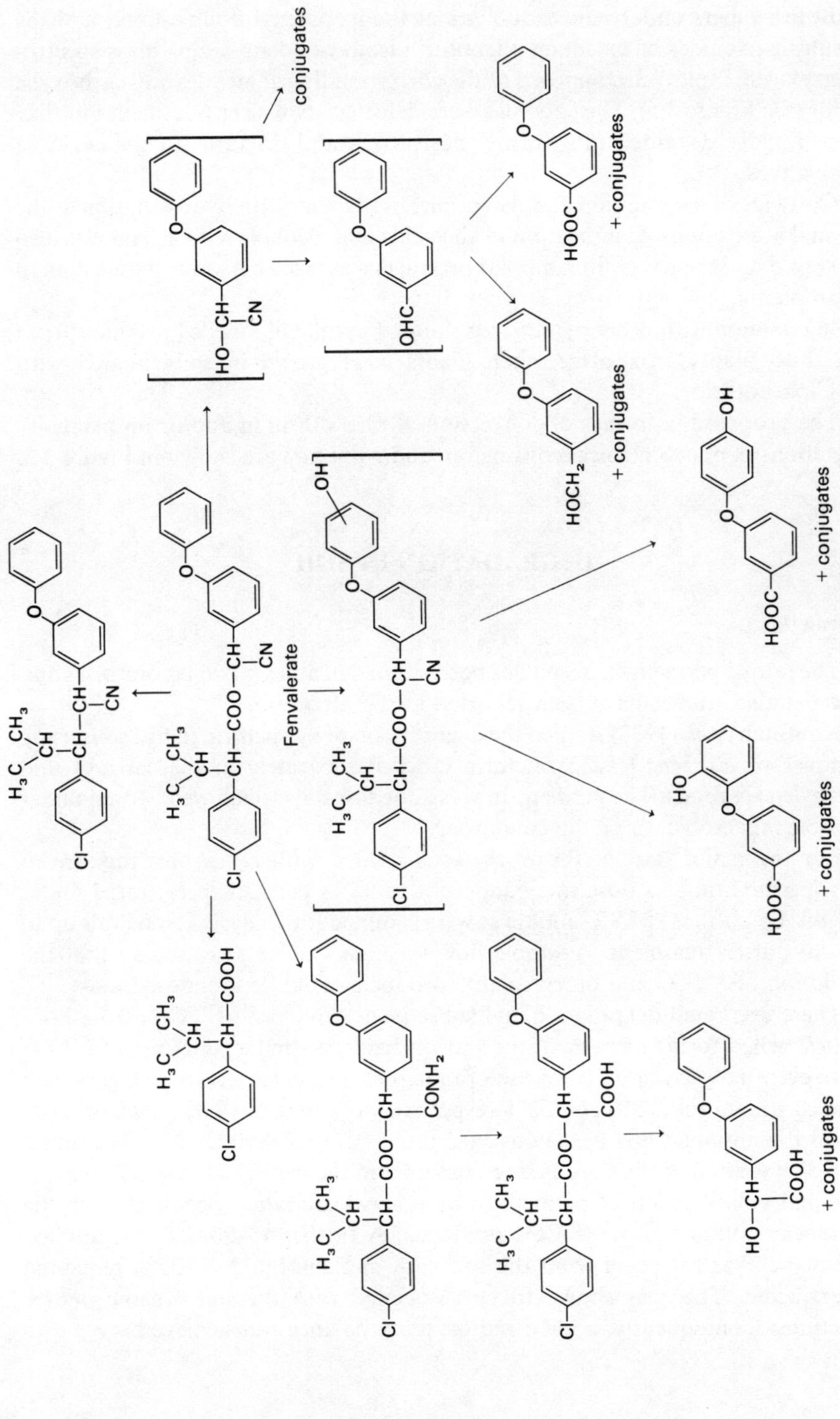

Figure 3.4 The metabolic pathway of fenvalerate in plants

Both isomers underwent ozonolysis at the isobutenyl double bond and the resulting ozonides of the intact phenothrin isomers (identified by mass spectrometry) were rapidly decomposed to the corresponding aldehydes and carboxylic acids (see Figure 3.5). The ozonides were detected soon after treatment but they were rapidly degraded into formyl derivatives and, in turn, to the carboxy derivatives.

Cleavage of the ester linkage also occurred together with hydroxylation at the 2'- and 4'-positions. Conjugation of the acids and alcohols with sugars was also observed and the formation of polar products was more extensive in rice than in bean plants.

In common with other pyrethroids, limited uptake of labelled products from soil into plants took place when plants were grown in soils treated with [^{14}C]phenothrin.

The proposed pathways of conversion of phenothrin in and/or on plants by metabolism, photochemical conversion, and oxidation are shown in Figure 3.5.

DEGRADATION IN SOIL

Permethrin

The fate of permethrin in soil has been studied in at least two laboratories and two detailed studies have been reported in the literature.

Kaufman et al. (1977) studied the degradation of permethrin in five soils using samples of *cis/trans* [^{14}C]permethrin labelled separately in the carboxy and methylene carbons. The permethrin was added to soils which were stored under aerobic, anaerobic, or sterile conditions.

For the major part of the work, soils treated with a *cis/trans* mixture of [^{14}C]permethrin at a dose rate equivalent to 0.2 lb per acre were stored under aerobic conditions at 25 °C and these were sampled for analysis at intervals up to 34 days after treatment. A simple flow-through system was used so that the evolution of $^{14}CO_2$ and other volatile products could be monitored.

There was rapid degradation and subsequent evolution of $^{14}CO_2$ from both radiolabelled forms of permethrin and on balance similar amounts of $^{14}CO_2$ were eventually given off from each radiolabel. For example, in a Hagerstown silty clay loam soil, 62% and 52% respectively of the methylene- and carboxyl-labelled compound had been converted into $^{14}CO_2$ within 27 days. The initial rate of evolution of $^{14}CO_2$ was more rapid from the methylene label. There was minimal volatilization of permethrin or other degradation products from the system in addition to the $^{14}CO_2$ produced. A further 15–20% of the applied radiolabel was extracted from the soil with methanol and 20–25% remained unextracted. This was shown to be associated with the soil organic matter fractions. Consequently, a good radioactivity balance was achieved.

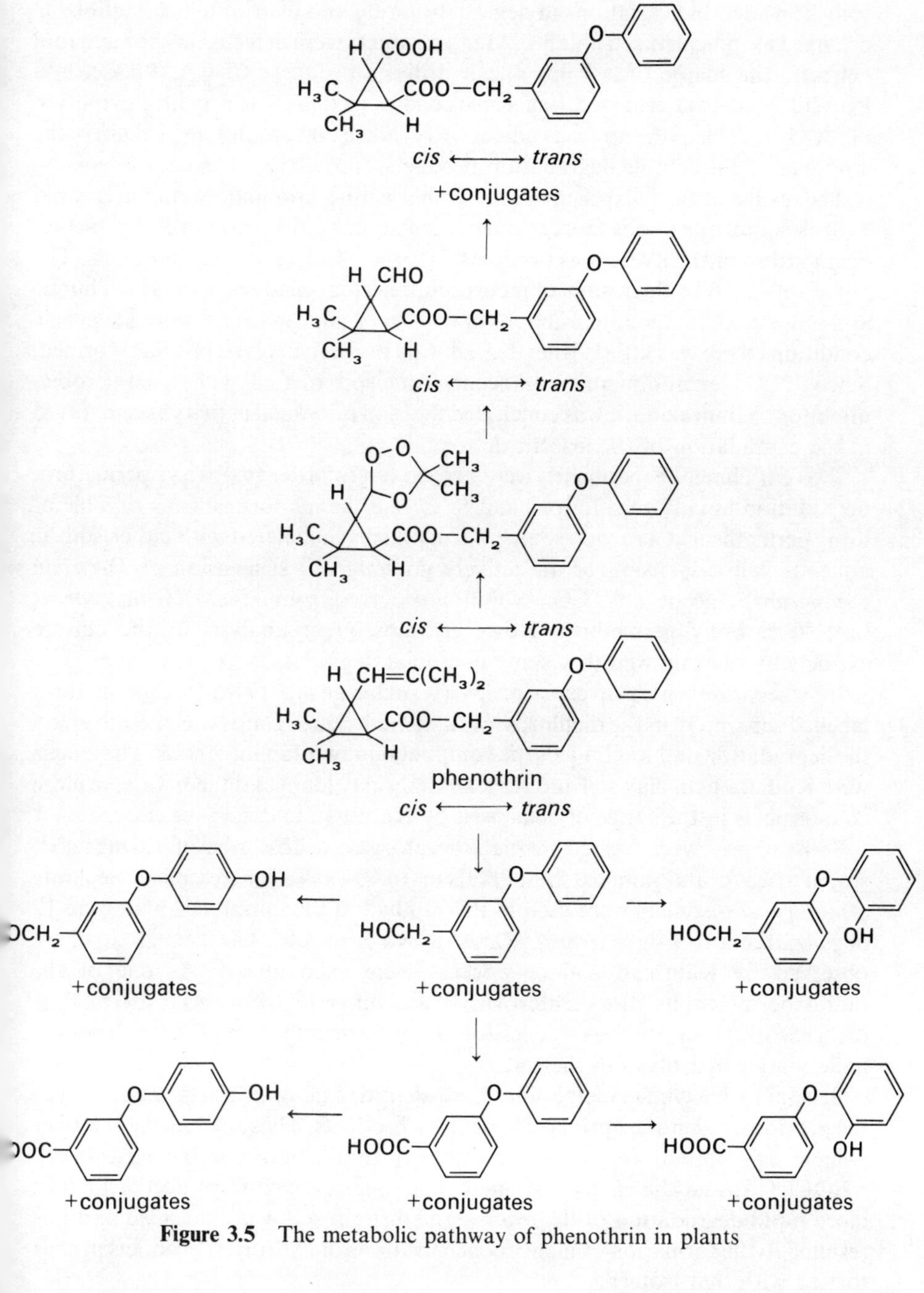

Figure 3.5 The metabolic pathway of phenothrin in plants

When the degradation of permethrin was examined using the range of five soils, considerable variations in degradation rate and distribution of radiolabel occurred as indicated in Table 3.2. Many products were detected in the methanol extracts, the major ones being the hydrolysis products Cl_2CA, PBAlc, and PBAcid. Under waterlogged, anaerobic conditions there was very little evolution of $^{14}CO_2$ (<1% of the applied radioactivity) and a corresponding increase in the amounts of extractable degradation products.

The results of these experiments show that permethrin underwent fairly rapid hydrolysis in most soils stored under aerobic conditions followed by further degradation of the hydrolysis products to form $^{14}CO_2$ and polar products. The latter appear to be the results of incorporation of labelled fragments into humic and fulvic acids and humin fractions of the soil organic matter. Under anaerobic conditions there was little further degradation of the hydrolysis products formed. Since $^{14}CO_2$ evolution did not occur from soil treated with the microbial inhibitor sodium azide, it was concluded that soil microbial activity was involved in the degradation of the insecticide.

Two enrichment experiments were carried out in order to observe permethrin degradation in culture solutions and to isolate the microorganisms capable of using permethrin as a carbon source. Using flasks inoculated with either soil, an aqueous soil suspension or an actively growing cell suspension of *Fusarium oxysporum* Schlecht, no $^{14}CO_2$ evolution occurred from either [^{14}C-methylene] or [^{14}C-carbonyl]permethrin. However, subsequent analysis of the culture extracts by t.l.c. showed that some microbial degradation had occurred.

In a separate series of experiments (Kaneko *et al.*, 1978) [^{14}C]permethrin labelled separately in the dichlorovinyl and methylene groups was used to study the degradation and leaching of the compound in two Japanese soils. These soils were Kodaira light clay soil and Azuchi sandy clay loam and there were no close counterparts in the range of soils used by Kaufman and co-workers.

Soil samples were treated in the laboratory at a dose rate of $1.0\,mg\,kg^{-1}$, stored at 25 °C and sampled at intervals up to 60 days after treatment. Separate *cis*- and *trans*-isomers were used in this study and the initial half-lives were 12 days (*cis*) and 6–9 days (*trans*). Rates of evolution of $^{14}CO_2$ similar to those observed by Kaufman and co-workers were encountered. As one of the radiolabel positions (the dichlorovinyl) was different from those used in the earlier work, this provides evidence for extensive degradation of the cyclopropyl moiety after hydrolysis in the soil.

Several hydroxylated metabolites were identified in soil extracts in addition to the previously identified primary hydrolysis products. These included 4′-HO-per (which was present in greater amounts from the more stable *cis*-isomer), *t*-HOCl$_2$CA, and the *cis*-hydroxymethyl lactones derived from each acid. The more rapid degradation of the *trans*-permethrin in soil was confirmed and this resulted in the formation of higher concentrations of hydrolysis products in soils treated with that isomer.

Table 3.2 Degradation of [^{14}C]permethrin in five soils (Kaufman *et al.*, 1977)

	Physical characteristics					Degradation products (% applied radiolabel)				
Soil	Sand (%)	Silt (%)	Clay (%)	OM (%)	pH	% moisture content	$^{14}CO_2$ released	Other volatiles	Extractable radioactivity	Unextractable radioactivity	Total
San Joaquin sandy loam	48.0	42.0	9.7	1.2	7.2	22.4	2.2	0.6	86.7	10.6	100.1
Dubbs fine sandy loam	48.8	44.0	7.2	1.0	5.9	23.7	46.0	0.7	17.1	38.7	102.5
Memphis silt loam	20.8	54.0	25.2	0.7	5.8	37.6	31.5	2.4	18.6	45.0	97.5
Hagerstown silty clay loam	17.0	50.6	32.4	2.3	7.5	32.6	51.0	0.4	22.5	26.0	99.9
Sharkey clay	20.8	32.0	47.2	6.1	5.9	45.5	31.1	0.3	40.7	28.5	100.6

OM = organic matter

A laboratory leaching study was also carried out using each of the Japanese soils, again incubated at $1.0\,\text{mg}\,\text{kg}^{-1}$. Elution with water was begun either immediately after treatment or after a 21 day ageing period during which degradation products would have formed. As expected, very little movement of permethrin itself occurred when leaching was started immediately. Even after 21 days' storage followed by leaching at least 80% of the applied radiolabel was still in the upper 5 cm of the column. However, radioactivity was also detected in lower sections and 0.3–2.6% was present in the column effluents. Compounds found in the lower soil layers included 4'-HO-per and PBAcid.

The current information on the degradation of permethrin in soil (Kaufman *et al.*, 1977; Kaneko *et al.*, 1978) is summarized in Figure 3.6.

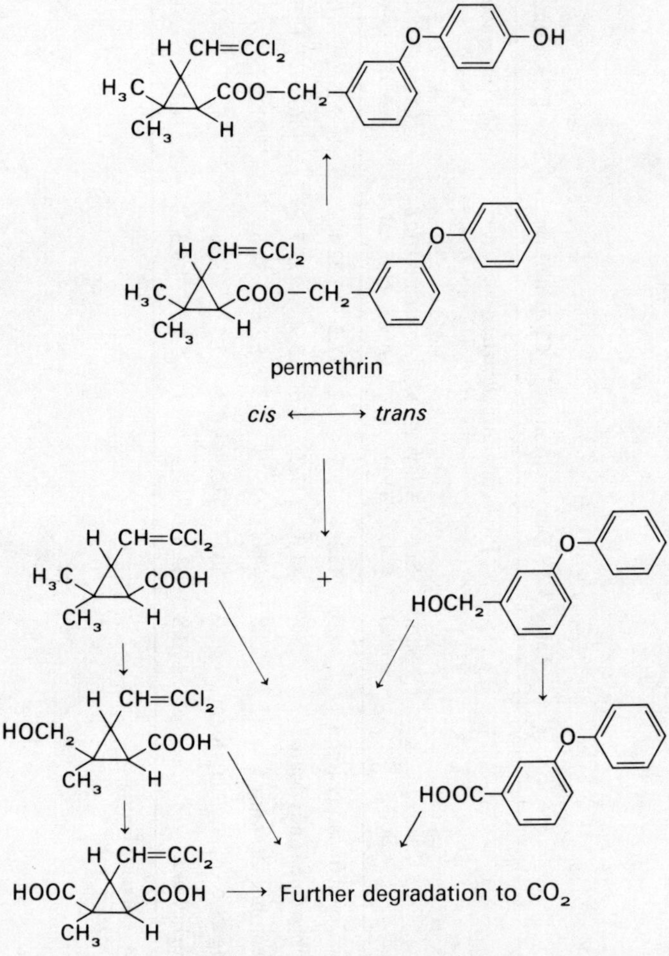

Figure 3.6 The degradation of permethrin in soil

Cypermethrin

The degradation of [^{14}C]cypermethrin has been studied in the laboratory in three soils. Two of these, a sandy clay and a clay, were taken from cotton growing areas in Spain and the third, a sandy loam, was taken from the UK (Roberts and Standen, 1977a). Separate *cis*- and *trans*-isomers (or in some cases a *cis/trans* mixture) were used labelled either in the C-1 position of the cyclopropyl ring or uniformly in the benzyl ring. Soils were treated at a dose rate of 2.5 mg kg^{-1} and were stored in the laboratory either in glass jars, in biometer flasks for collecting $^{14}CO_2$ released, or in a flow-through system for collecting all volatile products.

The initial half-life of the *cis*- and *trans*-isomers of cypermethrin was 4 and 2 weeks respectively in sandy clay and sandy loam soils although both isomers were more stable in the clay soil. As with permethrin, the major degradation route was hydrolysis leading to the formation of Cl_2CA and PBAcid. The initial hydrolysis product from the alcohol moiety would be the cyanohydrin which is unstable and would rapidly degrade and oxidize to form the PBAcid.

Under aerobic conditions further extensive degradation of the hydrolysis products occurred with the formation of $^{14}CO_2$. Bearing in mind that the positions of radiolabelling were in the cyclopropyl and benzyl rings (unlike the situation with permethrin studies), this $^{14}CO_2$ evolution is evidence for ring opening followed by degradation of the products formed. A minor additional degradation route of cypermethrin was oxidation to yield 4'-HO-cyper and 4'-HO-3-PBAcid.

When [^{14}C-benzyl]cypermethrin (*cis/trans* mixture) was incubated with waterlogged soil under nitrogen, hydrolysis occurred at a slightly slower rate than under aerobic conditions, but there was a build-up of PBAcid and little further metabolism. Another major difference was that less unextracted or 'bound' radioactivity remained in soils stored under anaerobic conditions. These differences are illustrated in Figure 3.7.

As the time intervals after treatment increased there were increases in the amounts of radioactivity not extracted by aqueous acetonitrile. As this increase paralleled the release of $^{14}CO_2$, it is likely to be the result of incorporation of degradation products (following ring opening) into soil organic matter fractions. This was confirmed by isolating organic matter fractions from soils containing 'bound' residues from which it was shown that fulvic acid, humic acid, and humin fractions all contained radioactivity. The fulvic acid fraction contained small amouts of PBAcid, 4'-HO-3-PBAcid, and Cl_2CA depending on the radiolabel used.

Soils containing 'bound' residues arising from cypermethrin applications were mixed with fresh soil and the evolution of $^{14}CO_2$ was monitored (Roberts and Standen 1980). Somewhat surprisingly, $^{14}CO_2$ was released at a slow but steady rate and between 21 and 37% of the radioactivity present as bound residues was mineralized over an 18 week period. These results suggest that the 'bound'

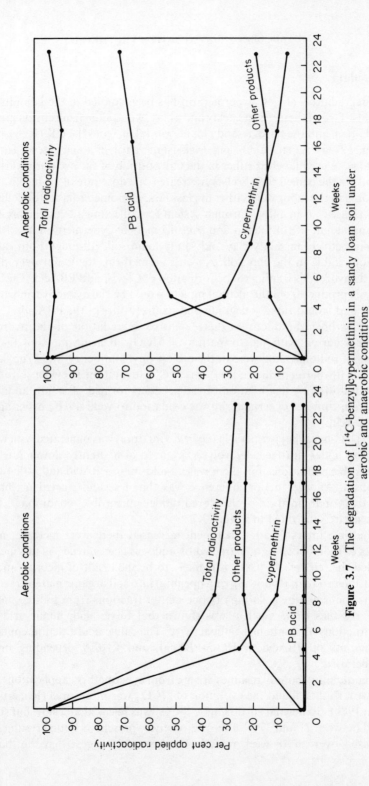

Figure 3.7 The degradation of [^{14}C-benzyl]cypermethrin in a sandy loam soil under aerobic and anaerobic conditions

residues were degrading further rather than accumulating and may be the result of incorporation of ^{14}C into the soil organic matter.

In subsequent work the cyclopropane dicarboxylic acid (16) was also identified by a combination of g.c./m.s. and synthesis of an authentic standard. This dicarboxylic acid was also formed from permethrin in soil (Leahey and Carpenter, 1980). It was also shown (Roberts and Standen, 1980) that when Cl_2CA itself was added to soil, it was converted initially into the hydroxymethyl analogue and then into (16).

In Figure 3.8 a degradation pathway of cypermethrin is proposed.

Fenpropathrin

Similar experiments to those described above for cypermethrin have been carried out on the structurally related compound fenpropathrin (Roberts and Standen, 1977b). Using [^{14}C]fenpropathrin labelled in the cyclopropyl and benzyl rings and the same three soils described previously (Roberts and Standen, 1977a), it was shown that fenpropathrin was somewhat more stable than cypermethrin, with an initial half-life in the range 4 to >16 weeks depending on the soil type.

Once again, hydrolytic cleavage of the ester bond predominated, followed under aerobic conditions by further degradation of the tetramethylcyclopropane carboxylic acid and PBAcid produced by ring opening and release of $^{14}CO_2$. In addition, hydrolysis at the cyano group was also observed and the resulting amide (25) was converted into the corresponding carboxylic acid (26). Oxidation of the phenyl ring was not observed with fenpropathrin.

As with cypermethrin, under waterlogged, anaerobic conditions the rate of hydrolysis of fenpropathrin was slower and the cyclopropane carboxylic acid and PBAcid were found to degrade slowly.

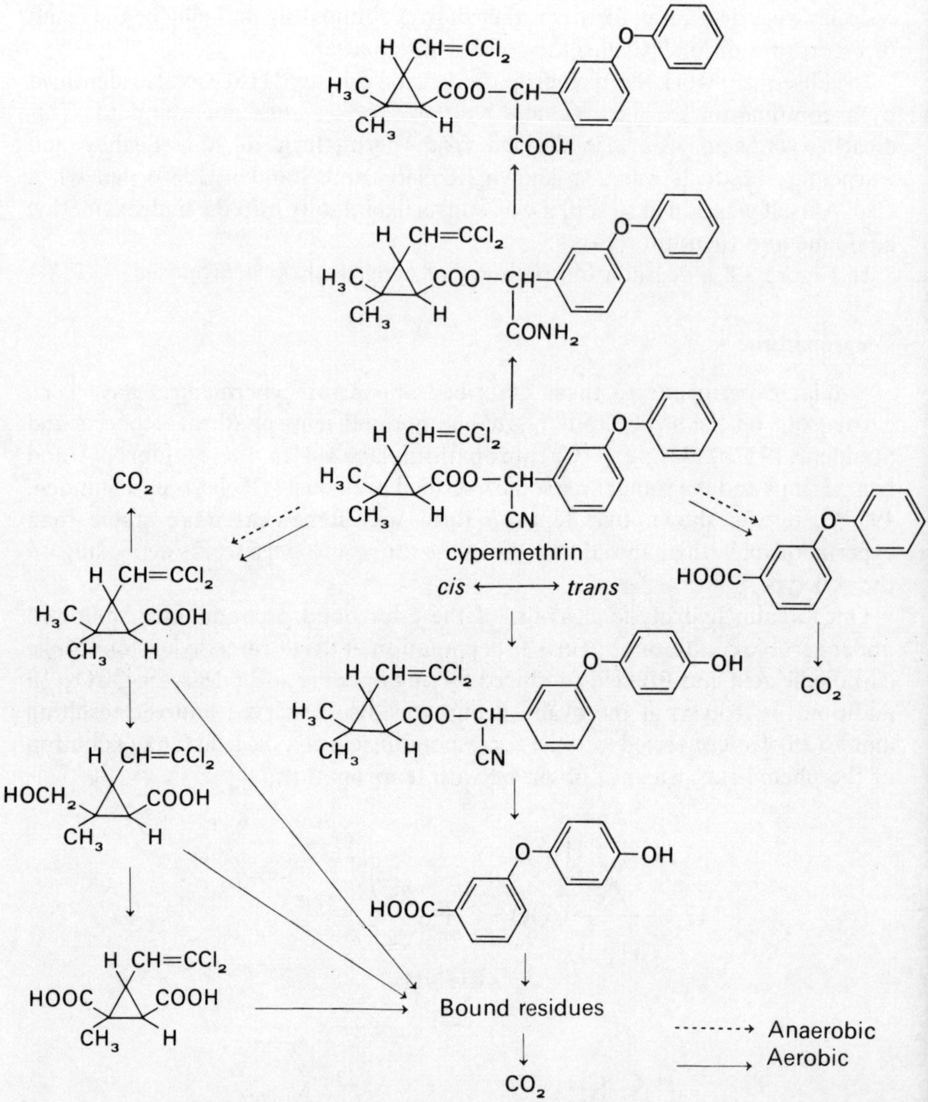

Figure 3.8 The degradation pathway of cypermethrin in soil

A proposed metabolic pathway for fenpropathrin in soil is shown in Figure 3.9.

Fenvalerate

Samples of [^{14}C]fenvalerate labelled separately in the carboxy and cyano groups have been used to study its degradation in four soils (Ohkawa et al., 1978)

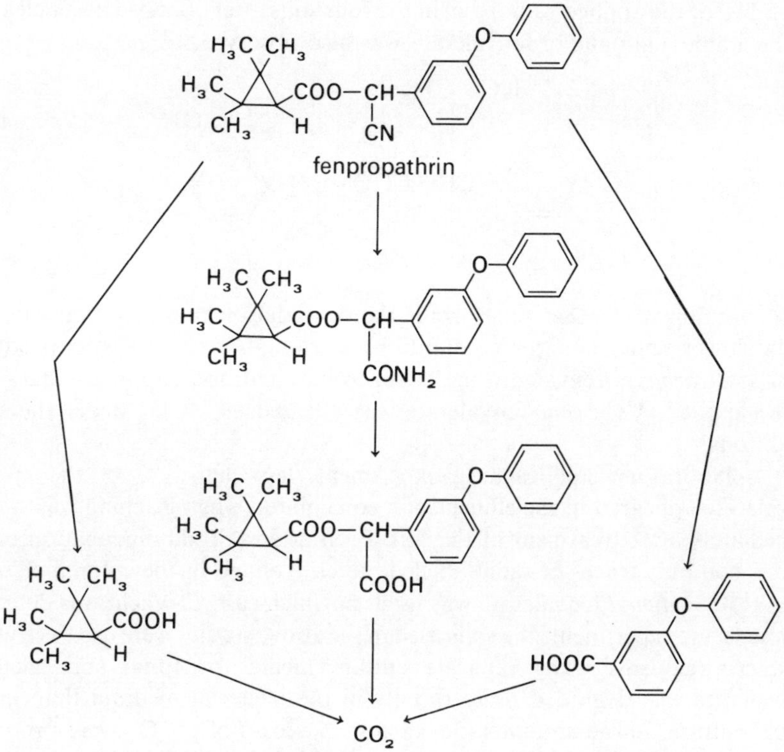

Figure 3.9 The degradation pathway of fenpropathrin in soil

including two of those used for permethrin degradation studies in the same laboratory (Kaneko et al., 1978). When [^{14}C]fenvalerate was added to soils at a dose rate of $1.0\,\text{mg}\,\text{kg}^{-1}$ and stored at $25\,°C$, the initial half-life ranged from 2 weeks to 3 months depending on the soil type.

The degradation products formed were analogous to those from the other pyrethroids discussed so far, and resulted from ester cleavage, ring hydroxylation in the 3-phenoxybenzyl moiety, and hydrolysis at the cyano group.

Thus, 2-(4-chlorophenyl)isovaleric acid (Cl-Vacid), 4-HO-fenvalerate, and $CONH_2$-fenvalerate were detected in extracts of all soils, although Cl-Vacid could not be detected when the cyano-labelled compound was used. Extensive $^{14}CO_2$ evolution occurred and the amounts were always greater when [^{14}C-*cyano*]fenvalerate was used. For example, 30 days after incubation of [^{14}C]fenvalerate in Katana sandy loam, 47.5% and 37.9% of the applied radiolabel had been released as $^{14}CO_2$.

In addition to these degradation pathways, another route not observed in studies with other pyrethroids was ether cleavage resulting in the formation of (27) which retains the ester intact. The amounts of (27) present ranged from

0.2–5.5% of the applied radiolabel in the four soils after 30 days. Ether cleavage of the amide analogue of fenvalerate was also observed.

$$Cl-\underset{}{\bigcirc}-\underset{\underset{CN}{|}}{CH}-COO-\underset{}{CH}-\underset{}{\bigcirc}-OH \quad \text{with } CH(CH_3)_2 \text{ group}$$

(27)

In common with other pyrethroids the degradation rate of fenvalerate was much slower under anaerobic conditions. Although no $^{14}CO_2$ was produced from [^{14}C-*carbonyl*]fenvalerate under anaerobic conditions, approximately 10% of the applied [^{14}C-*cyano*]fenvalerate was released as $^{14}CO_2$ under the same conditions.

In a laboratory soil leaching experiment, less than 1% of the applied fenvalerate appeared in the effluent of a soil column when leaching was started immediately after treatment of the soil. Even after a 30 day incubation of the treated soil only traces of radiolabelled material eluted in the column effluents when [^{14}C-*carbonyl*]fenvalerate was used, in which case Cl-Vacid was detected.

In separate experiments the soils used in the above studies were used as sources of microorganisms using separate culture media for fungi and bacteria. Fenvalerate was degraded more rapidly in the bacterial medium than in the fungal culture. Large amounts (35–42% in 2 weeks) of $^{14}CO_2$ were produced from both media when cyano-labelled fenvalerate was added compared with only 1.1–2.3% of $^{14}CO_2$ from the carbonyl label under the same conditions. In the latter case, Cl-Vacid was a major product, accounting for as much as 69% of the applied radiolabel. Essentially the same products were formed as in the soil studies but the carboxy analogue of fenvalerate (by conversion of the cyano group) was present in microbial solutions but not in soil extracts.

A summary of information available on fenvalerate degradation in soils and microbial culture is given in Figure 3.10.

Phenothrin

Two Japanese soils, Kodaira light clay soil and Katano sandy loam soil, were treated separately at dose rates of 1 mg kg^{-1} with the *cis*- and *trans*-isomers of [^{14}C]phenothrin labelled in the methylene of the benzyl group (Nambu *et al.*, 1980). Both isomers were rapidly degraded in soil with initial half-lives of 1–2 days. However, under waterlogged anaerobic conditions the degradation rate was much slower with half-lives of 2–4 weeks of the *trans*-isomer and 1–2 months for the *cis*-isomer.

After 6 months only 22–47% of the applied radioactivity was recovered from the soil samples and the major part of this was present as unextractable

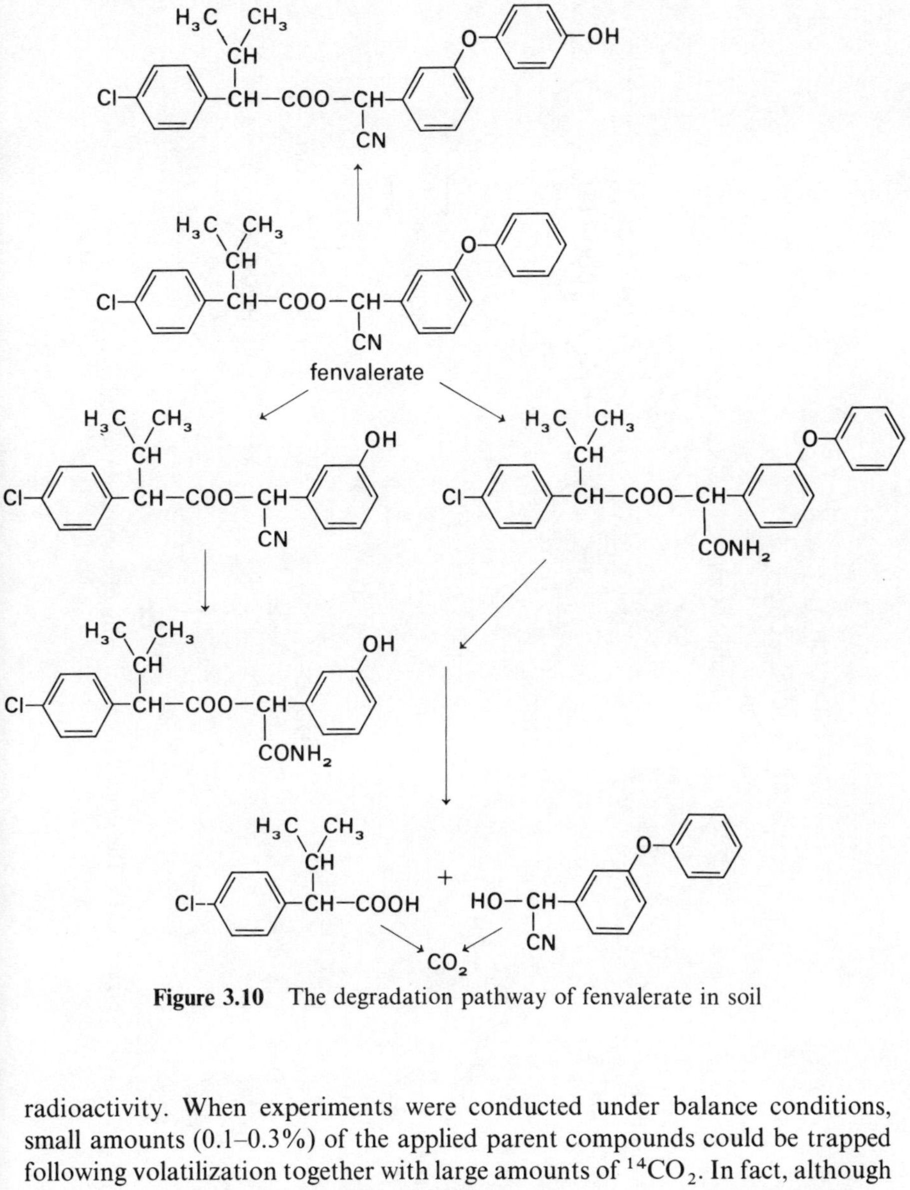

Figure 3.10 The degradation pathway of fenvalerate in soil

radioactivity. When experiments were conducted under balance conditions, small amounts (0.1–0.3%) of the applied parent compounds could be trapped following volatilization together with large amounts of $^{14}CO_2$. In fact, although the parent compounds had short half-lives, there was a steady, almost linear evolution of $^{14}CO_2$ over a 30 day period which must have been due to further degradation of the primary degradation products.

Amongst the degradation products identified during the course of the study were the hydrolysis products PBAlc and PBAcid, the 4′hydroxyderivatives of PBAlc, PBAcid, and of phenothrin itself. Ether cleavage was also observed to

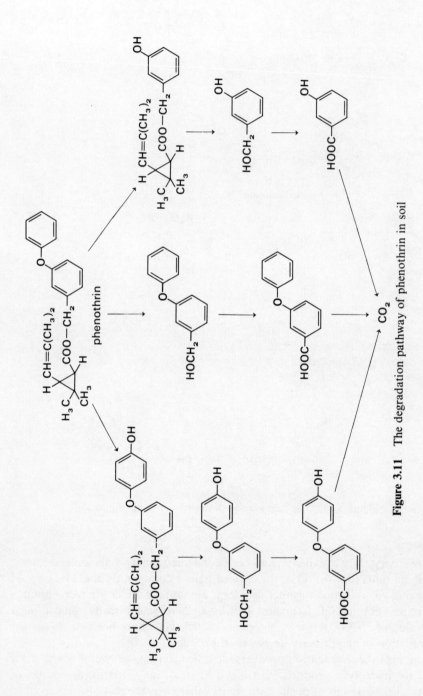

Figure 3.11 The degradation pathway of phenothrin in soil

produce the phenolic compounds shown in Figure 3.11 which also summarizes the proposed degradation pathway of phenothrin soils.

CONCLUSIONS

Information on plant metabolism and/or soil degradation is currently available on six pyrethroids and several common features are noteworthy. In all cases differences in rates of metabolism, and therefore in relative concentrations of products, occurred. The major primary routes of metabolism were hydrolysis at the ester bond and oxidation, of both aromatic rings and aliphatic groups in the 'acid' and 'alcohol' moieties.

In plants, the carboxylic acids and alcohols formed as hydrolysis products and the hydroxylated derivatives were usually converted into a variety of conjugates. Unlike the situation with the mammalian metabolism of pyrethroids in which a number of conjugates have been identified, information on the detailed nature of the plant conjugates is only beginning to emerge. This will no doubt be an area of further research in the future.

Pyrethroids are known to undergo photochemical decomposition (Holmstead et al., 1978a, b) and for this reason outdoor plant metabolism experiments would be particularly important so that the formation of photoproducts (e.g. on leaf surfaces) could be detected.

In soils, further degradation of the hydrolysis products to CO_2 was observed. Although some build-up of 'bound' (or unextracted) residues took place in the short term, in cases where this has been studied in further detail, it appears that turnover of the bound residues does occur.

Since publications of the fate of pyrethroids in plants and soils have emerged only in recent years, it is likely that the information given in this review, both on existing insecticides and on new compounds reaching the market, will be added to rapidly during the 1980s.

REFERENCES

Barlow, F., Hadaway, A. B., Flower, L. S., Grose, J. E. H., and Turner, C. R. (1977). 'Some laboratory investigations relevant to the possible use of new pyrethroids in control of mosquitoes and tsetse flies', *Pestic. Sci.* **8**, 291–300.

Elliott, M. (1977). 'Synthetic pyrethroids', in *Synthetic Pyrethroids* (Ed. R. F. Gould), ACS Symposia Series No. 42, pp. 1–28.

Elliott, M., and Janes, N. F. (1973). 'Chemistry of the natural pyrethrins', in *Pyrethrum, the Natural Insecticide* (Ed. J. E. Casida), pp. 56–100, Academic Press, New York.

Gaughan, L. C., and Casida, J. E. (1978). 'Degradation of *trans*- and *cis*-permethrin on cotton and bean plants', *J. Agr. Food Chem.* **26**, 525–528.

Gaughan, L. C., Unai, T., and Casida, J. E. (1977). 'Permethrin metabolism in rats and cows and in cotton and bean plants', in *Synthetic Pyrethroids* (Ed. R. F. Gould), ACS Symposia Series No. 42, pp. 186–193.

Hill, I. R., and Arnold, D. J. (1978). 'Transformations of pesticides in the environment—the experimental approach', in *Pesticide Microbiology* (Eds. I. R. Hill and S. J. L. Wright), p. 222, Academic Press, London.
Holmstead, R. L., Casida, J. E., Ruzo, L. O., and Fullmer, D. G. (1978a). 'Pyrethroid photochemistry: permethrin', *J. Agr. Food Chem.* **26**, 590–599.
Holmstead, R. L., Fullmer, D. G., and Ruzo, L. O. (1978b). 'Pyrethroid photochemistry-pydrin', *J. Agr. Food Chem.* **26**, 954–959.
Huff, R. K. (1980). *Pestic. Sci.* **11**, 290–293.
Hutson, D. H. (1979). 'The metabolic fate of synthetic pyrethroid insecticides in mammals', in *Progress in Drug Metabolism* (Eds. L. F. Chasseaud and J. W. Bridges), **3**, 215–252, Wiley, Chichester.
Kaneko, H., Ohkawa, H., and Miyamoto, J. (1978). 'Degradation and movement of permethrin isomers in soil', *J. Pestic. Sci.* **3**, 43–51.
Kaufman, D. D., Haynes, S. C., Jordan, E. G., and Kayser, A. J. (1977). 'Permethrin degradation in soil and microbial cultures', in *Synthetic Pyrethroids* (Ed. R. F. Gould), ACS Symposia Series No. 42, pp. 147–161.
Leahey, J. P., and Carpenter, P. K. (1980). 'The uptake of metabolites of permethrin by plants grown in ^{14}C-permethrin treated soil', *Pestic. Sci.* **11**, 279–289.
More, J. E., Roberts, T. R., and Wright, A. N. (1978). 'Studies of the metabolism of 3-phenoxybenzoic acid in plants', *Pestic. Biochem. Physiol.* **9**, 268.
Nambu, K., Ohkawa, H., and Miyamoto, J. (1980). 'Metabolic fate of phenothrin in plants and soils', *J. Pestic. Sci.* **5**, 177–197.
Ohkawa, H., Kaneko, H., and Miyamoto, J. (1977). 'Metabolism of permethrin in bean plants', *J. Pestic. Sci.* **2**, 67–76.
Ohkawa, H., Nambu, K., Inui, H., and Miyamoto, J. (1978). 'Metabolic fate of fenvalerate (SUMICIDIN) in soil and by soil microorganisms', *J. Pestic. Sci.* **3**, 129–141.
Ohkawa, H., Nambu, K., and Miyamoto, J. (1980). 'Metabolic fate of fenvalerate in bean plants', *J. Pestic. Sci.* **5**, 215–223.
Ohno, N., Fujimoto, K., Okuno, Y., Mizutani, T., Hirano, M., Itaya, N., Honda, T., and Yoshioka, H. (1974). 'A new class of pyrethroidal insecticides; α-substituted phenyl-acetic acid esters', *Agr. Biol. Chem.* **38**, 881–883.
Ohno, N., Fujimoto, K., Okuno, Y., Mizutani, T., Hirano, M., Itaya, N., Honda, T., and Yoshioka, H. (1976). '2-Arylalkanoates, a new group of synthetic pyrethroid esters not containing cyclopropanecarboxylates', *Pestic. Sci.* **7**, 241–246.
Roberts, T. R., and Dutton, A. J. (1980). Unpublished data.
Roberts, T. R., and Standen, M. E. (1977a). 'Degradation of the pyrethroid cypermethrin, NRDC 149, and the respective cis-(NRDC 160) and trans-(NRDC 159) isomers in soil', *Pestic. Sci.* **8**, 305–319.
Roberts, T. R., and Standen, M. E. (1977b). 'Degradation of the pyrethroid insecticide WL 41706 ($\pm$)α-cyano-3-phenoxybenzyl 2,2,3,3-tetramethyl-cyclopropane carboxylate in soils', *Pestic. Sci.* **8**, 600–610.
Roberts, T. R., and Standen, M. E. (1980). 'Further studies of the degradation of the pyrethroid insecticide cypermethrin in soils', *Pestic. Sci.* submitted for publication.
Ruzo, L. O., and Casida, J. E. (1979). 'Degradation of decamethrin on cotton plants', *J. Agr. Food Chem.* **27**, 572–575.
Seigler, D. G. (1977). in *Progress in Phytochemistry* (Eds. L. Reinhold, J. B. Harbourne, and T. Swain, **4**, p. 83, Pergamon, Oxford.
Wright, A. N., Roberts, T. R., Dutton, A. J., and Doig, M. V. (née Ford), (1980). 'The metabolism of cypermethrin in plants: the conjugation of the cyclopropyl moiety', *Pestic. Biochem. Physiol.* **13**, 71–80.

CHAPTER 4

The behaviour and mode of action of the phenoxyacetic acids in plants

J. B. Pillmoor and J. K. Gaunt

ABBREVIATIONS	148
GENERAL INTRODUCTION	148
UPTAKE OF HERBICIDE BY LEAVES	149
Introduction	149
Penetration of the Cuticle and Movement in the Apoplast	150
Entry to the Symplast	152
Factors Affecting Uptake to the Symplast	154
Summary	156
MOVEMENT WITHIN THE PLANT	156
Introduction	156
Local Movement in the Leaf	156
Polar Transport	158
Phloem Transport	159
Root Excretion	160
Factors Affecting Transport	160
The Effect of Herbicides on their own Transport	162
Summary	162
METABOLISM	163
Introduction	163
The Pathways of Metabolism	163
The Quantitative Assessment of Metabolism	169
The Kinetics of Metabolism	170
The Enzymes of Metabolism	172
The Biological Activity of Metabolites	174
Summary	176
INTERNAL LOCATION	176
Introduction	176
Subcellular Fractionation	177
Autoradiography	178
Efflux Analysis	179
The Use of Isolated Cells, Protoplasts, and Vacuoles	180
Summary	181
MODE OF ACTION	182
Introduction	182
Auxin Receptors	183
Biochemical Effects of Auxins	185
Do Plant Cells Respond to Auxin in Different Ways?	192

Toxic Action	192
Summary	194
SELECTIVITY	194
Introduction	194
The Assessment of Herbicide Sensitivity and Resistance	195
The Mechanisms of Selectivity	195
Summary	202
CONCLUDING REMARKS	202
REFERENCES	203

ABBREVIATIONS

ABA	abscisic acid
ATP	adenosine triphosphate
atrazine	2-chloro-4-ethylamino-6-isopropylamino-1,3,5-triazine
benzoylprop-ethyl	ethyl ester of N-benzoyl-N-(3,4-dichlorophenyl)-2-aminopropionic acid
benzoylprop-isopropyl	isopropyl ester of N-benzoyl-N-(3,4-dichlorophenyl)-2-aminopropionic acid
CoA	coenzyme A
2,4-D	2,4-dichlorophenoxyacetic acid
2,4-DB	4-(2,4-dichlorophenoxy)butyric acid
diphenamid	N,N-dimethyl-$\alpha\alpha$-diphenylacetamide
flamprop	N-benzoyl-N-(3-chloro-4-fluorophenyl)-2-aminopropionic acid
IAA	indole-3-acetic acid
MCPA	4-chloro-2-methylphenoxyacetic acid
MCPB	4-(4-chloro-2-methylphenoxy)butyric acid
monolinuron	N'-(4-chlorophenyl)-N-methoxy-N-methylurea
NAA	1-naphthylacetic acid
4-OH-2,5-D	2,5-dichloro-4-hydroxyphenoxyacetic acid
4-OH-POA	4-hydroxyphenoxyacetic acid
oxamyl	methyl N',N'-dimethyl-N-{(methylcarbamyl)oxy}-1-thio-oxamimidate
2,4,5-T	2,4,5-trichlorophenoxyacetic acid
2,4,6-T	2,4,6-trichlorophenoxyacetic acid
TCA	trichloroacetic acid
TIBA	2,3,5-tri-iodobenzoic acid
UDPG	uridine diphosphate glucose

GENERAL INTRODUCTION

The behaviour and the mode of action of the phenoxyacetic acids has been the subject of a vast amount of research over the past 30 years. Despite this, the biochemistry of their toxicity is not known, nor are the reasons for their selectivity completely understood. These problems have been reviewed many times, but the last major review (Loos, 1975) was written some time ago and it seems appropriate to reconsider the matter now. However, it is not our intention to simply average out all of the work that has been published on the phenoxyacetic acids over the past few years. Indeed we apologize in advance to any author who feels that his work has been inadequately presented. This is a statement of our personal opinion of the problem, which is biased by the results of investigations in this laboratory over the past decade.

This review is primarily concerned with the herbicidal phenoxyacetic acids, the most important of which are 2,4-D, MCPA, and 2,4,5-T. However, other phenoxyacetic acids and their derivatives will not be ignored. Where appropriate, mention will be made of the behaviour of the phenoxypropionic and phenoxybutyric acids. Since the phenoxyacetic acids are generally believed to act as herbicides because of their auxin activity, it is also appropriate to consider the behaviour and mode of action of IAA and of other synthetic auxins.

The approach to this subject is to consider the various aspects of behaviour of the phenoxyacetic acids in plants in an attempt to establish principles that underlie the processes of cuticle penetration, uptake into cells, internal distribution, movement between cells, long distance transport, and metabolic behaviour. The biochemistry and physiology of their mode of action as auxins and herbicides will next be considered. Finally, the different patterns of behaviour and activity in different plants will be compared in a search for the explanation for their selective action.

UPTAKE OF HERBICIDE BY LEAVES

Introduction

Since the normal application of the phenoxyacetic acids is to the leaves, only foliar uptake will be extensively considered here although many of the principles established will also be applicable to root uptake.

The first essential is to clarify what is meant by uptake. In most studies the term is applied rather generally to the movement of herbicide from the surface of a leaf to the interior. However, it must be remembered that after diffusion from the leaf surface the herbicide reaches both apoplast and symplast and quite different mechanisms of uptake are then involved. In this review, uptake into each of these compartments will be dealt with separately. Herbicide must first move from the leaf surface across the cuticle to epidermal cell walls and then perhaps directly to the rest of the interior apoplast. At some stage it must cross the plasma membrane and enter the symplast.

Studies of the kinetics of uptake of the phenoxyacetic acids have shown the presence of two distinct phases of uptake (Reinhold, 1954; Johnson and Bonner, 1956; Yamaguchi, 1965; Jenner *et al.*, 1968a). The first phase is unaffected by metabolic inhibitors and probably reflects simple diffusion through the cuticle and into the free space of the leaf (Poole and Thimann, 1964; Robertson and Kirkwood, 1969). The second phase is sensitive to metabolic inhibitors (Johnson and Bonner, 1956) and may only occur in the light (Sargent and Blackman, 1969). This phase is, therefore, an active process, dependent on metabolic energy. It probably represents movement across the plasmamembrane and into the symplast.

Penetration of the Cuticle and Movement in the Apoplast

At one time it was thought that the open stomata would provide the main route of entry for herbicides—at least for hydrophilic molecules. However, they are now generally considered to play only a supplementary role to the main pathway of uptake across the cuticle itself (Robertson and Kirkwood, 1969; Martin and Juniper, 1970). The structure, physiology, and role of the cuticle has been discussed by Van Overbeek (1956) and by Martin and Juniper (1970). Despite its lipophilic nature there are considered to be both aqueous and lipophilic routes for movement across this structure.

The aqueous route is believed to involve the ectodesmata (Sargent and Blackman, 1965; Robertson and Kirkwood, 1969). These are canals, or more precisely different regions, within the outer tangential walls of the epidermal cells (Martin and Juniper, 1970). Since these structures are concentrated in the region of the guard cells it has been suggested that these cells may play a major role in the uptake of polar herbicides (Franke, 1964; Sargent and Blackman, 1969; Kirkwood, 1972). However, it must be noted that the ectodesmata do not pass through the cuticle but only through the epidermal cell wall. Consequently they cannot provide a route for the initial movement of hydrophilic compounds across the external cuticle. The importance of the guard cells in uptake may be due to differences in the cuticle over these cells as compared to the rest of the leaf (Martin and Juniper, 1970), as well as their high content of ectodesmata.

The lipophilic route concerns the bulk of the cuticle. Its importance is suggested by the strong correlation that exists between the lipophilic nature of the herbicide and its ability to penetrate the cuticle (Norris and Freed, 1966; Richardson, 1977). This has been observed both for different herbicides and for different formulations of the same compound. It is likely that it also explains the faster uptake of the phenoxyacetic acids at low pH values, when acids will predominantly exist in the non-polar, un-ionized form (Szabo and Buckholtz, 1961; Baur et al., 1974; Richardson, 1977). There is a corollary to this, namely that the hydrophilic route across the cuticle is quantitatively of much less importance than the lipophilic pathway, at least as far as the phenoxyacetic acids are concerned.

Once a compound has diffused across the thickness of the cuticle it will reach the aqueous matrix of the epidermal cell walls. At this point a partition will occur.

The distribution of a herbicide between cuticle and cell wall will depend on its partition coefficient between the two media. Compounds that have considerable lipophilicity may be retained to a significant extent in the cuticle. This has been suggested for 2,4-DB (Loos, 1975) and for 2,4,5-T (Robertson and Kirkwood, 1969).

It is generally assumed that once a compound has entered the aqueous environment of the epidermal cell walls it is able to diffuse freely throughout the apoplast of the leaf. However, this may well be an oversimplification. Several factors may combine to limit the movement of the herbicide within the apoplast and may even confine it to the epidermis. These are as follows:

Adsorption on to cell wall components

Some workers have identified two phases in the process of movement of phenoxyacetic acids from leaf surface to the apoplast, one freely diffusible and the other exchangeable (Johnson and Bonner, 1956). These are both believed to occur within the 'free space' and the exchangeable part is suggested to be due to adsorption on to some component of either cuticle or cell wall. This idea remains to be substantiated but it is perhaps relevant that Zemskaya and Rakitin (1967) have suggested that 2,4-D can be bound to protein in the cell wall. Reinhold (1954) and Donaldson *et al.* (1973) have also reported an adsorptive component during uptake. Thus it is possible that some of the herbicide that penetrates the cuticle could be immobilized in the apoplast.

Local cell wall charges

Van Overbeek (1956) has drawn attention to another factor that may effect free movement in the apoplast. He pointed out that at the pH of the cell wall (between 5 and 6) the uronic acid residues of cell wall polysaccharides will be largely ionized (pK around 5). Phenoxyacetic acids in the cell wall will also be largely ionized (pK around 3). The resulting electrostatic repulsion may restrict movement in areas of the microenvironment of the cell wall. It is difficult to assess the importance of this. One of the reasons behind the proposal is that the phenoxyacetic acids are taken up by cells more readily at low pH values at which the ionization of both the herbicide and cell wall is less, as is their mutual repulsion. However, the model proposed by Rubery and Sheldrake (1973) to explain this observation seems more plausible and this will be discussed in more detail below.

Mass movement of water through the leaf apoplast

Such movement will be from the xylem towards the epidermis and the stomata. This will occur in any leaf that is actively transpiring and it will occur in exactly the opposite direction to the diffusion of herbicide. Again it is difficult to assess the quantitative importance of this factor.

Uptake to the symplast

It seems likely that the phenoxyacetic acids can be taken up by any leaf cell, provided it reaches that cell. However, following application to the cuticle it is by no means clear whether all the cells of the leaf have the same opportunity to take up the herbicide. Only the epidermal cells can be guaranteed access. If uptake to these cells is rapid, little herbicide may remain in the apoplast to penetrate directly by diffusion to the mesophyll or palisade cells.

Taking all of these factors into consideration it seems unlikely that very much herbicide moves extensively by diffusion within the apoplast. This conclusion agrees with the suggestion made by Crafts (1959, 1960) that movement of the phenoxyacetic acids is primarily in the symplast. Once uptake to the symplast occurs, movement between cells could either be via plasmodesmata or it could involve transfer across plasma membranes and intervening cell walls. Thus, even if there is very restricted diffusion of phenoxyacetic acids in the apoplast, they would none the less be expected to have access to apoplast and symplast throughout the leaf, unless immobilized in some way.

Entry to the Symplast

In early studies of the uptake of the phenoxyacetic acids to the symplast of plant cells, roots were used as the experimental material and the techniques applied were those that had long been used to follow ion absorption. The work of Smith and Epstein (1964) demonstrated that leaf slices could be used successfully to investigate the process of solute uptake by leaf tissue and since then this approach has been extensively applied to herbicides. More recently cell suspension cultures have also been used to great effect (e.g. Rubery and Sheldrake, 1974; Kurkdjian *et al.*, 1979).

In general, uptake of the phenoxyacetic acids to the symplast followed saturation kinetics. As herbicide concentration in the uptake medium is increased, the rate of uptake increases to a maximum. At this point further increases in concentration are without effect (Jenner *et al.*, 1968b; Rubery, 1977, 1978). This suggests the presence of either a saturable membrane carrier or an energy requirement or both of these.

The involvement of metabolic energy in the uptake process has been well documented (Reinhold, 1954; Johnson and Bonner, 1956; Sargent and Blackman, 1965; Yamaguchi, 1965; Franke, 1967; Moffit and Blackman, 1972; Que Hee and Sutherland, 1973). However, there is doubt that this is directly linked to the uptake mechanism, for example as an ATP-driven transport system. A view that is gaining wide acceptance is that energy is required to maintain a pH gradient between cell wall and cytoplasm and that herbicide entry is dependent upon such a gradient. It is clear that an energy-requiring proton pump is present in the plasma membrane of plant cells and is responsible for the proton gradient across this membrane (Marre, 1977; Smith and Raven, 1979). This can be linked to uptake of the phenoxyacetic acids in the two ways that are outlined below.

Diffusive uptake—the model of Rubery and Sheldrake (1973)

A cornerstone of this model is the fact that uptake of the phenoxyacetic acids by plant cells depends upon the pH of the medium. Uptake is very much greater at low pH than at pH values approaching neutrality. To explain this it is proposed that a phenoxyacetic acid can only diffuse across the plasma membrane in the un-ionized, lipophilic form and that no passive diffusion of the dissociated anion can occur. The phenoxyacetic acids are all weak acids and the degree of dissociation depends on pH of the medium. It will be relatively less at the pH of the cell wall (say 5–6) than at that of the cytoplasm (say 7). Diffusion of undissociated acid will occur across the plasma membrane into the cytoplasm down a concentration gradient. Within the cytoplasm dissociation of the acid will occur, reducing the concentration of un-ionized herbicide and permitting more diffusion. This process will continue until the concentration of undissociated herbicide in cell wall and in cytoplasm is the same. Since the pH is different in the two compartments, at equilibrium the amount of dissociated phenoxyacetic acid in each compartment will be different. There will be more total phenoxyacetic acid inside the cell than outside. The extent of the difference will depend entirely on the pH difference between the two compartments. The energy requirement of the uptake process is partly to maintain the pH gradient across the plasma membrane and partly to keep the cytoplasm pH constant via some form of biochemical pH stat (Smith and Raven, 1979).

Carrier mediated uptake

While the above model of diffusive uptake has been considered to explain all of the movement of auxins across the cell membrane (Raven, 1975) other work suggests that carrier proteins in the membrane are also involved in their transport (Rubery and Sheldrake, 1974; Rubery, 1977, 1978, 1979; Davies and Rubery, 1978). It was initially proposed that the anions of both IAA and 2,4-D could be transported into cells by an electroneutral, proton symporter. During the initial period of uptake this carrier will assist uptake into the cell, presumably driven by the gradients of both anion and proton across the plasma membrane. The relative importance of this carrier system in uptake, compared with diffusion of the undissociated auxin, is dependent upon the pH of the external medium. At the optimum pH (4.0) for carrier-mediated uptake of 2,4-D, the ratio of transport by the carrier to entry by diffusion is approximately 1 : 6.7. This ratio increases enormously as the pH is reduced. However, at pH 5, it has decreased to 1 : 2.5, suggesting that under normal conditions in an intact leaf both pathways may be important routes of entry for the phenoxyacetic acids.

While the anion carrier will initially mediate uptake of auxin into the cell, it may also be involved in efflux at a later stage. As uptake of auxin proceeds, by whatever route, the concentration of auxin anion inside the cell compared to that outside will become very large. For example, diffusive uptake alone would give more than 30 times more IAA or 2,4-D anion in the cytoplasm at pH 7.0 than in the cell wall at pH 5.5 at equilibrium. This concentration gradient for the auxin may be sufficient to overcome the proton gradient in the opposite direction. The result may be the export of anion by the same carrier that was originally involved with its uptake. The system would be at a dynamic equilibrium when the rate of entry by diffusion of the un-ionized auxin was equal to the rate of efflux by the anion carrier.

A third method of moving auxin across the plasma membrane has also been proposed. This is not linked to a proton gradient, although the mechanism is not clear. Furthermore it is involved solely in auxin efflux. However, it will be considered here since it completes the overall picture of auxin movements to and from a plant cell. It is proposed that a second auxin carrier exists (Davies and Rubery, 1978; Rubery, 1979). This carrier mediates the electrogenic export of auxin anion and is sensitive to TIBA. It is considered to have an asymmetrical distribution in the plasma membrane and to mediate polar transport. It has a greater specificity for IAA than the electroneutral symporter, but in at least one species it will also transport 2,4-D. The specificity of the electrogenic carrier from different species to various auxins and auxin herbicides apparently parallels the polar transport capacity of the tissue (Rubery, 1979).

Factors Affecting Uptake to the Symplast

Uptake of herbicide will be influenced in two ways. First, as initial uptake of the phenoxyacetic acids is largely controlled by the proton gradient across the plasma membrane, then any factor that affects the gradient will also affect uptake of the herbicide; for instance the pH of the uptake solution. Second, any process that affects the gradient of herbicide across the plasma membrane will affect the rate and extent of uptake. The following processes may therefore contribute to uptake by decreasing the concentration of phenoxyacetic acid in the cytoplasm.

(1) Movement of the herbicide to other cells within the leaf would be expected to increase uptake from the leaf surface. The principles of movement of the phenoxyacetic acids are considered in a later section. Similarly, transport out of the treated leaf has also been reported to increase uptake (McIntyre *et al.*, 1978; Phung-Hong-Thai and Field, 1979).

(2) Movement of herbicide to the vacuole or cell organelles. If the model of **Rubery and Sheldrake** is applied to the partition of phenoxyacetic acid between

cytoplasm, vacuole, and organelles, then the distribution will depend largely upon the pH in each compartment. Thus the vacuole, with its pH of say 5 will accommodate a lower concentration of phenoxyacetic acid than the cytoplasm. However, the relatively large volume of the vacuole means that the total amount of herbicide present may be greater there than in the cytoplasm at equilibrium. Among the organelles, the mitochondria could well accumulate high concentrations of phenoxyacetic acid due to their high pH. The same is true of the chloroplasts when photosynthesis is actively in progress and the stroma pH is around 8. This subject is dealt with more fully in a later section of this review.

(3) Metabolism of the phenoxyacetic acid will obviously reduce the cytoplasmic concentration, irrespective of the subsequent compartmentation of the metabolites. In species in which such metabolism is rapid and extensive, this may be the major means of maintaining continued cell uptake of phenoxyacetic acids over an extended period of time.

(4) Non-covalent binding of herbicide to protein or other macromolecules in the cytoplasm may also occur as is discussed later.

The consequence of each of these will be to increase uptake of phenoxyacetic acids from the cell wall.

It is apparent that uptake to the symplast may occur extensively and predominantly in the epidermis. This conclusion does not require the possession of any special properties by epidermal cells—merely that they are the first to be reached by herbicide on its passage from the leaf surface. However, it is worth mentioning that the guard cells have been implicated as a major entry point to the symplast (Franke, 1964; Sargent and Blackman, 1969; Kirkwood, 1972). Ectodesmata lead polar molecules to these cells; light is known to stimulate herbicide uptake (Sargent and Blackman, 1962, 1965, 1969) and the guard cells are the only photosynthetic cells of the epidermis. It is possible to imagine that they are especially active in uptake due to their photosynthetic function. This could provide ample energy to maintain the proton gradient across the plasma membrane and compartmentation of herbicide in chloroplasts could aid uptake still further.

A final point to consider with regard to uptake is the possibility that the phenoxyacetic acids may affect their own uptake. For instance, these herbicides are known to affect the permeability of the plasma membrane. Devlin (1974) and Zsoldos et al. (1978) have reported major alterations in this property following 2,4-D treatment of various tissues. Whatever the mechanisms behind such changes they could have a secondary effect on the flux of the herbicide across the cell membrane and thus upon its net uptake. It is also interesting to note that auxins are known to stimulate a plasma membrane bound proton pump (see later). As uptake of an auxin is dependent upon the proton gradient across the plasma membrane, an auxin could potentially increase its own uptake (Raven, 1975).

Summary

It would appear that the phenoxyacetic acids cross the cuticle readily by diffusion down a concentration gradient either using the lipophilic route (mainly for un-ionized molecules) or the hydrophilic route (mainly for the dissociated anion). At the inner boundary of the cuticle the herbicide will partition between the lipid phase of the cuticle and the aqueous medium of the epidermal cell wall and will tend to move by diffusion to the interior of the leaf through the apoplast. Such movement may be restricted by electrostatic considerations, by adsorption to cell wall components, or by mass flow of water in the cell wall in the opposite direction. Herbicide that reaches the plasma membrane of leaf cells will cross into the cytoplasm by both passive diffusion and via a specific carrier system, each of which is dependent upon the pH gradient across the plasma membrane. Initial entry to the symplast of the leaf is likely to occur primarily in the epidermal cells. Metabolism of herbicide, macromolecular binding, or compartmentation within the cell, as well as transport within the leaf and to the rest of the plant, can lead to continuous uptake to the symplast, further reducing movement in the apoplast.

MOVEMENT WITHIN THE PLANT

Introduction

Having discussed the initial uptake of herbicide into the symplast of the leaf it is now possible to consider its further movement within the plant. The overall patterns of movement of the phenoxyacetic acids in plants have been frequently described and reviewed (e.g. Eliasson, 1965; Crafts, 1967; Robertson and Kirkwood, 1970; Chkanikov *et al.*, 1971; Loos, 1975). It is not the intention here to repeat this. Less clearly understood are the mechanisms that underlie transport—both local and long distance. It is these aspects that will be considered here together with those factors that can influence the process and which help to explain the major differences found between species in the extent and pattern of movement.

Local Movement in the Leaf

Although the following discussion is applied to local movement from the epidermis of a leaf towards the vascular tissue, the principles that are introduced are applicable generally to any tissue and to movement during uptake by roots and after release from the phloem. Basically there are four pathways or processes that may be involved.

Movement in the apoplast by diffusion

This possibility has already been discussed with the conclusion that such movement is likely to be very restricted for the phenoxyacetic acids. Thus it is unlikely that these compounds will penetrate far into the mesophyll without entering the symplast. On the other hand, diffusion across the cell walls between the plasma membranes of adjacent cells is likely to occur readily because of the small distance involved.

Movement between cells via the plasmodesmata

Once within the cytoplasm a substance could move down a concentration gradient through the plasmodesmata to adjacent cells. However, it is not yet known whether these intercellular connections are generally available as channels for water and solute distribution. Thus their role in the movement of the phenoxyacetic acids remains imponderable.

Movement between cells by diffusion

Movement of un-ionized phenoxyacetic acid could occur by diffusion down concentration gradients from cytoplasm to cell wall to cytoplasm of an adjacent cell. The un-ionized molecule will readily cross cell membranes and so in theory will be able to move to all cells of a leaf. The main obstacle to such movement will be the fact that at the pH values found within a leaf, a very small proportion of the herbicide will be present in the un-ionized form. For example, in the case of 2,4-D ($pK_a = 2.8$), at pH 5.5 the ratio of dissociated to undissociated forms is 500:1 and at pH 7.0 it is 16,000:1. The very low concentration of free acid means that concentration gradients of this molecular form will be small and thus movement by diffusion through the leaf may be slow. It must be emphasized that this only applies to the undissociated form. The distribution of total 2,4-D between cytoplasm and cell wall if the cytoplasm pH is 7.0 and cell wall pH is 5.5 will be about 30:1 at equilibrium. However, this is only possible because the vast majority of the compound is ionized.

Movement between cells via membrane carrier proteins

Movement of the phenoxyacetate anion from cytoplasm to cell wall to cytoplasm may occur via the carrier protein systems discussed earlier. The proton symporter may only function in export when the concentration gradient of the anion across the plasma membrane reaches a threshold value. However, the carrier that is proposed to be important in polar transport will function in export all the time, provided it is present in a particular cell type and that the particular phenoxyacetic acid under consideration is able to interact. Once the herbicide has entered the cell wall, diffusion to the plasma membrane of an adjacent cell will allow uptake as described earlier. The net result will be to accelerate the local transport of herbicide away from the epidermis down a concentration gradient.

At first sight none of these four principles agrees with the suggestion by Crafts and Yamaguchi (1958) and Crafts (1959) that 2,4-D will be accumulated by a cell until it becomes 'saturated', only after which may herbicide be passed to a second cell. On the other hand, the data of Kirkwood et al. (1968) are consistent with these principles. They observed that the translocation of MCPA applied to the leaf surface of Vicia faba generally increased in proportion to dose up to the concentration at which damage was caused at the site of application. Perhaps the difference of opinion can be resolved by the following considerations.

(1) It may take a relatively long time for herbicide to penetrate to the phloem by diffusion processes if its distribution between apoplast and symplast is strongly in favour of the latter. The result of this would be an initial high concentration in the epidermis with a steep gradient to adjacent mesophyll cells which may be slow to diminish.

(2) It is likely that the cytoplasmic concentration of phenoxyacetic anion must reach a certain level before efflux occurs via the anion symporter.

Polar Transport

IAA is well known for its unique polar transport that results in movement of the molecule in stem segments in a morphologically basipetal direction, irrespective of the orientation of the segment or of the direction of the concentration gradient of the auxin (Goldsmith, 1977). This has been shown to occur in many tissues in a wide variety of plant species (Hertel and Leopold, 1963). The system is characterized also by its sensitivity to specific inhibitors such as TIBA (Hertel and Flory, 1968; Thomson et al., 1973). Polar transport will move auxin over short distances and has the potential to move it over long distances as well. For example, IAA applied externally to the apex of plants has been shown to be moved directly to the roots by a system that has all the characteristics of polar transport (Greenwood and Goldsmith, 1970). The exact location of the transport pathway is not clear. It has been reported to occur in phloem tissue, excluding the sieve tubes (Lepp and Peel, 1971; Morris and Kadir, 1972; Sheldrake, 1973; Wangermann, 1974), in parenchyma (Jacobs, 1967), in the cambial region (Morris and Thomas, 1978), and in the protocambium and protophloem of the undifferentiated region of the shoot apex (Greenwood and Goldsmith, 1970).

There are numerous reports that 2,4-D can also be transported in a polar fashion in stem segments, although at somewhat slower rates than IAA (McCready, 1963; McCready and Jacobs, 1963; Taylor and Warren, 1970; Wilkins and Wilkins, 1975) and with somewhat different kinetics (Gorter and Veen, 1966). It has been suggested that IAA and 2,4-D move by polar transport in different cells. For example, Jacobs (1967) found that IAA moved preferentially in cores of pith parenchyma while 2,4-D moved preferentially in vascular tissue cores.

While polar transport has been reported to occur in petioles, there is no record of its occurrence in the lamina of a leaf. Yet an observation made by Goldsmith *et al.* (1974) does suggest its presence. They found that TIBA could drastically inhibit the export of exogenous IAA after its application to a mature leaf. This inhibition did not appear to affect movement once the IAA was in the phloem. The implication is that the TIBA-sensitive system is involved in movement of IAA to the vascular system or in the loading of the phloem.

Overall, it seems likely that polar transport may have a wide distribution in the plant and that the underlying principles of movement could be applied to phenoxyacetic acid movement in many tissues. However, there is no direct evidence that polar transport of the phenoxyacetic acids occurs in intact plants. 2,4-D has been found not to move out of young, immature leaves (Crafts, 1959) and it has been reported that the compound can become restricted to the shoot apex (Eliasson, 1965). The question of polar transport of herbicide from these two tissues is particularly relevant when applied to the field situation where significant amounts of the herbicide will be absorbed or reach these areas and may not be exported in the phloem. Basically, while polar transport is an attractive possibility there is no reliable information on its contribution to the movement of the phenoxyacetic acids.

Phloem Transport

This is generally believed to account for the majority of phenoxyacetic acid movement from leaves. It has been extensively reviewed by Robertson and Kirkwood (1970). The reader is also referred to the work of Fensom (1972) and the review of Zimmermann and Milburn (1975) for further consideration of the translocation process itself. Both exogenously applied IAA as well as the phenoxyacetic acids move rapidly from mature leaves and with the characteristics of movement in the translocation stream (Radwan *et al.*, 1960; Pickering, 1965; Eschrich, 1968; Lepp and Peel, 1971; Goldsmith *et al.*, 1974). Such movement is dependent upon the simultaneous translocation of photosynthetic assimilates (Rohrbaugh and Rice, 1949; Crafts and Yamaguchi, 1958; Hoad *et al.*, 1971; Bonnemain and Bourbouloux, 1973; Radosevich and Bayer, 1979). Once in the phloem the transport of auxins appears to be determined by the flow of the translocation stream in a 'source to sink' manner. However, there are differences between the distribution of herbicide and assimilates (Crafts, 1960; Leonard *et al.*, 1967; Olunuga *et al.*, 1977) which suggest the existence of factors which specifically influence the behaviour of the herbicide. These include the excretion of the molecule from the roots of some species (Fites *et al.*, 1964; Basler *et al.*, 1970; Coble *et al.*, 1970).

There are no data available on the mechanisms of entry and egress of the phenoxyacetic acids to and from the phloem sieve tube cells. Loading has been suggested as a major barrier to translocation (Moorby, 1964; Wangermann,

1970). However, there is a very active proton pump in the phloem sieve tube plasma membrane. There is, therefore, a pH gradient across this membrane and any phenoxyacetic acid reaching the phloem should enter the sieve tubes passively according to the Rubery–Sheldrake model. Unloading presents a more difficult question that has yet to be resolved.

During translocation in the phloem it has been suggested that transfer to xylem can occur, resulting in even wider distribution of the herbicide (Eliasson, 1965; Long and Basler, 1973). Radwan et al. (1960) have also proposed that extensive transport of 2,4-D can occur in the xylem.

Root Excretion

This represents a special case of transport involving the movement of herbicide from the leaf to the roots and subsequent excretion (Fites *et al.*, 1964; Basler *et al.*, 1970; Coble *et al.*, 1970). It is a process that over a period of 2 to 3 weeks can eliminate a significant proportion of herbicide that enters the plant. At present, the mechanisms by which the herbicide is initially retained by the phloem and then excreted through the roots are unclear. From the principles of uptake outlined previously it is difficult to envisage how loss of herbicide from the roots could occur, However, the answer may be connected with the observations made by Blackman (1961). This author has reported that, while relatively high uptake of 2,4-D by roots was observed over a short time period, this was apparently followed by efflux of unchanged 2,4-D back to the uptake solution. This biphasic behaviour was only apparent for some species, others showing progressive accumulation with time. Similar behaviour has also been demonstrated for stem segments (Blackman, 1961; Saunders *et al.*, 1965a, b) and leaf slices (Neidermyer and Nalewaja, 1969). The mechanisms behind these observations are not known at present, but clearly warrant further research. It is also noted that Crafts and Yamaguchi (1960) reported that 2,4-D is absorbed only to a limited extent by roots. Most of the 2,4-D did not apparently enter the symplast but remained in the free space. These observations suggest that the behaviour of the phenoxyacetic acids in roots may be a specialized case and that the criteria, outlined previously for the local movement of these compounds in leaf tissue, may not be applicable here.

Factors Affecting Transport

In principle, the mechanisms described above will apply to the movement of the phenoxyacetic acids in all species. However, very large differences in mobility occur between species and in the same species under different conditions. In this section the possible reasons behind the differences will be considered.

Direct differences in the permeability of cell membranes to the parent phenoxyacetic acid

Any factor that alters the uptake, efflux, and subcellular distribution of herbicide will have consequences on cell to cell movement. Variation in the pH differences between cell wall, cytoplasm, vacuole, and organelles will certainly affect the passive distribution of herbicide between these compartments. The significance of this cannot, however, be estimated in the absence of detailed knowledge of the pH values in different plants and such information is not available. The role of the protein carrier systems in the plasma membrane (and perhaps other cell membranes) is still poorly understood. However, variations between species in the distribution, properties, or affinity of these proteins for the phenoxyacetic acids is likely to affect cellular distribution and mobility. Indeed Rubery (1979) has reported a striking difference between species in the specificity of the electrogenic carrier for 2,4-D. To date, however, there is insufficient data to judge the extent to which such variations occur.

Metabolism

The end products of metabolism of the phenoxyacetic acids are all more polar compounds than the parent herbicide. The majority are conjugated to either amino acids or sugars, the latter being no longer acidic. This will have a dramatic effect upon the mobility of the compound. Instead of having the ability to cross membranes by diffusion it is likely that specific carriers would be required. Very little direct information is available about metabolite mobility although efflux analysis (described later) should provide some answers. However, it has often been proposed that only the parent herbicide is mobile in the plant (Hallem, 1974). In studies of polar transport only free auxin is thought to be moved. During stem segment studies with IAA no metabolites have ever been found in receiver agar blocks while high levels of metabolites have been detected within the segments (Goldsmith and Thimann, 1962; McCready, 1963; Taylor and Warren, 1970; Wilkins and Wilkins, 1975).

For phloem transport the situation is less clear. The information obtained from aphid studies has yielded conflicting data. Chkanikov *et al.* (1972) found only 2,4-D present in the sieve tubes of soy bean (*Glycine max* var. primorskaya 494) with no evidence for the presence of amino acid or sugar conjugates. In the case of IAA, the studies of Eschrich (1968) and Lepp and Peel (1971) showed that free IAA was present in sieve tubes along with relatively small amounts of IAA aspartate and other metabolites. On the other hand, Hoad *et al.* (1971) found that the situation changed with time. They analysed the honeydew of aphids feeding on bark strips and found that at one hour after IAA treatment only parent IAA could be detected, whereas after 48 h the main component was IAA aspartate, with little or no IAA present. They found a similar situation in intact plants. These data contrast with the findings of Lepp and Peel (1971) that IAA aspartate applied directly to plants was relatively immobile. This situation needs clarification and a possible explanation is that the aspartate is actually synthesized within the sieve tubes and may then be unable to leave.

Further support for the idea that the parent herbicide is the only mobile form comes from studies of root excretion. No metabolite has yet been reported to be released from the roots by this process (Fites *et al.*, 1964; Basler *et al.*, 1970; Hallem, 1974; Williams, 1976).

The above observations combine to suggest that metabolism may be a form of immobilization. It does not mean that once metabolism has occurred the herbicide is 'fixed' since many metabolites may be metabolically unstable and able to regenerate the parent herbicide, a point that will be taken up again later. While Chkanikov *et al.* (1971) have reported a correlation between the extent of translocation and the level of free 2,4-D in the leaf for some species, they noted that this was not always the case. They therefore suggested that other factors, apart from metabolism, must contribute to immobilization of the herbicide. For example, protein and other macromolecular binding will also serve to immobilize a herbicide. Finally, compartmentation of the parent herbicide in the vacuole and other cell organelles may also contribute to immobilization.

The Effect of Herbicides on Their Own Transport

It is apparent that the phenoxyacetic acids can have an indirect effect upon their own movement. The consequence of their stimulation of proton pumps has already been mentioned. Other effects on energy metabolism or assimilate production could also affect the function of transport systems in a general way. The phenoxyacetic acids can also cause increased growth of specific areas of the stem. This will result in the development of new 'sinks' that will attract an increased flow of assimilates together with any herbicide moving in the phloem (Leonard *et al.*, 1966). Such a consideration may be relevant in the long term and could even help to explain the herbicida action of the compound by movement of more of it to a sensitive site.

Summary

The following principles appear to control the transport of phenoxyacetic acids in plants. Short-distance, intercellular movement may occur down concentration gradients. Distribution between different subcellular compartments will be determined by the pH of each region. The anion carrier in the plasma membrane may accelerate intercellular movement by enabling the efflux of the anion from a cell once the anion concentration increases beyond a certain level. Polar transport may participate in the movement of all auxins over short and long distances. The majority of long-distance movement from mature leaves probably takes place in the phloem. It is likely that only the parent herbicide is mobile and that metabolism leads to immobilization; although evidence for this is very limited. Other factors must also be involved in immobilization. The phenomenon of root excretion indicates that some mechanism exists in some species to determine the route and eventual destination of the phenoxyacetic acids. However, there is no information on the mechanism involved.

METABOLISM

Introduction

The pathways of metabolism of the phenoxyacetic acids have been relatively well documented and will not be discussed in detail here. This review concentrates upon less well understood aspects of metabolism. This includes the nature of sugar conjugates and bound residues, the technical difficulties of determining the level of metabolism, the enzymes and kinetics of metabolism and the toxicity and metabolic stability of metabolites.

The Pathways of Metabolism

Since the reviews of Robertson and Kirkwood (1970) and Loos (1975) there have been few additions to the list of metabolites and the reader is referred to these sources for an extensive treatment of the subject. A summary of present knowledge is shown for 2,4-D (Figure 4.1), MCPA (Figure 4.2), and 2,4,5-T (Figure 4.3). Processes of conjugation, hydroxylation, and side chain cleavage predominate. Conjugation of the parent phenoxyacetic acid to sugar residues has only been reported to involve glucose (e.g. Loos, 1975). However, it seems likely that other sugar residues may also be involved, particularly in the more polar conjugates (see later). Amino acid conjugation largely involves aspartic and glutamic acids, but the list of amino acids that have been found to be attached to 2,4-D is now long (e.g. Feung *et al.*, 1973). Side chain degradation probably occurs as a single stage oxidation to yield glycollic acid and the phenol corresponding to the parent herbicide (Fleeker, 1973; Williams, 1976). Subsequent metabolism of glycollic acid, perhaps via the photorespiratory pathway, yields CO_2 from the molecule. It is this reaction that has led to widespread reports of 'decarboxylation' of the phenoxyacetic acids. It seems probable that this is a misnomer since it is doubtful that any direct decarboxylation of the parent compound occurs. Another reaction affecting the side chain is the extension process reported by Linscott *et al.* (1968). However, the relative importance of this reaction remains to be established. Ring hydroxylation involving the familiar NIH shift (Guroff *et al.*, 1967) is a common feature of the metabolism of 2,4-D in many species (Fleeker and Steen, 1971). For 2,4,5-T the chlorine at position four is apparently lost during hydroxylation to give 4-OH-2,5-D (Hamilton *et al.*, 1971). However, none of the equivalent derivatives have

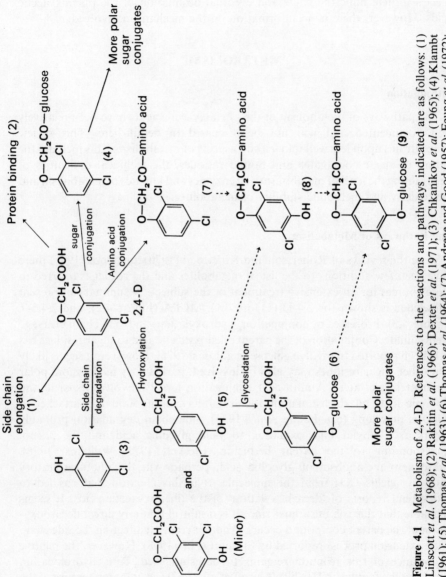

Figure 4.1 Metabolism of 2,4-D. References to the reactions and pathways indicated are as follows: (1) Linscott et al. (1968); (2) Rakitin et al. (1966); Dexter et al. (1971); (3) Chkanikov et al. (1965); (4) Klambt (1961); (5) Thomas et al. (1963); (6) Thomas et al. (1964); (7) Andreae and Good (1957); Feung et al. (1972); (8) Mumma and Hamilton (1975); (9) Chkanikov et al. (1977). ------> Routes yet to be substantiated

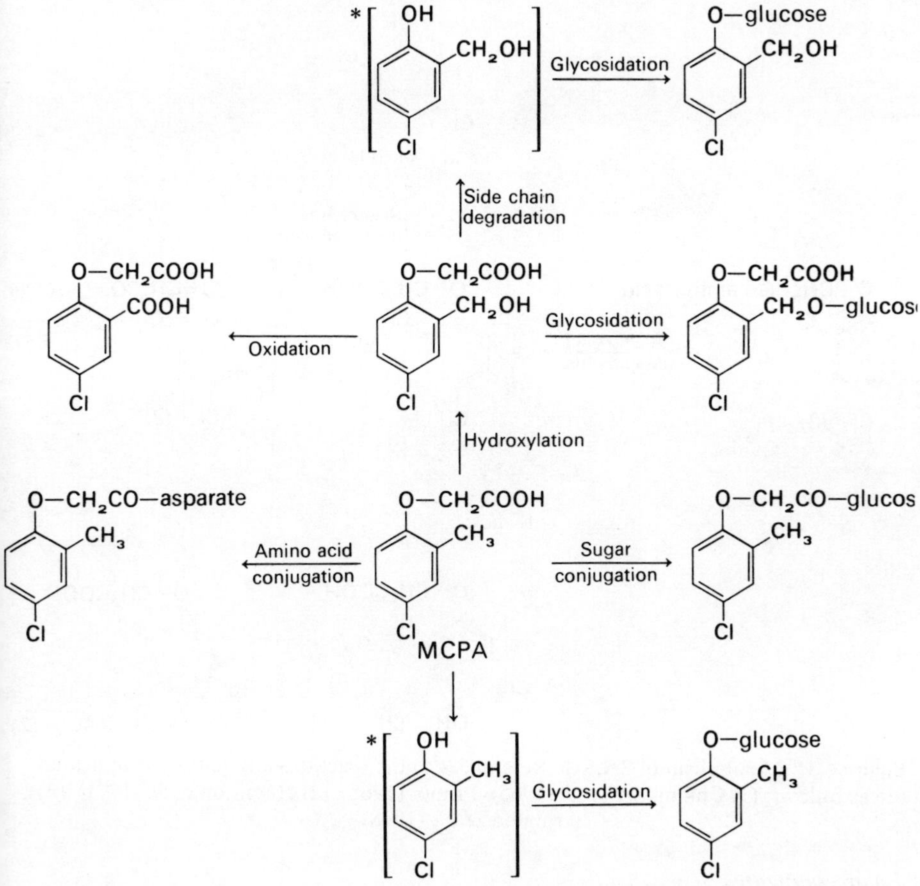

Figure 4.2 Metabolism of MCPA. * indicates not isolated from plant tissue in free form. (From Williams, 1976)

been found for MCPA. For this compound it is the 2-methyl group that is susceptible to oxidation and the hydroxymethyl derivative of MCPA has been found in all species examined (Collins and Gaunt, 1970, 1971; Williams, 1976). Subsequent oxidation of this to 4-chloro-2-carboxyphenoxyacetic acid has also been reported as a minor pathway in some species (Williams, 1976). Following the introduction of a hydroxyl group, either as a phenol or as a primary alcohol, glycoside formation appears to be rapid. Again, the only sugar reported to be involved in this type of conjugation is glucose. Apart from the relatively well-documented metabolic pathways mentioned above, three further areas of metabolism warrant further, more detailed discussion. These are the polar conjugates, bound residues, and protein binding.

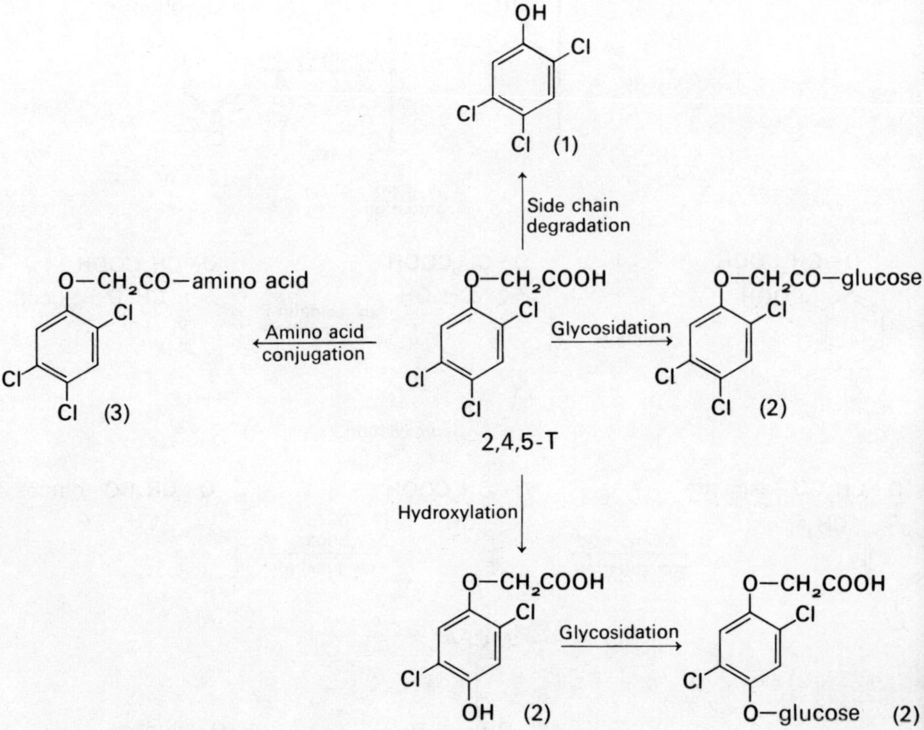

Figure 4.3 Metabolism of 2,4,5-T. References to the reactions and pathways indicated are as follows: (1) Chkanikov et al. (1965); Primer (1965); (2) Hamilton et al. (1971); (3) Arjmand et al. (1978)

Polar conjugates

An area of metabolism that is still poorly understood is the formation of conjugates that are far more polar than either the glucose esters of the phenoxyacetic acids or simple glycosides of hydroxylated metabolites. Work in this laboratory has shown the presence of several of these following MCPA application to a variety of plants (Williams, 1976; Pillmoor, unpublished data). This is particularly true in monocotyledonous species. So far their nature has not been established, although it seems likely that they consist of ester- and ether-linked glycosides with two or more sugar residues attached. Analogous derivatives have been found to be produced during metabolism of 3-phenoxybenzoic acid by More et al. (1978), who identified a glucosylarabinose ester as a metabolite in leaves of vine (*Vitis vinifera*). Metabolism of other types of herbicide for example flamprop, to yield a variety of sugar conjugates has also been reported (Roberts, 1977a, b). Progress in the characterization of these polar compounds is hampered mainly by technical difficulties in their separation and identification.

Bound residues

Following application of radiolabelled phenoxyacetic acids to plants some radioactivity is invariably found to be unextractable by the standard solvent systems used in metabolism studies. The amount involved varies with species but generally increases with time. Some of this is undoubtedly a consequence of the release of $^{14}CO_2$ during metabolism. The usual position of the radioactive label in the parent phenoxyacetic herbicide is at the C-1 or C-2 of the side chain. After oxidation of this unit, $^{14}CO_2$ may be released as was described earlier. Fixation of this CO_2 can lead to the appearance of ^{14}C in all plant products including the proteins, polysaccharides, and other polymers that form the bulk of the unextractable residues. Thus in species in which side chain degradation is extensive, large amounts of bound ^{14}C will be present. However, side chain degradation is only of minor quantitative importance in many species (Loos, 1975) and cannot explain the levels of bound residues often found. Such binding can be considerable, especially in the long term. Even so, this phenomenon has led people to overlook the fact that the herbicide itself, and its metabolites, can become covalently bound to components of the insoluble fraction. The existence of herbicide and metabolite binding has been established largely by the use of ring-labelled phenoxyacetic acids, often using the radioactive isotope of chlorine, ^{36}Cl. For example, application of 2,4-D(2-^{14}C;^{36}Cl) to *Ribes sativum* indicated that over two-thirds of the radioactivity associated with the ethanol-insoluble material could have come from an intact herbicide molecule, with only one-third due to re-incorporation of $^{14}CO_2$ released by side chain degradation (Fleeker, 1973). Steen (1972) has reported that a range of monocotyledonous species contained from 9% to 39% of the absorbed radioactivity from 2,4-D(^{36}Cl) as ethanol-insoluble material. In contrast, the dicotyledonous species investigated contained only 2–5% of the absorbed radioactivity in this fraction. Alternatively, ^{14}C ring labelled herbicide can be used for investigation of bound residues. Significant levels of incorporation of ring-[^{14}C]2,4-DB into insoluble material in white clover (*Trifolium repens*) cell suspensions have been reported (Smith, 1979). Association of herbicide and metabolites with insoluble residues creates a particular problem for the herbicide residue analyst concerned with crop plants at harvest, many months after herbicide application (e.g. Meagher, 1966).

The nature of the insoluble components to which herbicide is bound is not at all clear. One of the first problems encountered with this type of work is the exact definition of what constitutes an insoluble residue. The amount of radioactivity associated with the insoluble residue fraction will be dependent on the final extraction system employed (Wieneke, 1975). For example, analysis of the metabolites of MCPA present in the straw of treated wheat (*Triticum aestivum*), grown to maturity, indicated that some 33% of the total radioactivity was insoluble to extraction with acetone/water at room temperature. However, over 50% of this 'insoluble' fraction was released by washing with phosphate buffer (pH 6.9) at room temperature or, alternatively, 60% could be released by boiling with 1% (w/v) sodium chloride (NaCl). The insoluble radioactivity remaining after both acetone/water extraction and boiling with NaCl, however, could only be solubilized by harsh chemical treatment (unpublished data). While this final fraction might be considered as the true insoluble material, the NaCl extract would contain the water-soluble pectic polysaccharides of the cell wall (Harborne, 1973; Albersheim, 1976; Masuda and Sugawora, 1978) and any associated herbicide residues.

Most reports implicate cell wall structural polymers as the likely site of binding. Hutber *et al.* (1978) demonstrated that 4-OH-POA was associated with a cell wall fraction suggested to be lignin; Meagher (1966) reported that 2,4-D became bound to a pectic polymer in the peel of citrus fruit. IAA has been shown to be bound to a high molecular weight, cellulose glucan through an ester linkage (Bandurski and Piskornik, 1973). However, the routine separation and characterization of bound residues present serious technical difficulties. Separation of cell wall fractions by chemical solubilization techniques has been used, but with only partial success due to cross-contamination of fractions, artefact production, and difficulties in the identification of components in the fractions isolated (Still *et al.*, 1975; Verloop, 1975; Haque *et al.*, 1976). Another approach has been to use selective enzymic hydrolysis to solubilize specific cell wall fractions (Frear and Swanson, 1972; Smith, 1979) but again with relatively little success to date.

It is apparent that much work is still needed to elucidate techniques that will permit the identification of the polymers to which the herbicide is bound. An associated problem is to release the bound phenoxyacetic acid or metabolite by procedures that do not chemically modify the substance. This is an essential preliminary to their characterization.

Protein binding

While binding to cell wall polymers is believed to involve covalent bonds, the phenoxyacetic acids are also reported to bind to proteins by non-covalent interactions. This is not strictly a metabolic process, but it does involve both parent herbicide and metabolites and probably necessitates the synthesis of specific binding proteins (Evans, 1973) and so will be considered in this section.

In some species, protein binding is reported to be very extensive and is considered to involve the major proportion of the phenoxyacetic acid that enters a plant (Rakitin *et al.*, 1966; Zemskaya and Rakitin, 1967; Zemskaya *et al.*, 1971, 1973). However, other workers have been unable to find such high levels of binding (Dexter *et al.*, 1971; Hallem and Eliasson, 1972; Evans, 1973). For example, while Rakitin *et al.* (1966) found that over 72% of the 2,4-D applied to oats (*Avena satium*) was bound to protein within 24 h, Dexter *et al.* (1971) could find only 30% binding using the same technique. Furthermore, the technique used in each case involved preparation of proteins by trichloroacetic acid precipitation. The use of any precipitation procedure in this work is extremely hazardous because of the coprecipitation of low molecular weight materials. This was demonstrated clearly by Zenk (1964) who showed that IAA could be coprecipitated with proteins, thus giving totally spurious binding data. It is also relevant that Dexter *et al.* (1971) have reported that much of the radioactivity associated with TCA-insoluble material was bound through covalent bonds and these authors suggested that the use of such a harsh precipitation procedure would have broken down many non-covalent herbicide/macromolecular associations.

However, not all studies of protein binding have involved precipitation procedures. Evans (1973), for example, made an extensive study of the binding of MCPA to proteins. He separated protein-bound herbicide from free MCPA by dialysis or gel filtration and concluded that binding did take place and that it appeared to involve special proteins, the synthesis of which was apparently induced by the presence of MCPA. However, under his experimental conditions, only about 3% of the MCPA entering the leaves of oat plants (*Avena sativum*) was found to be protein bound in the first 24 h after application. This figure has recently been confirmed for 2,4-D in the same species by work in this laboratory (Soriano, unpublished data). She too was unable to repeat the findings of the Russian workers. Thus, while protein binding undoubtedly occurs, its quantitative importance must remain in some doubt.

The Quantitative Assessment of Metabolism

It has been suggested that all species show the same qualitative pattern of metabolism (Feung *et al.*, 1975). However, it is recognized that there are tremendous quantitative differences between species in both the extent and pattern of metabolism. Quantitative variation is also found between different tissues and organs of the same plant (Canny and Markus, 1960; Norris and Freed, 1966; Davidonis *et al.*, 1977; Feung *et al.*, 1978) and between intact tissue and isolated tissue from the same plant (Schultz and Tweedy, 1971). There also appear to be variations with age (Whitehead and Switzer, 1963; Evans, 1973; Chkanikov *et al.*, 1977; Davidonis *et al.*, 1978). The physiological significance of such quantitative differences in herbicide metabolism is uncertain. Of prime importance in this type of investigation is the availability of reliable data. It is essential to be able to measure accurately the level of each metabolite and of free herbicide in a plant. Unfortunately this is not an easy objective. While some metabolites are relatively stable, others are easily broken down by the extraction or separation techniques used for their isolation. This is particularly true of the sugar esters of the phenoxyacetic acids. Hydrolysis of these conjugates can lead to gross errors in estimates of the levels of free herbicide and metabolites present in a tissue. The following data taken from recent experiments in this laboratory serve to illustrate this point well (Oropeza, unpublished data). Leaves of

Sorghum vulgare where treated with [^{14}C]2,4-D for 24 h after which different procedures were used to estimate the pattern and level of metabolism. The use of ethanol to extract the tissue led to the formation of the ethyl ester of 2,4-D. Up to 40% of the 2,4-D entering the plant was in this form. Of the remainder some 10% appeared as glycosides and the rest was free 2,4-D. The use of isopropanol as extractant gave some 2% of the isopropyl ester of 2,4-D, about 10% glycosides and the remainder as free 2,4-D. The use of acetone as extractant gave 60% of glycosides and only 40% as free 2,4-D. Two problems are apparent here—transesterification and simple hydrolysis of polar metabolites. Transesterification of sugar ester conjugates has been reported for IAA (Stowe *et al.*, 1968) and ABA (Milborrow and Mallaby, 1975). The latter authors observed that extensive transesterification only occurred under slightly alkaline conditions. We have confirmed that with flamprop the use of dilute sodium bicarbonate to wash leaves prior to extraction with boiling ethanol leads to extensive transesterification of the sugar conjugates of this herbicide with the ethanol. Little or no transesterification occurred if the bicarbonate wash was omitted (unpublished data). We have also found that hydrolysis of the sugar ester conjugates of both MCPA and flamprop can occur under very slightly alkaline conditions. Even the use of dilute sodium bicarbonate to separate acidic and neutral components from plant extracts can cause extensive hydrolysis of these conjugates to release the parent herbicides.

Extraction procedures using very low pH values may cause breakdown of sugar esters of 2,4-D (Bristol *et al.*, 1977). It seems likely that macromolecular conjugates involving ester linkages will also be susceptible to hydrolysis, if extracted under the conditions outlined above.

Protein–herbicide complexes involving non-covalent links are also very unstable in the presence of organic solvents (Brian, 1958; Winter and Thimann, 1966; Evans, 1973). Thus, during normal extraction procedures, protein-bound phenoxyacetic acid will appear as free herbicide, another factor which is given little attention is quantitative metabolism studies. Consequently, many of the methods that have been widely used in quantitative studies of herbicide metabolism are vulnerable to major errors and may, in particular, grossly overestimate the amount of free, parent herbicide present.

A further complication of this issue is that conditions of application of herbicide can affect the extent of metabolism (Feung *et al.*, 1978). This means that it may not be valid to compare quantitative data from different sources. An example is provided by the findings of Haque *et al.* (1978) that application of monolinuron-β-glucose by stem injection or via uptake by excised leaves resulted in major differences in metabolism.

The Kinetics of Metabolism

In order to understand fully the behaviour of a herbicide in a plant it is not enough merely to describe the levels and distribution of metabolites and parent herbicide at a given time after application. Data are required of the rate of metabolism, the metabolic pathways involved, and the turnover of metabolites. Ideally such information is needed from samples taken the first hour or so after application and from samples taken throughout the subsequent life of the plant. The obvious approach to this is to analyse the pattern of metabolism at different times after treatment. It is worth emphasizing that this type of study demands dependable quantitative methods of analysis. From the results of such a survey it is possible to propose sequences of metabolism and to get an idea of the metabolic stability of the parent herbicide and metabolites. Detailed studies of turnover can then be initiated.

At this point it is worth noting that changes in the metabolites present with time could be due to several distinct processes.

(1) There may be several phases of metabolism of the parent herbicide. For example, Montgomery et al. (1971) have reported that corn (*Zea mays*) and bean (*Phaseolus vulgaris*) plants form glycosides of both parent 2,4-D and its hydroxylated derivative. However, with increasing exposure times, the percentage of the conjugates with 2,4-D as the aglycone apparently increased, indicating a change in the metabolic capabilities of the tissue with time.
(2) Changes in the metabolites present in an extract from a whole plant with time could reflect movement of herbicide into different parts of the plant, with different metabolic capabilities.
(3) Changes in the metabolites present with time could be due to actual interconversion of metabolites in specific precursor/product relationships.

In the first few hours after application, a lag period is often apparent before metabolism begins, which could be explained by the activation or synthesis of the enzymes involved in metabolism. This is discussed in more detail later in this section.

The first metabolites formed after application of 2,4-D to soya bean (*Glycine max*) are amino acid conjugates (Chkanikov et al., 1972; Feung et al., 1972). 2,4-D glutamate appears to be produced initially. However, this compound is reported to be converted subsequently to other metabolites including 2,4-D aspartate (Feung et al., 1972; Mumma and Hamilton, 1975) and hydroxylated derivatives (Feung et al., 1973). This metabolic sequence is supported by the finding that hydroxylated derivatives of 2,4-D can be conjugated to amino acids (Mumma and Hamilton, 1975). The glycoside of 4-OH-2,5-D can also apparently occur conjugated to an amino acid (Chkanikov et al., 1977). Other studies have led to the suggestion that amino acid conjugates may be slowly converted to sugar conjugates over longer periods of time (Chkanikov et al., 1972).

In contrast, Bristol et al. (1977) from studies with wheat callus (from *Triticum monococcum*), consider that hydroxylation of 2,4-D and subsequent sugar conjugation occur first and that this is initially the main metabolic pathway.

However, this pathway appears to become less active with time, whereas formation of the amino acid conjugates continues in a slow, linear fashion. This observation is similar to that made by Montgomery *et al.* (1977), mentioned previously, as both reports suggest that the high rate of hydroxylation initially seen decreases with time and is replaced by direct amino acid or sugar conjugation of the parent herbicide. However, for *Coleus* explants (*Coleus rhenaltianus*) Veen (1966) has suggested that sugar ester conjugates of IAA are initially formed, followed by subsequent transfer to amino acid conjugates. A cautionary note is needed since reliable quantitative data are rarely available for the sugar ester conjugates due to their chemical instability. For example, the extraction procedure used by Bristol *et al.* (1977) could have led to the destruction of such compounds. Although these authors do note that any breakdown of the carboxylic glycosides would not have introduced significant error into their results for wheat (*Triticum monococcum*), the same will not necessarily be true if other species were investigated by the same procedure. Furthermore, it is important to distinguish between sugar ester conjugates and ether-linked glycosides since only the former will involve the parent phenoxyacetic acid molecule.

Monosaccharide conjugates are probably precursors of the more polar sugar conjugates and possibly of macromolecular covalent complexes (Hallem and Eliasson, 1972). Little data are available for the phenoxyacetic acids but this route has been found for other compounds. In this laboratory we have found that flamprop is initially metabolized to the glucose ester in oats (*Avena sativum*). This metabolite reaches a maximum level after one day and then decreases concomitantly with the appearance of more polar sugar conjugates (unpublished data). Phenol conjugation with glucose has been shown to precede the formation of phenol gentiobiosides (Yamaha and Cardini, 1960; Frear, 1975). The compound oxamyl is initially glucosylated prior to conversion to more polar compounds (Harvey *et al.*, 1978). A glycoside of 2,4-DB has been suggested to be incorporated into the cell wall structural material (Smith, 1979) of clover callus tissue (*Trifolium repens*). A similar progression has been proposed for the metabolism of diphenamid (Hodgson *et al.*, 1974).

The results presented above all indicate that initial herbicide metabolites are unstable within the plant. It seems probable that upon entry to a plant tissue there is rapid and extensive metabolism to what may be termed an 'emergency metabolite'. This is followed in the longer term by conversion to what are presumably more stable, less toxic, or more easily stored metabolites. However, much more research is needed to establish the biochemical relationships between metabolites before such a proposal can be substantiated.

The Enzymes of Metabolism

Very little is known of the enzymes responsible for the metabolism of the phenoxyacetic acids since very few successful cell-free studies have been made.

One of the more interesting observations is that conjugation of applied phenoxyacetic acid, IAA, or NAA with aspartic acid appears to involve inducible enzymes. A lag period of several hours occurs between application and the detection of aspartate conjugation (Zenk, 1961; Sudi, 1966; Veen, 1966; Venis, 1972; Rekoslavskaya and Gamburg, 1977). Inhibitor studies have shown that both RNA and protein synthesis are required during the lag period (Venis, 1964, 1972). The lag period can only apparently be abolished by pretreatment with active auxins (Sudi, 1964; Venis, 1972). The case of IAA aspartate formation in blackberry (*Rubus* sp.) tissue cultures is additionally interesting since this tissue is autonomous for auxin and contains relatively high levels of endogenous IAA (Rekoslavskaya and Gamburg, 1977). Despite this the aspartate conjugation system only appears to be induced by the addition of exogenous IAA. It is possible that the induction system shows a concentration dependence and that aspartate conjugation is a specialized response of the tissue to deal with auxin concentrations above normal hormone levels.

In contrast to the above, conjugation of IAA and NAA with sugar does not show a lag period (Zenk, 1961; Veen, 1966). However, there is evidence that sugar conjugation of 2,4-D is markedly increased by pretreatment of the plant (Oropeza, personal communication). A similar situation was found by Makeev *et al.* (1977) for the hydroxylation of 2,4-D in cucumber (*Cucumis sativus*) and pea (*Pisum sativum*) plants. The capacity for hydroxylation by cell-free extracts was increased three or four times by pretreatment of intact tissue with 2,4-D. In each of these cases there is an apparent stimulation in the capacity of a particular metabolic reaction by the herbicide above an existing background level. This differs from the aspartate conjugation system in which the process does not occur in the absence of the herbicide.

The situation for protein binding of MCPA appears to resemble an inducible system (Evans, 1973). The increase in binding proteins in response to the presence of herbicide showed a lag period and was blocked by RNA and protein synthesis inhibitors.

There have been only one or two cell-free demonstrations of the metabolism of phenoxyacetic acids. Makeev *et al.* (1977) found hydroxylation of 2,4-D to be catalysed by microsomal systems. The enzyme preparation could also catalyse the hydroxylation of cinnamic acid and appears to resemble the system involved in phenol synthesis, which has been described by Russell (1971). It is interesting to note that it has much in common with the well described cytochrome P450 system found in animals. This result would suggest that hydroxylation of the phenoxyacetic acids involves an enzyme with a role in normal intermediary metabolism. In contrast, Hamilton *et al.* (1971) have reported that a microsomal preparation from pea (*Pisum sativum*), capable of converting *trans*-cinnamic acid to *p*-coumaric acid, did not convert 2,4-D to its hydroxylated derivatives. Finally an enzyme has been isolated from cucumber (*Cucumis sativum*) that catalyses the glucosylation of 4-OH-2,5-D (Makoveichuk *et al.*, 1978). This enzyme appears to be a soluble protein and to use UDPG as the glucose donor. A number of reports have been made of cell-free glycosylation of plant phenols, which also use UDPG as the glucosyl donor (Yamaha and Cardini, 1960; Feingold et al., 1964; Kleinhofs et al., 1967). The specificity of such systems does not appear to have been widely studied, although Hutber et al. (1978) have reported that an enzyme from wheat germ, which was able to catalyse quinol glucoside formation, had no effect on 4-OH-POA. It is evident from these results that much more characterization of the enzymes involved in both hydroxylation and glycosylation of the phenoxyacetic acids is required before it can be decided whether herbicide metabolism is mediated by enzymes with a role in normal intermediary metabolism or whether special enzymes are required with, perhaps, a specific defence role in the plant.

To date, there do not appear to have been any cell-free demonstrations of the conjugation of the phenoxyacetic acids with glucose or amino acids. For IAA, Kopcewicz et al. (1974) have reported enzymic esterification to glucose by an enzyme preparation from immature kernels of Zea mays which was dependent on CoA and ATP. However, these authors do note that only the 2-O, 4-O, and 6-O esters of IAA were formed by their system and that formation of the 1-O esters may occur through a mechanistically different reaction, possibly involving UDPG.

A very interesting recent finding is that the internal concentration of free 2,4-D is apparently regulated by metabolism in soy bean callus (Davidonis et al., 1977). This implies the presence of rather precise control mechanisms capable of detecting the level of 2,4-D and of regulating the activity of the enzymes involved in its metabolism. Furthermore the level at which 2,4-D is maintained varies with root and leaf tissue. However, the physiological significance of these observations cannot yet be evaluated. Whereas it would be expected that mechanisms exist to regulate the concentration of IAA, it is difficult to imagine that enzymes controlling IAA metabolism are the same as those that control 2,4-D breakdown.

The subcellular location of metabolism is not known. It is often assumed that it will occur within the cytoplasm and the association of a 2,4-D hydroxylation system with microsomal membranes is in agreement with this. However, there have been suggestions that some types of metabolism occur outside the cell in the apoplast (Crafts, 1960; Szabo, 1963; Zemskaya and Rakitin, 1967; Robertson and Kirkwood, 1970).

The Biological Activity of Metabolites

In order to assess the physiological significance of metabolism it is essential to know whether the metabolites have any herbicidal or hormonal activity. This question has been difficult to answer directly since there has been no *in vitro* method of determining auxin activity. Thus standard auxin bioassays have been used. These suffer serious drawbacks in providing comparative data between parent phenoxyacetic acid and its metabolites. The drawbacks stem from (1) differences in physico-chemical properties which lead to different uptake and distribution in the test tissue; and (2) differences in metabolic stability within the tissue. In only a few cases has any attempt been made to measure these variables in a bioassay and thus allow proper interpretation of the results.

In general, the sugar ester and amino acid conjugates of auxins show activity in auxin bioassays at levels that are comparable with the parent compound (Zenk, 1961; Feung *et al.*, 1977). However, it is thought that both of these types of metabolites are unstable in the plant and may be readily hydrolysed (Evans, 1973; Davidonis *et al.*, 1979). It is therefore believed that they are inactive *per se*. A similar conclusion was reached by Hiraga *et al.* (1974) for the glucosyl esters of gibberellins. The observation of Davidonis *et al.* (1979) that the aspartic and glutamic acid derivatives of 2,4,5-T have very low auxin activity is also in agreement with this. However, it is as yet unexplained why the same conjugates of 2,4-D are highly active in the same assay system (Feung *et al.*, 1977), although this may be due to differences in the stability of the conjugates in the assay system.

Hydroxylation of the phenoxyacetic acids, either in the aromatic ring (2,4-D and 2,4,5-T) or at the methyl group of MCPA, seems to inactivate the molecule (Fleeker and Steen, 1971; Hamilton *et al.*, 1971; Collins, 1972; Bristol *et al.*, 1977). Similarly the loss of the side chain will eliminate auxin activity, although there have been suggestions that the resulting phenols could still be toxic (Luckwill and Lloyd-Jones, 1960b; Dubovoi *et al.*, 1973). If this is so, then subsequent glycosylation of the phenols (Williams, 1976) may be required before the molecule completely loses biological activity.

Recently two developments have made it possible to examine directly the auxin activity of metabolites. The first of these is the discovery of receptor proteins that bind auxins *in vitro*. These are discussed in a later section. So far the phenoxyacetic acid metabolites have not been tested in competition experiments that would indicate whether they could be expected to have any biological activity.

The second development is the finding in this laboratory that the enzyme tocopherol oxidase is controlled *in vitro* by auxins, apparently with high specificity (Gaunt *et al.*, 1980). This is again discussed in more detail later in this review. So far a few metabolites of MCPA have been tested in this system (Gaunt and Plumpton, unpublished data) including the hydroxymethyl derivative and two amino acid conjugates of the parent herbicide. None of these compounds showed auxin activity, supporting the view mentioned above. It is worth recording that in this system neither indolebutyric acid nor 2,4-DB was active as an auxin, confirming the belief that these compounds are only active after β-oxidation (Wain, 1955a, b, c). However, even if the toxicity of the metabolites can be reliably determined, the action they will exert in the *in vivo* situation will be dependent on their intracellular location. To date, little information is available on this.

From the above discussion it is apparent that metabolic reactions can be divided into two distinct classes:

(1) Those reactions that irreversibly inactivate the herbicide, such as hydroxylation and side chain loss or alteration;
(2) Those reactions that inactivate the molecule but produce a derivative that is metabolically unstable and which may subsequently release the free herbicide. This is true of sugar ester and amino acid conjugation as well as protein and macromolecular complexing. It has already been noted that these reaction products are converted to other compounds in the long term, but it is worth remembering that in the short term they represent a pool of potential herbicide within the plant.

Summary

The phenoxyacetic acids are readily metabolized in most plants, although there is considerable variation in the quantitative pattern of metabolism between species. Metabolites are generally believed to be inactive as auxins and their formation can be thought of as a detoxication mechanism. However, some metabolites may regenerate the parent phenoxyacetic acid upon hydrolysis and could thus serve as a reservoir of potential herbicide. There is evidence that after a period of rapid metabolism there is a slow, progressive change and the metabolites that are formed first may undergo further metabolic alteration. The enzymes involved in metabolism are poorly understood, but some appear to be induced by the presence of herbicide. Protein binding can also occur and reduce the levels of free herbicide, although there is some dispute about the quantitative significance of this process.

INTERNAL LOCATION

Introduction

The lack of data concerning the internal location of the phenoxyacetic acids and their metabolites is referred to several times in this review. If we are to understand the different ways in which plants respond to these compounds it is essential that the following information is available:

(1) their distribution in the various tissues within the plant;
(2) their subcellular distribution in each cell;
(3) their concentration within each compartment;
(4) the permeability of different membranes to each compound;
   and the fluxes that occur across each membrane.

It is important to emphasize that these questions are asked of metabolites as well as of the parent herbicide. This is necessary since many metabolites may be converted back into the parent herbicide and their metabolic stability may vary in different compartments of the cell. For example, herbicide conjugates may be quite stable in the vacuole but readily hydrolysed by enzymes in the cytoplasm. A further point worth noting is that a description of the internal location of the herbicide and its metabolites at a single time period after treatment is only of limited value. What is really required is a detailed study of their internal location over at least the first few weeks of treatment.

This information is required at two quite different levels—tissue and subcellular. While there is some information on the distribution of phenoxyacetic acids in different parts of the plant after leaf application, this does not distinguish between different cell types within each region. Even in a leaf given a surface application of herbicide there is little knowledge of the distribution between epidermis, mesophyll, vascular, and other tissues. At the subcellular level, a few studies have been made of phenoxyacetic acid distribution with very variable results. We are left with almost no direct information about the four points raised above. The major reasons for this are the severe experimental difficulties encountered in this type of work. Consequently, the following discussion will largely be about the methods available. At the outset it is worth emphasizing that all of the methods to be discussed have at least one drawback, either in the interpretation of results or in actual technical difficulties. It is evident, therefore, that to begin to understand the internal location of herbicides it is necessary to use several approaches. Only when all methods suggest the same conclusion can one feel confident about that conclusion. Most of the discussion will be focused upon subcellular localization. Four main approaches are available, namely subcellular fractionation, autoradiography in thin sections, efflux analysis, and studies with isolated cells, protoplasts, and vacuoles.

Subcellular Fractionation

The accumulation or association of a herbicide or metabolite in a particular organelle or membrane may be demonstrated directly by isolation of particulate fractions from homogenates of tissue previously exposed to the herbicide. Techniques of subcellular fractionation by differential or isopycnic centrifugation procedures have been developed to permit the separation of organelles and membranes in relatively undamaged condition and relatively pure. Most procedures use aqueous media, but this allows the diffusion of soluble compounds out of organelles during their isolation. In an attempt to overcome this problem, non-aqueous isolation procedures have been used. The difference in results from these procedures is dramatic, as is illustrated by the work of Hallam and Sargent (1970) on the subcellular distribution of 2,4-D after application to bean leaf discs. Aqueous fractionation procedures led to the presence of the majority of the herbicide in the soluble supernatant, not associated with any particulate fraction. In contrast, the use of a non-aqueous isolation procedure showed the majority of the 2,4-D to be localized in the chloroplasts. An identical situation has also been reported for the subcellular distribution of 2,4,5-T in *Rubus procerus* (Bretherton and Hallam, 1979). Such results agree with those of Zemskaya and Rakitin (1967) and Evans (1973) for the aqueous procedure and of Bertagnolli and Nadakavukaren (1974) for the non-aqueous procedure. Although the aqueous method is certain to give inaccurate data the reliability of the non-aqueous procedure is uncertain. However, we have already argued that a high herbicide concentration in the chloroplasts may be expected on the grounds of the pH gradient between cytoplasm and chloroplast stroma; the results of the non-aqueous procedure support this argument. Further support for the non-aqueous procedure is evident from the results of Hallam and Sargent (1970), Bertagnolli and Nadakavukaren (1974), and Bretherton and Hallam (1979), in that all three groups of authors have observed either morphological or physiological changes in the chloroplasts after treatment, suggesting that herbicide might be expected to be associated with these structures.

Even if procedures are developed that allow the isolation of pure membranes and organelle components from a tissue, complete with all of their content, there would still be a major problem in distinguishing between three subcellular compartments. Soluble compounds in the cell wall, cytosol, and vacuole will all appear together after sedimentation of all particulate components. Consequently this general approach to understanding herbicide distribution has a limited value.

Autoradiography

The use of whole plant autoradiography for investigation of the distribution of herbicides throughout the plant gives useful qualitative data and has been extensively used. There have also been several studies of distribution between the various cell types in a thin section. However, there are several technical problems here. The main danger is the leakage and redistribution of water-soluble compounds during sectioning and processing of plant material after treatment with radioactive herbicide. This is normally overcome by the rapid freezing of tissue and the sectioning of the frozen material. Suitable methods have been discussed by Chayen *et al.* (1960), Radwan *et al.* (1960), Appleton (1964), Gahan *et al.* (1967), and Sanderson (1972). Apart from technical difficulties, the method has the additional drawback that it does not allow discrimination between parent herbicide and metabolites, nor does it give accurate quantitative data. None the less there have been attempts at localizing 2,4-D at both the intra- and intercellular level. Radwan *et al.* (1960) demonstrated the presence of radioactive 2,4-D in the phloem and xylem of petioles and stems of bean plants. Pickering

(1965) showed the accumulation of 2,4-D in the chlorenchyma cells of bean leaves. However, a more thorough investigation of the distribution of herbicide with time in the various cell types of the leaf after foliar application is still required. An extension of the use of this technique to look at subcellular distribution is an attractive possibility. Pickering (1965) felt able to distinguish between radioactivity located in the cytoplasm and in the vacuole. He reported no evidence for extensive vacuolar concentration of radioactivity. Liao and Hamilton (1966) claimed that both 2,4-D and IAA were associated with cytoplasm and nuclei in root tip squashes.

In general this approach has not yet received the attention it deserves in the herbicide field. We must look to future developments that should permit the routine preparation of sections suitable for both autoradiography and for the electron microscope.

Efflux Analysis

Efflux analysis was originally developed for the investigation of the subcellular compartmentation of inorganic ions by MacRobbie and Dainty (1958a, b) and Pitman (1963). The technique has also been employed by other authors including Hope (1963)., Cram (1968, 1973), Davies and Higinbotham (1976), and has been reviewed by MacRobbie (1971). The procedure basically involves measurement of the efflux of a radioactively labelled compound from tissue slices that have been loaded previously with that compound. A semilogarithmic plot of the radioactivity in the tissue at different stages of efflux is constructed. The presence of a straight-line component in the graph is taken to indicate first order diffusion kinetics, limited by movement across a single membrane. By 'curve-peeling' techniques several kinetically distinct phases of efflux can usually be distinguished. Three such phases have often been identified for inorganic ions and these have been tentatively equated to free space, cytoplasm, and vacuole. However, the cytoplasm will only be distinguished from the vacuole provided that the flux of the compound under study across the tonoplast is slower than its flux across the plasmalemma and also provided that the vacuole is linked in series through the cytoplasm to the free space.

From the data that can be obtained from an efflux experiment the concentration of compound in each compartment and the fluxes between compartments can be calculated. The equations used to calculate the various fluxes, for a variety of different situations, have been summarized by Walker and Pitman (1976). It must be recorded that several assumptions have to be made about the particular system in order to apply these equations. For example, it is assumed that all intracellular pools of the labelled compound have the same specific activity and that the system has reached a quasi-steady state at the end of the loading period. While it is clear that the interpretation of efflux analysis data requires caution, the technique provides a valuable approach to the problem of subcellular localization of a variety of compounds. Although initially used with inorganic ions it has more recently been applied to organic compounds. For instance, Osmond and Laties (1969) and Kluge and Heininger (1973) have used it to study malate distribution; Dela Fuente and Leopold (1972) applied a type of efflux analysis to investigate the transport of IAA in sunflower stem sections; Drake (1979) has recently applied efflux analysis to study gibberellin compartmentation. This extension of efflux analysis is not without the introduction of a major new problem, namely that the compound under study may be metabolized in the tissue. This has two consequences, first, at the end of the loading period the system may not be in a quasi-steady state, contrary to one of the basic assumptions of the method, and second, the distribution of radioactivity between parent compound and metabolite must be ascertained at each stage of the efflux. As an alternative to this very time-consuming approach, other workers have used model compounds that are not metabolized in the tissue under study (Price, 1973; Bridges and Farrington, 1974).

In the case of the phenoxyacetic acids in plant tissues in which metabolism is rapid and extensive, efflux analysis may have limited value. At best it can only give a guide to the distribution and mobility of the parent compound and of the major metabolites, but even this information would be valuable bearing in mind the present lack of data. The procedure has been used to study 2,4-D distribution in barley roots (Shone *et al.*, 1974) and three, kinetically distinct components were observed. However, in addition to the problems already mentioned the phenoxyacetic acid herbicides present additional difficulties. These compounds can directly affect membrane permeability (Devlin, 1974; Zsoldos, *et al.*, 1978), which may alter fluxes during analysis. The biphasic pattern of herbicide uptake followed by export (Blackman, 1961) that has been mentioned earlier also implies that a change in membrane permeability can occur following treatment. Thus, the time of exposure of the tissue to a phenoxyacetic acid must be carefully determined during an efflux analysis experiment to try to avoid such alterations during the course of the experiment.

The Use of Isolated Cells, Protoplasts, and Vacuoles

The use of isolated cells and protoplasts as a research tool for investigation of general metabolism of plant cells, as well as for the biochemical mode of action of herbicides, has gained much support in recent years. One of the main attractions of these systems is that the extracellular environment can be so much more easily defined than when using an intact plant or even with leaf discs or tissue slice experiments (e.g. Gnanam and Kulandaivelu, 1969; Rehfeld and Jensen, 1973; Boulware and Camper, 1972, 1973; Porter and Bartels, 1977; Pavlenko *et al.*, 1978).

By using single cell suspension cultures direct information on the distribution of the phenoxyacetic acids between the cell and the external medium has been obtained by Rubery (1977, 1978) as has been discussed earlier in this review. Valuable data on the permeability of the plasma membrane can thus be determined. The same approach has been used to relate the internal concentration of 2,4-D to the physiological behaviour of *Acer* cells in suspension culture (Leguay and Guern, 1977). Working with cells in which cell division is dependent upon an external supply of auxin they observed that as long as the 2,4-D concentration within the cells was above 3×10^{-7}M, cell division continued. As soon as the internal concentration fell below this threshold value, cell division ceased. This is a very important advance in knowledge and for the first time discussion about auxin action can be focused upon a rather precise value for the hormone content of the cell.

Although invaluable information may be obtained from experiments with isolated cells, they still do not provide detail of the subcellular distribution of herbicide. The possibility of working with isolated vacuoles does permit an advance in this direction. Two approaches are theoretically possible. First, the isolation of vacuoles from plants previously treated with herbicide may permit the direct analysis of either parent compound or metabolites within this compartment. This would only be possible if the tonoplast retained the compound, preventing leakage during isolation. However, if efflux analysis demonstrates the presence of an immobile pool of radioactivity, or possibly a very slowly exchanging pool, there would be a good chance of directly determining by isolation whether it was located in the vacuole. Such an approach has been successfully employed to demonstrate the storage of phenol glycosides in the vacuole (Saunders and Conn, 1978; Grob and Matile, 1979; Loffelhardt *et al.*, 1979). The second approach is to measure the uptake and efflux of herbicide and metabolites by isolated vacuoles and thus obtain direct data on the permeability of the tonoplast. However, at present it seems doubtful that isolated vacuoles have sufficient stability to allow the manipulations needed for this.

Finally, the isolation of homogeneous preparations of intact organelles such as mitochondria and chloroplasts is relatively easy. A good deal of data on the permeability of the membranes surrounding these organelles to both herbicide and metabolites should be a straightforward task.

Summary

Very little is clearly understood of the subcellular distribution of the phenoxyacetic acids and their metabolites. Four experimental approaches are available that could be used to study this. While each has problems of a technical nature or of interpretation, together they should provide some badly needed information.

MODE OF ACTION

Introduction

The question of the mode of action of the phenoxyacetic acid herbicides is ambiguous. It usually refers to the reasons underlying their toxicity, but confused with this is the fact that these compounds are all auxins. As such they regulate an extremely large number of physiological activities in plants. It is likely that few of these are potentially damaging to a cell or tissue. So from the outset it is necessary to recognize that many plant responses to the phenoxyacetic acids do not lead to injury or death. To understand their mode of action, however, demands a biochemical account of each and every one of the observed physiological effects. Each must be explained at the molecular level. The sum of the answers will describe the mode of action of the phenoxyacetic acids. Amongst the answers must lie the explanation of their phytotoxicity.

Let us first consider how an auxin, or indeed any other plant hormone, might be expected to function, that is, their theoretical mode of action at the cellular level. The following events must take place.

(1) The hormone must be perceived and recognized at some receptor site or sites in the cell.
(2) As an immediate consequence of hormone interaction with its receptor, one or more specific reactions will be affected. These will be primary or direct reactions. They require the presence of auxin and it should be possible to demonstrate the reactions and their responses to hormone in cell-free systems.
(3) Consequential upon the primary reactions will be alterations in the rates of other reactions. These will be secondary or indirect responses. They are not regulated directly by auxin but are dependent on whatever changes were initiated by the primary reactions. It is possible to envisage level after level of secondary reactions, each responding to changes wrought by the preceding level. As conditions within the cell alter, so the number of secondary effects is multiplied until eventually one can imagine that the entire biochemistry of the cell will have altered. It is not difficult to draw up a scheme whereby a single primary reaction can lead to this. For example, consider the consequences following the stimulation by auxin of the activity of an ATP-requiring proton pump in the plasmalemma—a plausible possibility. The following sequence of events could occur:

(a) increased proton gradient across the plasma membrane leading to (i) an altered distribution of weak acids, such as auxins, between cell wall and cytoplasm (see earlier discussion); (ii) increased fluxes of other solutes across the plasma membrane via proton cotransport systems; (iii) an altered membrane potential with subsequent effects on ion fluxes across the plasma membrane.

(b) increased cytoplasmic pH leading to (i) altered activities of cytoplasmic enzymes; (ii) increased acid production to reduce cytoplasmic pH.
(c) decreased cell wall pH which will lead to (i) altered activities of cell wall enzymes; (ii) possible changes in the interactions between structural components of the cell wall.
(d) increased potassium uptake to maintain electrical neutrality in the cytoplasm. This could also involve other ion movements.
(e) increased mitochondrial respiration rate to provide the ATP required for the proton pump, leading to (i) increased rate of glycolysis and/or the pentose phosphate pathway; (ii) increased hydrolysis of energy reserves such as sucrose and starch.
(f) changes in cellular composition may even lead to an altered pattern of gene expression.

The 'knock-on' effect of a single change is extensive and dramatic. Considered in this light it is not really surprising that auxins have been found to have so many biological effects in plants.

The sum of the primary and secondary reactions will actually describe the mode of action of auxins, at least in biochemical terms. The answer will be different in each tissue and may vary with species. Existing observations on the biochemical effects of the phenoxyacetic acids must be fitted into the framework of the above theoretical outline. It is necessary to consider which reactions are primary and which are likely to be secondary responses and what is the approximate hierarchy among them. Finally, we must ask which reactions to the phenoxyacetic acids are responsible for cellular and tissue damage and plant death. In the following discussion each of these points will be considered.

Auxin Receptors

There have been occasional reports that auxins can interact with cell components such as phospholipids with biochemically significant consequences (Kennedy and Harvey, 1972; Veen, 1974; Parups and Miller, 1978). However, there is no evidence that such interactions are specific or sensitive enough to be involved in the primary role of auxin action. The weight of evidence suggests that proteins fulfil this role. There has been an increasing number of reports of the existence of proteins that bind auxins with high affinity. These have been reviewed by Venis (1977) and Lamb (1978). Auxin-binding proteins have been reported to be soluble (Wardrop and Polya, 1977) or to be associated with various membrane fractions, identified as the plasma membrane (Lembi et al., 1971; Jacobs and Hertel, 1978), the endoplasmic reticulum (Ray, 1977; Dohrmann et al., 1978; Cross and Briggs, 1979), and the tonoplast (Dohrmann et al., 1978). There may be at least two different receptor proteins in a single cell (Batt and Venis, 1976; Dohrmann et al., 1978). The membrane-bound receptor has often been solubilized without loss of auxin-binding capacity (Cross and Briggs, 1979).

Not all proteins that bind auxins can be considered as receptor molecules involved in auxin function. Even bovine serum albumin can bind auxins (Matlib *et al.*, 1971; Murphy, 1979) which can hardly be of physiological significance to plants. Furthermore, there are some phenoxyacetic acid binding proteins in plants which possibly have a defensive role (Chkanikov *et al.*, 1971; Zemskaya *et al.*, 1971; Evans, 1973). These have already been discussed. In order for a binding protein to be credible as a receptor it must not only show a high affinity for auxin, it must also show specificity and should interact only with auxins, anti-auxins or with compounds that inhibit polar auxin transport. Several of the putative auxin receptors that have been studied appear to lack the high degree of specificity demanded for their role (Ray *et al.*, 1977b) despite the fact that there is a generally good correlation between binding and biological activity. At first sight this is disturbing and requires explanation. It is possible that extraction of these proteins from membranes or an alteration in their ionic environment could affect protein conformation and thus specificity (Poovaiah and Leopold, 1976). Alternatively there may be specific small molecular weight compounds present in cells which modify the interaction between auxin and receptor (Ray *et al.*, 1977a; Venis and Watson, 1978). More work is needed to resolve this point. Meanwhile, caution should be exercised before considering a protein that binds auxins as a receptor (Murphy, 1979). Nevertheless, it seems certain that auxin receptor proteins do exist, although the number and location of these in a cell is still in doubt. Their function is also unknown and there has been no demonstration of any biological activity for these proteins beyond their ability to bind auxins. At present one can but speculate on their cellular role(s). A number of basic possibilities are likely.

(1) They are enzymes.
(2) They are membrane proteins responsible for transport of one or more solutes between subcellular compartments. This could include auxin itself and there have been suggestions that some auxin-binding proteins have the properties required of an auxin transport carrier (Jacobs and Hertel, 1978).
(3) They interact with chromatin to alter its template availability.
(4) They have a purely regulatory function. That is, they work by interaction with a second protein that is either an enzyme, a transport protein or a gene repressor. This, for example, is the way in which hormone receptor proteins appear to modulate the action of adenylate cyclase in the plasma membrane of many animal cells (Cuatrecasas, 1974).

In each of the above the receptor is seen as an allosteric protein. In the simplest model it will exist in two conformational states, depending on the presence or absence of auxin. One conformation will permit the primary reaction to occur, the other will not.

Biochemical Effects of Auxins

The literature records literally hundreds of different effects of auxins upon cellular biochemistry. These range from minor changes in the level of some cell component to an altered pattern of gene transcription. They have been listed and reviewed on several earlier occasions (Penner and Ashton, 1966; Ashton and Crafts, 1973; Loos, 1975) and no useful purpose would be gained by cataloguing them again or even by extensively updating the list. In this section we wish to review those areas of cellular metabolism that have been consistently shown to be affected by auxins. In particular, attention will be focussed upon candidates for primary reactions of auxins. All such reactions should have the following properties:

(1) Specificity towards a range of auxins;
(2) reversal of the auxin effect by auxin antagonists;
(3) sensitivity to auxins at concentrations in the range likely to be of physiological significance—say 10^{-4} to 10^{-7} M;
(4) immediate response to an application of the auxin;
(5) direct regulation by auxin that should be demonstrable *in vitro*.

The general areas that will be considered are gene expression, mitochondrial and chloroplast function, membrane transport, cell wall changes, interactions with other hormones, and regulation of specific enzymes.

Control of gene expression

The activity of numerous enzymes has been reported to be altered during auxin action. This must either be due to the activation or inhibition of an existing protein or to an altered pattern of gene expression. There is ample evidence to show that the latter is true in many cases (Venis, 1977). This comes from several types of experimental observation.

(1) Auxin-induced changes may be blocked by inhibitors of RNA transcription and protein synthesis (Venis, 1964; Fan and Maclachlan, 1966; Venis, 1972; Rutherford and Deacon, 1973; Bates and Cleland, 1979).
(2) The level of polysomes has been shown to increase after auxin treatment, reflecting an increased rate of protein synthesis (Trewavas, 1968; Davies and Larkins, 1972).
(3) Auxin may cause a change in the pattern of RNA synthesis (Ingle and Key, 1965; Key *et al.*, 1966).

From the above examples and similar reports it is clear that auxin-induced changes involve altered transcription and/or translation. However, the means by which this is achieved has yet to be explained. One way could be via changes in RNA polymerase activity (Guilfoyle *et al.*, 1975). While this might account for gross changes in the pattern of RNA synthesis, such as the several-fold stimulation of ribosomal RNA, it is unlikely to explain the minor alterations in messenger RNA synthesis that appear to lie behind some changes in protein synthesis. A more plausible model would involve auxin in the alteration of the template availability of nuclear chromatin, so permitting very selective changes in protein synthesis. Some observations support such a proposal (Teissere et al., 1973; Likholat and Pospelov, 1974). This could be a consequence of changes in chromosomal proteins (Chen et al., 1973; Murray and Key, 1978) such as the phosphorylation of non-histone proteins.

In the above suggestions the question of how auxin can cause the changes proposed remains unresolved. There have been one or two reports that auxins can directly stimulate RNA synthesis by nuclear preparations in vitro (Matthysse and Phillips, 1969; Mondal et al., 1972), but there has been no substantiation of these findings and it is possible that control of RNA synthesis is indirect. There have been demonstrations that auxin leads to the release or activation of proteins that in turn regulate transcription. The primary action of the auxin would seem, therefore, to occur at a site remote from the nucleus (Hardin et al., 1972).

There remains the possibility that auxin may regulate at some stage of post-transcriptional processing of premessenger RNA or of translation. These complex processes are still poorly understood but it is clear that they are susceptible to very precise regulation in eukaryotic cells (Revel and Groner, 1978).

The kinetics of auxin-induced changes in transcription and protein synthesis have been studied and some remarkably rapid responses claimed. For example, Masuda and Kamisaka (1969) reported changes in RNA synthesis within 10 min of auxin treatment. In general, however, it is felt that no convincing changes in RNA and protein synthesis occur until at least 30 min after hormone application (Venis, 1977). It is known that auxin can elicit some much more rapid responses in plants which cannot, therefore, be explained in terms of transcriptional control. Thus, even if auxin does directly regulate gene expression it must also have some other primary action.

Control of mitochondrial and chloroplast function

During the action of auxins in the promotion of stem section elongation there is an increased respiration rate, reflecting increased mitochondrial activity, as would be expected of the energy-dependent growth process. This effect cannot be demonstrated in vitro and auxins have not been found to stimulate the respiration rate of isolated mitochondria (Key et al., 1960; Baxter and Hanson, 1968). However, there have been reports of the inhibition of both respiration rate and oxidative phosphorylation by the phenoxyacetic acids (Stenlid and Saddik, 1962; Lotlikar et al., 1968). These have led to suggestions that this might be responsible for the toxicity of the phenoxyacetic acid herbicides. However, inhibitory effects are generally found only at relatively high herbicide concentrations (10^{-4} to

10^{-3} M) which may not be reached within the plant. Furthermore, inhibition is not specific. The non-herbicidal compound 2,4,6-T is as effective as 2,4,5-T, and 2,4-DB and MCPB, which are unlikely to have auxin activity *per se* (see later), are more effective than 2,4-D and MCPA (Lotlikar *et al.*, 1968). It thus seems very unlikely that the phenoxyacetic acids owe their toxicity to any direct action upon mitochondria unless they are applied at abnormally high levels.

The effect of auxins on chloroplast function has also been investigated, again with variable results. Application to intact tissue or to isolated cells can cause increased photosynthetic activity at low concentrations or inhibition at higher levels (Kulandaivelu and Gnanam, 1975; Ashton *et al.*, 1977; Paul *et al.*, 1979). Inhibitory effects may be partly due to structural damage to chloroplasts that has been found in some species (Hallam, 1970; Bretherton and Hallam, 1979). They may also be a consequence of direct inhibition of photosynthetic reactions. For instance, Moreland and Hill (1962) found 2,4,5-T to inhibit the Hill reaction in isolated chloroplasts, albeit at rather high concentrations. But in another study Robinson *et al.* (1978) could find no effect of IAA on photosynthetic electron transport or phosphorylation in isolated chloroplasts. It is thus difficult to generalize about auxin effects on chloroplasts. Two variables may explain the differences of opinion put forward in the above reports, (1) the concentration of auxin used and (2) the plant species involved. At high enough concentrations all chloroplast function may be inhibited by auxins, but there may be differences of sensitivity between species. Stimulation of photosynthetic activity has not been reported in isolated chloroplasts and may be an indirect consequence of auxin action in intact cells.

Membrane effects

There has been considerable interest in the effects of auxins on cell membranes, in particular on the ion transport properties of the plasma membrane. Auxins cause rapid stimulation of proton extrusion from a cell (Rayle, 1973). Such an effect is believed to be responsible for the rapid increase in growth rate that is caused by auxins. It is a very specific effect and is regulated at physiologically significant concentrations of auxins. Therefore, it becomes a good candidate for a primary action of auxin. It is generally considered that an ATP-driven proton pump is responsible for the proton efflux and that this is directly stimulated by auxin (Marre, 1977). However, as yet there has been no report of such a reaction *in vitro*. A word of caution must be introduced at this point since it is still not proven that an ATPase is involved in proton extrusion (Cross *et al.*, 1978) and so a search for such a system may be futile. Indeed, Cross *et al.* (1978) found no ATPase activity to be associated with auxin receptor proteins of maize membranes. Thus the mechanism of auxin involvement in the proton pump is still uncertain.

Stimulation of a plasma membrane-associated proton pump does have several consequences. These include an increase in dark CO_2 fixation and an alteration in the flux of several organic and inorganic compounds across the plasma membrane. As proton extrusion occurs the cytoplasmic pH rises. This is believed to cause increased CO_2 fixation by phosphoenolpyruvate carboxylase, leading to malate synthesis and thus the restoration of the normal cytoplasmic pH (Stout et al., 1978; Smith et al., 1979). Alterations in the flux of inorganic ions may occur, particularly of potassium and chloride which restore electrical neutrality to the cell (Bentrup et al., 1973; Pfruner and Bentrup, 1978; Stout et al., 1978; Pike and Richardson, 1979). The increased proton gradient across the plasma membrane may also have secondary effects upon proton-mediated cotransport systems and thus account for alterations in the flux of several inorganic ions or organic molecules. Other effects that are probably associated with the altered proton flux are rapid changes in membrane potential during auxin action (Cleland et al., 1977; Goring et al., 1979).

In addition to the above events that appear to be directly attributable to altered activity of the proton pump, auxins have been found to promote direct changes in membrane structure (Morre and Bracker, 1976) and in physical properties (Helgerson et al., 1976) although the significance of these is not known.

Cell wall changes

The cell extension caused by auxins is now generally accepted as being due to an increase in cell wall plasticity. This is probably brought about by the pH change in the cell wall that is a consequence of the increased activity of the proton pump discussed above. Exactly how a decrease in pH leads to increased plasticity is not clear. However, much interest has been focused upon cell wall components and their biochemistry during auxin action. Changes in polysaccharide composition have been found (Sakurai and Masuda, 1978a, b), as have alterations in the activity of various enzymes involved in cell wall metabolism (Masuda and Yamamoto, 1970; Johnson et al., 1974; Goldberg, 1977; Likholat and Druzhinina, 1977). None the less, the relationship between these changes and cell wall rigidity is not known.

Hormonal interactions

It has long been recognized that auxins stimulate ethylene production (Morgan and Hall, 1962). It has been suggested that many of the effects observed following auxin treatment of tissues are a consequence of this (Maxie and Crane, 1967). While this may be true for certain long term effects, it appears not to be so in the short term since auxin-induced ethylene production appears dependent upon protein synthesis (Yu et al., 1979). Furthermore, several of the symptoms of herbicide damage are quite unlike the damage caused by ethylene. The conclusion is that while ethylene produced after phenoxyacetic acid treatment may contribute to toxicity it is not sufficient to be the sole cause of plant death (Loos, 1975).

The phenoxyacetic acids have also been reported to stimulate reactions that could lead to an alteration in levels of endogenous IAA. This might occur in two ways:

(1) by causing increases in phenolic compounds that act as inhibitors of IAA oxidase (Volynets and Pal'chenko, 1977);
(2) via induction of enzymic conjugation of auxins to amino acids (Sudi, 1964; Venis, 1972).

The former would increase endogenous auxin levels, the latter would have the opposite effect. However, any change brought about by these means would almost certainly be small by comparison with the amount of auxin needed to be added to bring it about. It is highly unlikely that either response is of physiological significance to herbicide action.

Specific enzymic effects

There have been literally hundreds of reports of changed levels of activity of a wide range of enzymes following auxin treatment. How may they be brought about?

Measurements of enzyme activity that are carried out in properly conducted enzyme assays *in vitro* indicate the concentration of the active form of that particular protein in the tissue under investigation. Thus a change in enzyme activity after auxin treatment should reflect an altered enzyme population. This could be a consequence of transcriptional or translation regulation as has already been considered. However, another possibility is that auxin causes the modification of an existing protein to a form with altered activity. For example, the dramatic increase in invertase levels in chicory root tissue following 2,4-D treatment appears to be due to the dissociation of a high molecular weight protein with low invertase activity into low molecular weight subunits which have very high activity (Gordon and Flood, 1979). The mechanism involved is not known, but auxin does not directly cause the interconversion *in vitro*.

It is worth pointing out that if an enzyme assay is conducted using crude preparations containing many other plant constituents there is a danger that activity may be influenced by some component of the system and may therefore not be a true guide to enzyme population. In this case it is possible to imagine that auxin treatment could lead to an alteration in an activator or inhibitor of the enzyme under investigation. The resulting assay data would lead to an erroneous conclusion about the effects of auxin.

If auxin directly regulates enzyme activity then the enzyme levels determined by assays made before and after auxin treatment will not, of course, show any change. The only way of finding such an action is by directly investigating the effect of auxin *in vitro* on enzyme activity. Clearly it is not practical to study all enzymes for auxin effects and there is no way of knowing how many have been studied with negative results. However, of the many that must have been checked, a very few have been reported to be affected by auxins. Since these obviously represent primary biochemical actions of auxin they are extremely interesting. Unfortunately few of them look very convincing. They include the following examples.

(1) Citrate synthetase was found to be activated slightly by IAA at 10^{-12} M (Sarkissian, 1970). However, others have not been able to repeat this observation (Zenk and Nissl, 1968; Brock and Fletcher, 1969).
(2) A glucan synthetase in the membranes from onion stems was reported to show a slight stimulation by 2,4-D (Van der Woude *et al.*, 1972). However, Ray (1973) could not repeat this finding.
(3) A membrane-bound ATPase from mung-bean hypocotyls was reported to give 150% stimulation by IAA (Kasamo and Yamaki, 1974). However, the same effect was given over the range from 10^{-5} to 10^{-13} M which suggests that the system is of dubious physiological significance.
(4) A mitochondrial NADH-dehydrogenase has been reported to be completely inhibited by 2,4-D and 2,4,5-T at 2 mM (Manella and Bonner, 1978). However, this concentration is very high and IAA had no effect on the enzyme suggesting that if this enzyme response is significant, it is only likely to be so in the toxic action of the phenoxyacetic acid herbicides and not as part of the normal mode of action of auxins.
(5) A calcium-stimulated K^+-ATPase from rice roots has been found to be promoted by IAA and 2,4-D in the range 10^{-8} to 10^{-10} M (Erdei *et al.*, 1979). However, the level of stimulation was relatively low and the specificity of the system has not been studied.
(6) The enzyme tocopherol oxidase has recently been shown to be regulated by the plant hormones, including the auxins, in this laboratory (Gaunt *et al.*, 1980). This enzyme has striking regulatory properties and is controlled by light and photoperiod as well as by the hormones (Gaunt and Plumpton, 1978, 1980). Regulation by auxins is shown in two ways:
(a) by inhibition of enzyme activity in the light. Under defined conditions this can be complete at 10^{-7} M and it has been shown to be very specific for auxins. Furthermore, anti-auxins reverse auxin-induced inhibition of the enzyme, again with high specificity.
(b) by reversal of dark inhibition of the enzyme. This response is found at 10^{-8} M under defined conditions. It is worth repeating that neither 2,4-DB, indolebutyric acid or MCPA aspartate are effective against this system

(Gaunt and Plumpton, unpublished data) which leads us to believe that they are only effective as auxins after conversion to 2,4-D, IAA, and MCPA respectively. All of the control functions of this enzyme system depend upon the presence of cell membranes. Furthermore, the substrate of the reaction is also a membrane component. It seems likely that tocopherol oxidase is involved with some important membrane activity although the nature of this is not known. Thus, while it seems likely that the system will prove to be physiologically important *in vivo,* its role is a matter for speculation.

Of the above examples it can be seen that only the tocopherol oxidase system appears to obey the criteria required of a primary action of auxin. However, this has yet to be repeated in other laboratories and a cellular role must be demonstrated, before its acceptance can be general. None the less, the principle that auxin has a primary role in modulating enzyme action would seem to be established.

How many primary reactions exist for auxins?

From the data presented so far in the section it seems likely that auxins have more than one primary reaction. The evidence for this is as follows.

(1) There are several auxin receptors with different affinities for auxins and with different distributions in cells. Their different properties could reflect different functions.
(2) There is some evidence for the direct regulation of transcription by auxins, perhaps mediated by a binding protein. This possibly offers an attractive explanation for many of the enzyme changes elicited by auxins. However, a clear demonstration of a primary action here is still awaited.
(3) There is strong evidence for a direct role in enzyme regulation, although whether this involves an auxin–enzyme interaction or is mediated by a receptor protein with a regulatory role is not known.

Our present view, as yet unsupported by direct experimental evidence, is that the auxin receptor proteins have a purely regulatory function and mediate between the hormone and some catalytic or functional molecule in the cell. Membrane-bound receptors float in the lipid matrix of the membrane and their interactions with other proteins of the membrane determine the activity of the latter. There is no immediately apparent reason why a single receptor should not interact with several different catalytic or transport proteins in a membrane. Certain receptors could also be imagined to leave the plasma membrane and migrate to the nucleus, there causing changes in the pattern of transcription by direct interaction with either DNA itself or with the chromatin proteins involved in gene repression/depression. Such a model is very broad and would easily account for all of the membrane and enzymic consequences of auxin action described above.

Do Plant Cells Respond to Auxin in Different Ways?

Any textbook of plant physiology lists a frightening array of the different effects that auxins have upon plants. It is apparent that the majority, if not all, of the different cell types in a plant are able to respond to auxins, ranging from root cells to stem and leaf cells, from meristematic cells to differentiated cells, from guard cells to pith cells, from reproductive cells to vegatative cells, from transport cells to storage cells. Most of the effects of auxins have only been described in physiological terms and it is easily seen that different cell types do not necessarily respond in the same way nor to the same concentrations of auxin. While some responses may be similar, the overall response is usually different. Unfortunately, biochemical investigations of auxin action have been conducted very spasmodically and serious attention has been paid to only a few tissues, such as the coleoptile. Thus it is not possible to begin to classify auxin responses in detail in order to see if there are reactions or receptors common to all or several cell types. Is there such a thing as a universal auxin response by a plant cell? The question of the biochemistry of the response of different cells will not be answered quickly. This uncertainty creates serious difficulties when it comes to the integration of all of the observed auxin effects into a model. The data recorded in the literature represent responses of all possible tissues in a wide range of species. It is a daunting task to try to correlate such biochemical data with physiological observations in order to explain mode of action.

Toxic Action

The consequences of leaf application of a phenoxyacetic acid herbicide to a plant are very varied and depend on the amount applied. Tolerant species may show no visible effects, although this is rare. There may be a slight stimulation of growth that is beneficial to crop performance (Wiedman and Appleby, 1972). In some plants early symptoms of damage, such as epinastic curvature of stem and petiole, are temporary and are followed by more or less complete recovery. In others a sequence of events occurs that culminates in plant death. Anatomical and morphological studies of plants have shown the following symptoms to be associated with toxicity:

(1) leaf chlorosis, often accompanied by severe chloroplast damage (Tukey et al., 1945; Hallam, 1970; Nadakavukaren and McCracken, 1977; Bretherton and Hallam, 1979);
(2) altered stomatal function (Pemadasa and Jeyaseelan, 1976; Rao et al., 1977; Pemadasa, 1979);
(3) stem tissue proliferation affecting various tissues in different species (Beal, 1945; Eames, 1950; Sun, 1955; Key et al., 1966; Cardenas et al., 1968);
(4) root initiation in stem tissue (Beal, 1945);
(5) disintegration of root tissues (Coble and Slife, 1971);

(6) abnormal apical growth (Key *et al.*, 1966; Cardenas *et al.*, 1968; Sharman, 1978).

How many of these effects can be considered to be the cause of death of a plant? The root decay reported by Coble and Slife (1971) appeared to kill honey vine milkweed, but this does not seem to be a general toxic effect in plants. The early view of Eames (1950) that the massive cell division in the stem leads to phloem collapse and plant death by strangulation has received widespread support. However, there is doubt that the tissue proliferation completely blocks the phloem in all cases (Sun, 1955; Whitworth and Muzik, 1967). Leaf death following chloroplast disruption could also contribute to plant death in those species showing such damage (Hallam, 1970). On the other hand, abnormal behaviour of the apical meristem may result in stunted and deformed plants, but may not be lethal.

At a cellular level, the symptoms described above reflect two basic modes of action:

(1) direct toxicity, likely to result in cell death. For example, chloroplast disintegration will doubtless lead to cell damage and death. The relationship of this to tissue and plant death is clear.
(2) indirect toxicity, in which the cell that is responding to the herbicide is not directly damaged although its behaviour becomes abnormal. This describes the majority of herbicidal effects, such as the initiation of cell division or of abnormal cell differentiation. In these cases the death of the plant is not easily explained. It follows from this that it is impossible even to speculate upon the biochemistry of toxicity.

It is apparent that our knowledge of the mode of action of the phenoxyacetic acids has not advanced greatly since their introduction. They were discovered as compounds able to emulate IAA. The early view of their herbicidal action was that they caused an increase in the auxin content of the plant and that the resulting hormonal imbalance led to abnormal cellular behaviour. This general explanation remains the most accurate description of their action in the majority of plants.

One further point that deserves attention here is the size of the auxin overdose that is needed to cause abnormal cellular behaviour. This is not yet known but may be much lower than is generally believed. First we must consider the normal level of free auxin in a cell. Reliable data are very difficult to find. Most information has been obtained from bioassays and is thus open to criticism. Careful chemical analyses of IAA levels in plants have been undertaken, but such figures do not distinguish the physiologically important concentration in different subcellular compartments and rarely distinguish between different tissues. However, such data do give an idea of the total auxin content. For example, McDougall and Hillman (1978) found up to 3.8 µg IAA/kg fresh weight in shoots of *Phaseolus vulgaris*. If we assume that the shoot of a young plant weighs 10 g, then the total shoot content is around 40 ng of IAA per plant. How does this compare with a herbicidal dose? The amount of a phenoxyacetic acid herbicide reaching a plant during a spray application at normal dose rates is almost certainly in excess of 100 µg. While accepting the approximate nature of the figures it is still obvious that the plant is receiving a massive overdose of auxin—at least one thousand times more than is already present. One consequence of such a calculation is the implication that if only a very small proportion of the herbicide entering a plant reaches sensitive cells the hormonal balance of those cells will be seriously altered, although it remains unknown by how much it is necessary to disturb the endogenous auxin concentration of a cell in order to upset cellular behaviour.

Summary

The phenoxyacetic acid herbicides are all auxins. Their application leads to a serious hormonal imbalance within a plant. This results in abnormal cellular behaviour which eventually causes death of sensitive species, although the precise cause of death is poorly understood. Despite much information on the biochemical responses of cells to auxins their mode of action is not known.

SELECTIVITY

Introduction

One of the most fascinating aspects of the behaviour of the phenoxyacetic acid herbicides is their selective action. A few micrograms applied to one species will kill it within two weeks or so while a milligram applied to another may have no obvious effect. Although selectivity is sometimes viewed in 'all or nothing' terms—a plant either dies or it is not affected—this is an oversimplified and inaccurate view. It is likely that all plants will be affected in some way by the phenoxyacetic acids even though this may not be obvious (Brian, 1958). Agronomically advantageous growth effects have even been described for resistant cereals (Wiedman and Appleby, 1972). If the dose rate is high enough, even so-called resistant plants show toxic symptoms. For example, Boyle (1954) showed that 2,4-D application could produce the same response in oats (*Avena sativum*) and bean (*Phaseolus vulgaris*), although it required five times the dose to produce the response in oats. There is also tremendous variation in the sensitivity of a species with age. Thus cucumbers (*Cucumis sativus*) are sensitive to 2,4-D as very young seedlings and must have reached a certain physiological age before they become resistant (Chkanikov et al., 1977) and wild carrot (*Daucus carota*) is also only resistant to 2,4-D after the cotyledon stage (Whitehead and Switzer, 1963). Similar observations have also been reported for other herbicides

(Paterson, 1977; Miller *et al.*, 1978). Other factors can also influence the sensitivity of a plant to herbicides. For example, the degree of plant competition has been found to show a direct correlation with the toxic effects of MCPA (Fogelfors, 1977). A similar situation has been observed with other herbicides (Jeffcoat and Sampson, 1973; Hill and Stobbe, 1978).

The Assessment of Herbicide Sensitivity and Resistance

Sensitivity and resistance are seen as relative terms. All levels of response can be found between the extremes of death and no visible change. The position of a species on the scale changes with dose, age, and environmental conditions. It is really necessary to quantitate the toxic symptoms shown by a plant for the purposes of correlation of toxicity with aspects of herbicide behaviour. This is a very difficult task that has not been resolved. The use of a scale with grades ranging from 'very susceptible' to 'highly resistant' is only useful in a very general sense. We need to know much more about the response of a plant—exactly what symptoms appear, over what time scale, and whether recovery or death occurs in the long term. A morphological description of symptoms alone is inadequate and a knowledge of both anatomical (at cellular and subcellular levels) and biochemical effects are required before an objective comparison can be made between the behaviour of the plant and behaviour of the herbicide. It is also worth pointing out that estimates of herbicide behaviour are usually made under laboratory conditions following spot application to a single leaf, while estimates of plant response are made after field or glasshouse spray treatment. It is dangerous to draw correlations between such disparate systems.

Turning to consider the behaviour of the herbicide, this has many facets, as have been shown in this review. A complete picture of all of these is required for a species before we can begin to piece together an explanation for plant responses. This includes knowledge of tissue distribution, subcellular distribution, transport, metabolism, protein binding, and root excretion. Although there have been a great many studies that have attempted to explain resistance there has not been one that has succeeded in matching all of these points of herbicide and plant behaviour satisfactorily.

The Mechanisms of Selectivity

It seems probable that selectivity of the phenoxyacetic acids is a consequence of several facets of herbicide and plant behaviour and that no single aspect predominates. This must complicate the search for an explanation. However, let us consider what is known of the problem. At the outset it must be said that any toxic action can only be a consequence of the accumulation of a toxic concentration of the toxic molecule at the particular site of action. There are thus

two possible ways of explaining the selective action of a herbicide:
(1) by differences in the sensitivity of the site(s) of action;
(2) by differences in the concentration of herbicide that reaches the site(s) of action.

Each of these is considered below.

What is the sensitivity of the active site in different species?

To answer this question we need to know where the active site is as well as having a method of directly assessing its response to the herbicide. This presents major problems. As has been discussed in the previous section it is likely that all plant cells respond to auxins, that different cell types respond in different ways and that only one or two are responsible for the symptoms of toxicity that accompany auxin action. Since these have not been unambiguously defined it is difficult to pursue the matter further. However, we can perhaps limit the discussion to cells in petiole and stem that seem to be responsible for the abnormal growth and tissue proliferation that typify phenoxyacetic acid toxicity. Do these tissues show differences in sensitivity to the phenoxyacetic acids in different species? This is not known. It is recognized that in auxin bioassays using different tissues the response to these compounds is similar. For example, *Avena* coleoptile and pea third internode assays show similar levels of sensitivity to the same concentration of MCPA, yet the intact plants respond very differently with oats being very resistant and peas of moderate sensitivity (Collins, 1972). This implies that in each species the site of action has the same sensitivity. But it should be pointed out that bioassays are conducted on very young seedlings that differ markedly in their sensitivity to the herbicide from the older plants to which the resistance data refers. Other drawbacks to the interpretation of bioassay data have already been mentioned.

Today there are two direct approaches available that could be used to compare directly the sensitivity of active sites in different species to the phenoxyacetic acid herbicides. These are (1) a study of the affinity of isolated receptor proteins for the compounds, and (2) a study of the sensitivity of tocopherol oxidase to control by different auxins *in vitro*. So far neither of these approaches has been used.

Thus the active site question remains open. It is worth remembering that it has recently been convincingly demonstrated that resistance of some plants to the triazines is due to differences in active site sensitivity (Pfister *et al.*, 1979). These compounds exert their toxic action through inhibition of photosynthesis, caused by binding to a protein that has a role in photosystem II. Several 'biotypes' of groundsel (*Senecio vulgaris*) have been found to be highly resistant to atrazine. For these biotypes it has been shown that the herbicide will no longer bind to the protein in the photosystem II complex. There is no obvious reason why an analogous situation should not also apply for the phenoxyacetic acids.

Does the herbicide concentration at the active site vary between species?

An answer to this question depends upon identification of the herbicide-sensitive tissue. Although this is still uncertain, apical meristems, stems, and roots are obvious candidates and it should be a relatively straightforward matter to analyse the level of herbicide in these areas after application to the plant. This has been done in some investigations, for example by Chkanikov *et al.* (1971), in which the level of radioactivity in shoot, apex, stem, and roots was measured after leaf application of 2,4-D to several species of varying resistance. Their data showed few correlations. In the sensitive sunflower (*Helianthus annus*) 48% of the absorbed radioactivity was accumulated in the apex after 72 h, while in the equally sensitive mustard (*Sinapsis alba*) there was only 11% in the apex after the same time period. Higher levels of radioactivity were found in stems of both these sensitive species by comparison with resistant plants. Little accumulation was found in the roots of any species. The most consistent correlation appeared to be between susceptibility and the rate of translocation of herbicide out of the treated leaves. Even this was not without its exception and high translocation was found in the semiresistant buckwheat (*Fagopyrum sagittatum*).

From the foregoing it is apparent that although analytical data can be obtained, interpretation of the data is difficult. Inability to obtain meaningful correlations in the above work may be partly because no account of herbicide metabolism, binding, and inter, and intracellular distribution was made for the apex or stems; presumably because the latter two factors, in particular, are very difficult to resolve. Expression of results solely as a percentage of the absorbed radioactivity in the sensitive area under study gives no indication of the actual amount present as the toxic molecule. Even if the amount of toxic molecule present can be determined this may be a gross overestimation of that available for physiological action, due to compartmentation and binding. For example, of the total herbicide found in the stems only a small fraction may be available for physiological action simply because the majority is in the vascular system rather than in the sensitive cells of the region.

Another factor, often overlooked, is that the concentration of herbicide needed to elicit toxic responses in sensitive tissue may be very low and that it is quite unnecessary for the herbicide to be accumulated to high levels. All that is required for a plant to suffer damage is that a concentration be maintained at a certain value for a period of time. This concept has been applied to explain the selective toxicity of 2,4-DB (Hawf and Behrens, 1974). They suggest that 2,4-DB may be metabolized to 2,4-D by β-oxidation in all plants and that the 2,4-D produced is subsequently degraded. The relative rates of β-oxidation and 2,4-D degradation will determine the concentration of 2,4-D (the toxic molecule) in the plant. If the balance between these two processes differs with species then the free-2,4-D concentration will also differ and this could account for the selective action of 2,4-DB. The same suggestion is made by Hill and Stobbe (1978) from the results of experiments with the herbicide benzoylprop-ethyl. They found a direct correlation between the amount of conjugated metabolites of benzoylprop and the degree of injury—quite the contrary to what might have been expected. However, formation of the conjugates is dependent upon initial hydrolysis of the applied ester. High levels of conjugate formation simply reflect a rapid rate of hydrolysis to the free acid. The degree of injury depends on the concentration of free acid—which is determined by the relative rates of hydrolysis of benzoylprop-ethyl and the subsequent removal of free herbicide by conjugation or perhaps other mechanisms.

While the phenoxyacetic acids require no activation by metabolism (except in the case of ester formulations) they must be moved to their site of action. Thus the concentration in the sensitive tissue will be a balance between its import from the site of application and its removal by any process that reduces its effective concentration—such as export, metabolism, binding, or compartmentation. Species variation in any one of these factors could lead to selectivity. This is an attractive proposal that has not received adequate experimental consideration. It requires a careful quantitative study of the concentration of herbicide in tissues that are known to show toxic responses in a particular species. The threshold concentration that must be reached in order to give toxic symptoms may be low and represent only a very small proportion of the total herbicide applied to or entering the plant. An important part of this model is that the sensitive tissue must be exposed to the herbicide for sufficient time to permit lethal damage to occur. How long this requires is not known. Symptoms of toxicity develop quickly after application of the phenoxyacetic acids, with typical stem and petiole bending detectable within a few hours. But many plants recover completely after showing these symptoms and develop no lasting damage. For example, although resistant and susceptible strains of wild carrot (*Daucus carota*) showed the same degree of injury after 1 week, the resistant strains subsequently 'grew out' of this initial injury and showed no lasting damage (Whitehead and Switzer, 1963). Either the concentration in the sensitive tissue reaches only a sublethal level, or, what is more likely, it is not sustained for a long enough period. From this discussion it will be apparent that we do not yet have direct data about the concentration of free herbicide at the putative active sites in the plant. On the other hand there have been many investigations of herbicide behaviour that are relevant to this matter. They show that variations in concentration would be expected due to differences between species in almost every aspect of herbicide behaviour that has been studied. Let us move on to consider these studies. The majority have been directed towards elucidating the mechanism(s) of resistance. A variety of approaches have been used:

(1) a comparison of the behaviour of a particular herbicide in a range of species of different sensitivity to the compound. This is perhaps the commonest approach with many examples in the literature (Morgan and Wayne, 1963;

Rakitin *et al.*, 1966; Chkanikov *et al.*, 1971; Dexter *et al.*, 1971; Fleeker and Steen, 1971; Williams, 1976; Wyrill and Burnside, 1976).
(2) a comparison of the behaviour of a herbicide in plants of the same species but which differ markedly in their resistance. The variation in response could be due to age or variety or 'biotype' (e.g. Whitehead and Switzer, 1963; Whitworth and Muzik, 1967; Chkanikov, *et al.*, 1977).
(3) a comparison of the behaviour of two closely related phenoxyacetic acids in a single species which responds differently to each compound (e.g. Leafe, 1962; Slife *et al.*, 1962; Sanad, 1971).

Any of these approaches could throw light upon the question of resistance mechanisms if applied comprehensively. Each could establish the level of free herbicide in sensitive tissues and explain how it was obtained, but rarely is this objective reached. Far too many studies have been confined to only one or two aspects of herbicide behaviour, usually transport and metabolism, with no consideration given to the complete story. Furthermore, most studies deal with the general situation in the plant or perhaps with the 'treated leaf' and do not consider that the important tissue for analysis may behave in a special way. Unfortunately an incomplete survey is of little value. Despite this criticism a lot of information is now available of the different patterns of behaviour of the phenoxyacetic acids in plants and of how this relates (albeit superficially) to plant resistance.

Uptake

It is apparent that if the same dose of a phenoxyacetic acid is applied to the leaves of different species, the amount of herbicide entering each is not the same. The reasons for this probably relate to the composition and structure of the cuticle (Robertson and Kirkwood, 1969; Baker and Bukovac, 1971; Norris, 1974). The cuticle may retain herbicide within it and thus restrict entry to the leaf symplast. There have been suggestions that these differences in uptake can explain selectivity (Wathana *et al.*, 1972) but on the other hand studies have shown no general correlation between resistance and uptake (Chkanikov *et al.*, 1971; Dexter *et al.*, 1971).

Transport

Many studies have demonstrated the differences that exist between species in the mobility of the phenoxyacetic acids. Since these compounds have to be transported from treated leaves in order to be toxic this can obviously contribute to selectivity. Indeed a general correlation has sometimes been reported between the sensitivity of a plant and the mobility of the herbicide (Rakitin *et al.*, 1966; Chkanikov *et al.*, 1971; Dexter *et al.*, 1971; Wathana *et al.*, 1972; Hallem, 1974,

1975). The differential toxicity of several phenoxyacetic acids to the same plant has also been suggested to be due to differences in mobility (Slife *et al.*, 1962; Sanad, 1971). For perennial weeds it has often been suggested that insufficient herbicide is translocated to the roots, thereby allowing regrowth to occur after the initial destruction of the apical part of the plant (Richardson, 1976; McIntyre *et al.*, 1978). However, there are many exceptions (Whitworth and Muzik, 1967; Chkanikov *et al.*, 1971) indicating that while immobilization undoubtedly plays an important role in selectivity other factors must also be taken into account. Immobilization in the treated leaf could occur by metabolism to non-mobile metabolites, as was discussed before, although Chkanikov *et al.* (1971) have suggested that other mechanisms must also be involved.

Root excretion

This is a special case of transport. There is enormous variation between species but root excretion only appears to reach significant proportions in a very few plants and then only over long periods of time— say 2 or 3 weeks. It will clearly reduce the content of herbicide within the plant and must be considered as a potential defence mechanism, as has been suggested by Fites *et al.* (1964), Basler *et al.* (1970), and Coble *et al.* (1970). When the mechanism underlying this process is understood we will be in a better position to judge its importance. It is possible to imagine a role in resistance if it diverts transported herbicide from sensitive tissue to the root from which the molecule is excreted, but how it could do this is by no means clear.

Metabolism

The pattern and extent of metabolism is very variable between species and even within a single plant, as has already been discussed. Since it is likely that all metabolites are non-herbicidal and immobile, metabolism would seem to be an excellent defence mechanism. It should be pointed out that not everyone considers this to apply to all metabolites. In particular Bristol *et al.* (1977) consider that amino acid conjugates cannot be thought of as a detoxification mechanism *per se* and Chkanikov *et al.* (1977) suggest that, while amino acid conjugates are immobile and glycosides of the hydroxylated derivatives are physiologically inactive the reverse is not necessarily true. None the less we adhere to our opinion for the reasons given earlier. Several investigations have shown correlations between resistance and metabolism. However, many of these surveys have been inadequate. Some have only considered a single metabolic reaction—for example 'decarboxylation' (Luckwill and Lloyd-Jones, 1960a, b) or hydroxylation (Fleeker and Steen, 1971). Of major importance in this work is the need for reliable quantitative data—which is very difficult to obtain unless the instability of certain metabolites is recognized and overcome. In the studies of

Rakitin *et al.* (1966), Montgomery *et al.* (1971), and Bristol *et al.* (1977) attempts have been made to correlate resistance and metabolism. However, all three have potentially serious flaws in the methodology used, as has been more fully discussed in a previous section. Thus, Rakitin *et al.* (1966) employ an extraction procedure involving TCA precipitation, which must be considered unreliable due to coprecipitation of free herbicide and/or destruction of some metabolites. The extraction procedure of Montgomery *et al.*(1971), involving homogenization with ethanol and heating for 1 h in the presence of sodium bicarbonate ($NaHCO_3$), would be expected to cause extensive breakdown of any sugar ester conjugates present. Bristol *et al.* (1977) also employ an extraction procedure involving ethanol and dilute $NaHCO_3$. In both of these latter studies it is also noted that no attempt was made to estimate protein binding. Nevertheless, it must be noted that all three studies reported that resistant monocots contain very low levels of free 2,4-D as compared to susceptible dicots.

An unknown quantity in this area is the significance of phenoxyacetic acid conjugates as a reservoir of potential herbicide. These metabolites are rapidly formed in most species and are known to 'turn over' in the plant. If turnover involves hydrolysis to the parent herbicide, then a continuing supply of herbicide will be available. This would give a new significance to the model introduced earlier in this discussion in which toxicity may depend upon the maintenance of a low concentration of herbicide over a long period of time.

Protein binding

There is so much disagreement about the level of protein binding that it is impossible to judge its significance in reducing the level of free herbicide in the plant. The Russian workers have reported major differences in the level of binding between resistant and susceptible species and consider the process to be an important defence mechanism (Rakitin *et al.*, 1966; Zemskaya and Rakitin, 1967; Zemskaya *et al.*, 1971, 1973). These data look very impressive. None the less the failure of others to repeat their findings raises a query over the general significance of binding. Evans (1973) also found a poor correlation between binding and resistance.

Compartmentation

There is almost nothing to be said about the relative subcellular distribution of the phenoxyacetic acids in different species. Thus, even if the amount of herbicide present in susceptible tissue was reliably determined, there would still be uncertainty of its concentration at the active site in the cell. One can speculate that the free herbicide may be compartmented in the vacuole. Alternatively it may be actively excluded from the cytoplasm of cells. These possibilities are supported by the work of Backman (1961) and Saunders *et al.* (1965a, b). Indeed the latter authors reported a correlation between biphasic uptake in stem segments and susceptibility, in that resistant species showed only progressive accumulation of 2,4-D with time in contrast to susceptible species which showed subsequent efflux after an initial period of uptake. However, Neidermyer and Nalewaja (1969) were unable to repeat this correlation for leaf slices and found biphasic uptake in all species studied.

In conclusion it can be said that there is still no satisfactory explanation to account for the differential response of plants to the phenoxyacetic acids. It seems likely that in resistant species the effective herbicidal concentration in sensitive tissues of the plant is kept below the toxic level by a combination of factors. Although superficial correlations have been found between certain specific reactions and plant behaviour it is unlikely that any single process is entirely responsible for selectivity.

Summary

The behaviour of the phenoxyacetic herbicides in different species is very variable. It includes different patterns of movement, metabolism, binding, and root excretion. There are technical problems which must be overcome before a complete quantitative picture can be established. These apply particularly to metabolism, protein binding, and subcellular compartmentation. Until these are resolved it is impossible to relate herbicide behaviour and plant response, which must be the objective in a search for the mechanism of resistance. Data are required over a long period of time and should include information of metabolite turnover. Finally a much more comprehensive assessment of resistance and susceptibility is also required.

In the face of inadequate data it is only possible to conclude very generally that it is doubtful that any single aspect of herbicide behaviour will explain selectivity.

CONCLUDING REMARKS

Since the introduction and widespread use of the phenoxyacetic acids as selective herbicides a vast amount of research effort has been devoted to them. Yet there is still no clear biochemical explanation for their mode of action or for their selectivity. In this review we have summarized existing knowledge and have discussed critically the experimental approaches that are being used to further this knowledge. It is apparent that not only is there a lack of information in many areas, there is also a great deal of disagreement about some of the available data. This is partly due to the many technical problems that remain to be resolved.

A brief summary of each of the major aspects of research on these herbicides has already been presented in the text and there is no purpose in repeating these conclusions. Suffice to say that we feel that we are still only scratching the surface of the problem of the complex interaction between plant and herbicide. Our own experimental approach is to try to build up a comprehensive picture of the behaviour of a single compound in a single species. Only when we have a complete and accurate account of each aspect of its behaviour can we hope to understand the way the plant responds to herbicide application.

Many of the possible fates that can befall a herbicide molecule in a plant appear to be detoxication reactions, for example, metabolism, macromolecular binding, and root excretion. Plants seem well equipped to defend themselves against xenobiotics that may be toxic. What is the evolutionary origin of these defensive reactions? The history of herbicides is far too short for plants to have developed defence mechanisms that can specifically counteract this chemical warfare devised by man. It seems likely that the various mechanisms have actually evolved to protect plants from chemical warfare waged by the plant pathogens. Many pathogenic bacteria and fungi produce exotoxins that have a role in infection. Plants are generally resistant to the multitude of pathogenic organisms in existence but relatively little is known of the biochemistry of such resistance. Studies on defence against herbicides will perhaps provide insight into an area which has far wider significance to the majority of plants than is at first apparent.

REFERENCES

Albersheim, P. (1976). 'The primary cell wall', in *Plant Biochemistry* (Eds. J. Bonner and J. E. Varner), 3rd edition, p. 264, Academic Press, New York.
Andreae, W. A., and Good, N. E. (1957). 'Studies on 3-indoleacetic acid metabolism. IV. Conjugation with aspartic acid and ammonia as processes in the metabolism of carboxylic acids', *Pl. Physiol.* **32**, 566–572.
Appleton, T. C. (1964). 'Autoradiography of soluble labelled compounds', *J. Roy. Micr. Soc.* **83**, 277–281.
Arjmand, M., Hamilton, R. H., and Mumma, R. O. (1978). 'Metabolism of 2,4,5-T. Evidence for amino acid conjugates in soybean callus tissue', *J. Agr. Food Chem.* **26**, 1125–1128.
Ashton, F. M., and Crafts, A. S. (1973). *Mode of Action of Herbicides*, pp. 69–109, 276–284, Wiley, New York.
Ashton, F. M., DeVilliers, O. T., Glenn, R. K., and Duke, W. B. (1977). Localization of metabolic sites of action of herbicides', *Pestic. Biochem. Physiol.* **7**, 122–141.
Baker, E. A., and Bukovac, M. J. (1971). 'Characterisation of the components of plant cuticles in relation to penetration of 2,4-D', *Ann. Appl. Biol.* **67**, 243–253.
Bandurski, R. S., and Piskornik, Z. (1973). 'An indole-3-acetic acid ester of a cellulosic glucan', in *Biogenesis of Plant Cell Wall Polysaccharides* (Ed. F. Loewus), p. 297, Academic Press, New York.
Basler, E., Slife, F. W., and Long, J. W. (1970). 'Some effects of humidity on the translocation of 2,4,5-T in bean plants', *Weed Sci.* **18**, 396–398.
Bates, G. W., and Cleland, R. E. (1979). 'Protein synthesis and auxin-induced growth: inhibitor studies', *Planta* **145**, 437–442.
Batt, S., and Venis, M. A. (1976). 'Separation and localisation of two classes of auxin binding sites in corn coleoptile membranes', *Planta* **130**, 15–21.

Baur, J. R., Bovey, R. W., and Riley, I. (1974). 'Effect of pH on foliar uptake of 2,4,5-T-1^{14}C', *Weed Sci.* **22**, 481–486.

Baxter, R., and Hanson, J. B. (1968). 'The effects of 2,4-D upon the metabolism and composition of soybean hypocotyl mitochondria', *Planta* **82**, 246–260.

Beal, J. M. (1945). 'Histological reactions of bean plants to certain of the substituted phenoxy compounds', *Bot. Gaz.* **107**, 200–217.

Bentrup, F. W., Pfruner, H., and Wagner, C. (1973). 'Evidence for differential action of IAA upon ion fluxes in single cells of *Petroselinum sativum*', *Planta* **110**, 369–372.

Bertagnolli, B. L., and Nadakavukaren, M. J. (1974). 'Some physiological responses of *Chlorella pyrenoidosa* to 2,4-dichlorophenoxyacetic acid', *J. Exp Bot.* **25**, 180–188.

Blackman, G. E. (1961). 'A new physiological approach to the selective action of 2,4-D', in *Proceedings of the 4th International Conference on Plant Growth Regulation*, pp. 233–245, Iowa State University Press, Iowa, USA.

Bonnemain, J. L., and Bourbouloux, A. (1973). 'The transport and metabolism of ^{14}C-IAA in intact plants', in *Transactions of the 3rd Symposium on the Accumulation and Translocation of Nutrients and Regulators in Plant Organisms* (Ed. R. Antoszewski), pp. 207–214, Proceedings of the Research Institute of Pomology, Skierniewice, Poland.

Boulware, M. A., and Camper, N. D. (1972). 'Effects of selected herbicides on plant protoplasts', *Physiol. Plant.* **26**, 313–317.

Boulware, M. A., and Camper, N. D. (1973). 'Sorption of some ^{14}C-herbicides by isolated plant cells and protoplasts', *Weed Sci.* **21**, 145–149.

Boyle, F. P. (1954). 'Physiology and chemistry of 2,4-dichlorophenoxyacetic acid action on resistant and non-resistant plants', in *Congr. internat. bot., Paris. Rapps. et communs.* **8**, sect. 11/12, pp. 184–185.

Bretherton, G., and Hallam, N. D. (1979). 'The movement of 2,4,5-trichlorophenoxyacetic acid into the leaves of *Rubus procerus* P. J. Muell and its effect on chloroplast ultrastructure', *Weed Res.* **19**, 307–313.

Brian, R. C. (1958). 'On the action of plant growth regulators. II. Adsorption of MCPA to plant components', *Pl. Physiol.* **33**, 431–439.

Bridges, R. C., and Farrington, J. A. (1974). 'Compartmental computer model of foliar uptake of pesticides', *Pestic. Sci.* **5**, 365–381.

Bristol, D. W., Ghanuni, A. M., and Oleson, A. E. (1977). 'Metabolism of 2,4-D by wheat cell suspension cultures', *J. Agr. Food Chem.* **25**, 1308–1314.

Brock, B. L. W., and Fletcher, R. A. (1969). 'Activation of citrate synthetase by indoleacetic acid', *Nature (London)* **224**, 184–185.

Canny, M. J., and Markus, K. (1960). 'The breakdown of 2,4-dichlorophenoxyacetic acid in shoots and roots', *Aust. J. Biol. Sci.* **13**, 486–500.

Cardenas, J., Slife, F. W., Hanson, J. B., and Butler, H. (1968). 'Physiological changes accompanying the death of cocklebur plants treated with 2,4-D', *Weed Sci.* **16**, 96–100.

Chayen, J., Cunningham, G. J., Gahan, P. B., and Silcox, A. A. (1960). 'Life-like preservation of cytoplasmic detail in plant cells', *Nature (London)* **186**, 1068–1069.

Chen, L. G., Ali, A., Fletcher, R. A., Switzer, C. M., and Stephenson, G. R. (1973). 'Effects of auxin-like herbicides on nucleohistones in cucumber and wheat roots', *Weed Sci.* **21**, 181–184.

Chkanikov, D. I., Makeev, A. M., Pavlova, N. N., and Dubovoi, V. P. (1971). 'Behaviour of 2,4-D in plants of different 2,4-D sensitivity', *Sov. Pl. Physiol.* **18**, 1067–1072.

Chkanikov, D. I., Makeev, A. M., Pavlova, N. N., and Dubovoi, V. P. (1972). '*N*-(2,4-dichlorophenoxyacetyl)-L-glutamic acid: a new metabolite of 2,4-D', *Sov. Pl. Physiol.* **19**, 364–369.

Chkanikov, D. I., Pavlova, N. N., and Gortsuskii, D. F. (1965). 'Production of halophenols from halophenoxyacetic acids in plants', *Khim. v. Sel'sk. Khoz.* **3**, 56–60.

Chkanikov, D. I., Pavlova, N. N., Makeev, A. M., Nazarova, T. A., and Makoveichuk, A. Y. (1977). 'Paths of detoxification and immobilisation of 2,4-D in cucumber plants', *Sov. Pl. Physiol.* **24**, 457–463.

Cleland, R. E., Prins, H. B. A., Harper, J. R., and Higinbotham, N. (1977). 'Rapid hormone-induced hyperpolarisation of the oat coleoptile transmembrane potential', *Pl. Physiol.* **59**, 395–397.

Coble, H. D., and Slife, F. W. (1971). 'Root disfunction in honeyvine milkweed caused by 2,4-D', *Weed Sci.* **19**, 1–3.

Coble, H. D., Slife, F. W., and Butler, H. S. (1970). 'Absorption, metabolism and translocation of 2,4-D by honeyvine milkweed', *Weed Sci.* **18**, 653–656.

Collins, D. J. (1972). 'Studies on the metabolism of the phenoxyacetic acids in plants', Ph.D. Thesis, Univ. Wales.

Collins, D. J., and Gaunt, J. K. (1970). 'The metabolic fate of 4-chloro-2-methyl-phenoxyacetic acid in peas', *Biochem. J.* **118**, 54P.

Collins, D. J., and Gaunt, J. K. (1971). 'The metabolism of 4-chloro-2-methyl-phenoxyacetic acid in plants', *Biochem. J.* **124**, 9P.

Crafts, A. S. (1959). 'Further studies on comparative mobility of labelled herbicides', *Pl. Physiol.* **34**, 613–620.

Crafts, A. S. (1960). 'Evidence for hydrolysis of esters of 2,4-D during absorption by plants', *Weeds* **8**, 19–25.

Crafts, A. S. (1967). 'Absorption and translocation of labelled tracers', *Ann. N.Y. Acad. Sci.* **144**, 357–361.

Crafts, A. S., and Yamaguchi, S. (1958). 'Comparative test on the uptake and distribution of labelled herbicides by *Zebrina pendula* and *Tradescantia flaminensis*', *Hilgardia* **27**, 421–454.

Crafts, A. S., and Yamaguchi, S. (1960). 'Absorption of herbicides by roots', *Amer. J. Bot.* **47**, 248–255.

Cram, W. J. (1968). 'Compartmentation and exchange of chloride in carrot root tissue', *Biochim. Biophys. Acta* **163**, 339–353.

Cram, W. J. (1973). 'Chloride fluxes in cells of the isolated root cortex of *Zea mays*', *Aust. J. Biol. Sci.* **26**, 757–779.

Cross, J. W., and Briggs, W. R. (1979). 'Solubilised auxin-binding protein: sub-cellular localisation and regulation by a soluble factor from homogenates of corn shoots', *Planta* **146**, 263–270.

Cross, J. W., Briggs, W. R., Dohrmann, U. C., and Ray, P. M. (1978). 'Auxin receptors of maize coleoptile membranes do not have ATPase activity', *Pl. Physiol.* **61**, 581–584.

Cuatrecasas, P. (1974). 'Membrane receptors', *Ann. Rev. Biochem.* **43**, 169–214.

Davidonis, G. H., Arjmand, M., Hamilton, R. H., and Mumma, R. O. (1979). 'Biological properties of amino acid conjugates of 2,4,5-trichlorophenoxyacetic acid', *J. Agr. Food Chem.* **27**, 1086–1088.

Davidonis, G. H., Mumma, R. O., and Hamilton, R. H. (1977). 'Metabolism of 2,4-D in soybean: Evidence for a 2,4-D saturation level at physiological concentrations', *Pl. Physiol. Suppl.* **59**, 619.

Davidonis, G. H., Mumma, R. O., and Hamilton, R. H. (1978). 'Metabolism of 2,4-dichlorophenoxyacetic acid in soybean root callus and differentiated soybean root cultures as a function of concentration and tissue age', *Pl. Physiol.* **62**, 80–83.

Davies, E., and Larkins, B. A. (1972). 'Polyribosomes from peas. II. Polyribosome metabolism during normal and hormone-induced growth', *Pl. Physiol.* **52**, 339–345.

Davies, P. J., and Rubery, P. H. (1978). 'Components of auxin transport in stem segments of *Pisum sativum* L.', *Planta* **142**, 211–219.

Davies, R. F., and Higinbotham, N. (1976). 'Electrochemical gradients and K^+ and Cl^- fluxes in excised corn roots', *Pl. Physiol.* **57**, 129–136.

Dela Fuente, R. K., and Leopold, A. C. (1972). 'Two components of auxin transport', *Pl. Physiol.* **50**, 491–495.

Devlin, R. M. (1974). 'Influence of plant growth regulators on the uptake of Naptalam by *Potomogeton*', *Proc. N.E. Weed Sci. Soc., Philadelphia* **28**, 99–105.

Dexter, A. G., Slife, F. W., and Butler, H. S. (1971). 'Detoxification of 2,4-D by several plant species', *Weed Sci.* **19**, 721–726.

Dohrmann, U., Hertel, R., and Kowalik, H. (1978). 'Properties of auxin binding sites in different sub-cellular fractions from maize coleoptiles', *Planta* **140**, 97–106.

Donaldson, T. W., Bayer, D. E., and Leonard, O. A. (1973). 'Absorption of 2,4-dichlorophenoxyacetic acid and 3-(p-chlorophenyl)-1,1-dimethylurea (monuron) by barley roots', *Pl. Physiol.* **52**, 638–645.

Drake, G. A. (1979). 'Flux studies and compartmentation analyses of gibberellin A_1 in oat coleoptiles', *J. Exp. Bot.* **30**, 429–437.

Dubovoi, V. P. Chkanikov, D. I., and Makeev, A. M. (1973). 'Certain characteristics of 2,4-D decarboxylation in plants', *Sov. Pl. Physiol.* **20**, 1073–1078.

Eames, A. J. (1950). 'Destruction of phloem in young bean plants after treatment with 2,4-D', *Amer. J. Bot.* **37**, 840–847.

Eliasson, L. (1965). 'Interference of the transpiration stream with the basipetal translocation of leaf-applied chlorophenoxy herbicides in aspen (*Populus tremula* L.)', *Physiol. Plant.* **18**, 506–515.

Erdei, L., Toth, I., and Zsoldos, F. (1979). 'Hormonal regulation of Ca^{2+} stimulated K^+ influx and Ca^{2+}, K^+-ATPase in rice roots: *in vivo* and *in vitro* effects of auxins and reconstitution of the ATPase', *Physiol. Plant.* **45**, 448–452.

Eschrich, W. (1968). 'Translocation of labelled indolyl-3-acetic acid in sieve tubes of *Vicia faba*', *Planta* **78**, 144–157.

Evans, W. K. (1973). 'Studies on the selective herbicidal activity of MCPA (4-chloro-2-methylphenoxyacetic acid)', Ph.D. Thesis, Univ. Wales.

Fan, D. F., and Maclachlan, G. A. (1966). 'Control of cellulase activity by indoleacetic acid', *Can. J. Bot.* **44**, 1025–1034.

Feingold, D. S., Neufeld, E. F., and Hassid, W. Z. (1964). 'Enzymes of carbohydrate synthesis', *Modern Methods of Plant Analysis* (Eds. K. Paech and M. V. Tracey), pp. 474–519, Springer-Verlag, Berlin.

Fensom, D. S. (1972). 'A theory of translocation in phloem of *Heracleum* by contractile protein microfibrillar material', *Can. J. Bot.* **50**, 479–497.

Feung, C. S., Hamilton, R. H., and Mumma, R. O. (1973). 'Metabolism of 2,4-dichlorophenoxyacetic acid. V. Identification of metabolites in soybean callus tissue cultures', *J. Agr. Food Chem.* **21**, 637–640.

Feung, C. S., Hamilton, R. H., and Mumma, R. O. (1975). 'Metabolism of 2,4-dichlorophenoxyacetic acid. VII. Comparison of metabolites from five species of plant callus tissue cultures', *J. Agr. Food Chem.* **23**, 373–376.

Feung, C. S., Hamilton, R. H., and Mumma, R. O. (1977). 'Metabolism of 2,4-dichlorophenoxyacetic acid. XI. Herbicidal properties of amino acid conjugates', *J. Agr. Food Chem.* **25**, 898–900.

Feung, C. S., Hamilton, R. H., Witham, F. H., and Mumma, R. O. (1972). 'The relative amounts and identification of some 2,4-dichlorophenoxyacetic acid metabolites isolated from soybean cotyledon callus cultures', *Pl. Physiol.* **50**, 80–86.

Feung, C. S., Loerch, S. L., Hamilton, R. H., and Mumma, R. O. (1978). 'Comparative metabolic fate of 2,4-dichlorophenoxyacetic acid in plants and plant tissue culture', *J. Agr. Food Chem.* **26**, 1064–1067.

Fites, R. C., Slife, F. W., and Hanson, J. B. (1964). 'Translocation and metabolism of radioactive 2,4-D in Jimson weed', *Weeds* **12**, 180–183.
Fleeker, J. R. (1973). 'Removal of the acetate-moiety of 2,4-dichlorophenoxyacetic acid in *Ribes sativum*', *Phytochem.* **12**, 757–762.
Fleeker, J., and Steen, R. (1971). 'Hydroxylation of 2,4-D in several weed species', *Weed Sci.* **19**, 507–510.
Fogelfors, H. (1977). 'The competition between barley and five weed species as influenced by MCPA treatment', *Swed. J. Agr. Res.* **7**, 147–151.
Franke, W. (1964). 'Role of guard cells in foliar absorption', *Nature (London)* **202**, 1236–1237.
Franke, W. (1967). 'Mechanisms of foliar penetration of solutions', *Ann. Rev. Pl. Physiol.* **18**, 281–300.
Frear, D. S. (1975). 'Pesticide conjugates—glycosides', in *Bound and Conjugated Pesticide Residues* (Eds. D. D. Kaufman, G. G. Still, G. D. Paulson, and S. K. Bandal), pp. 35–54, Amer. Chem. Soc. Symposium series no. 29.
Frear, D. S., and Swanson, H. R. (1972). 'New metabolites of monuron in excised cotton leaves', *Phytochem.* **11**, 1919–1929.
Gahan, P. B., McLean, J., Kalina, M., and Sharma, W. (1967). 'Freeze-sectioning of plant tissue: the technique and its use in plant histochemistry', *J. Exp. Bot.* **18**, 151–159.
Gaunt, J. K., Matthews, G. M., and Plumpton, E. S. (1980). 'Control *in vitro* of tocopherol oxidase by light and by auxins, kinetin, gibberellic acid, abscisic acid and ethylene', *Biochem. Soc. Trans.* **8**, 186–187.
Gaunt, J. K., and Plumpton, E. S. (1978). 'Control *in vitro* of tocopherol oxidase by light in extracts from leaves of *Xanthium strumarium* L.', *Biochem. Soc. Trans.* **6**, 143–145.
Gaunt, J. K., and Plumpton, E. S. (1980). 'Photoperiodic control *in vivo* and *in vitro* of tocopherol oxidase in *Xanthium strumarium* L.', *Biochem. Soc. Trans.* **8**, 187–188.
Gnanam, A., and Kulandaivelu, G. (1969). 'Photosynthetic studies with leaf cell suspensions from higher plants', *Pl. Physiol.* **44**, 1451–1456.
Goldberg, R. (1977). 'On possible connections between auxin-induced growth and cell-wall glucanase activities', *Pl. Sci. Lett.* **8**, 233–242.
Goldsmith, M. H. M. (1977). 'The polar transport of auxins', *Ann. Rev. Pl. Physiol.* **28**, 439–478.
Goldsmith, M. H. M., Cataldo, D. A., Karn, J., Brenneman, T., and Trip, P. (1974). 'The rapid non-polar transport of auxin in the phloem of intact *Coleus* plants', *Planta* **116**, 301–317.
Goldsmith, M. H. M., and Thimann, K. V. (1962). 'Some characteristics of movement of indoleacetic acid in coleoptiles of *Avena*. I. Uptake destruction, immobilisation and distribution of IAA during basipetal translocation', *Pl. Physiol.* **37**, 492–505.
Gordon, A. J., and Flood, A. E. (1979). 'Effect of 2,4-dichlorophenoxyacetic acid on invertases in chicory root', *Phytochem.* **18**, 405–408.
Goring, H., Polevoy, U. V., Stahlberg, R., and Stumpe, G. (1979). 'Depolarisation of transmembrane potential of corn and wheat coleoptiles under reduced water potential and after IAA application', *Plant Cell Physiol.* **20**, 649–656.
Gorter, C. J., and Veen, H. (1966). 'Auxin transport in explants of *Coleus*', *Pl. Physiol.* **41**, 83–86.
Greenwood, M. S., and Goldsmith, M. H. M. (1970) 'Polar transport and accumulation of indole-3-acetic acid during root regeneration by *Pinus lambertiana* embryos', *Planta* **95**, 297–313.
Grob, K., and Matile, P. H. (1979). 'Vacuolar location of glucosinolates in horseradish root callus', *Pl. Sci. Lett.* **14**, 327–335.

Guilfoyle, T. J., Lin, C. Y., Chen, Y. M., Nagao, R. T., and Key, J. L. (1975). 'Enhancement of soybean RNA polymerase I by auxin', *Proc. Natl. Acad. Sci. US* **72**, 69–72.

Guroff, G., Daly, J. W., Jerina, D. M., Renson, J., Witkop, B., and Udenfriend, S. (1967). 'Hydroxylation-induced migration. The NIH shift', *Science* **157**, 1524–1530.

Hallam, N. D. (1970). 'The effect of 2,4-D and related compounds on the fine structure of the primary leaves of *Phaseolus vulgaris*', *J. Exp. Bot.* **21**, 1031–1038.

Hallam, N. D., and Sargent, J. A. (1970). 'The localisation of 2,4-D in leaf tissue', *Planta* **94**, 291–295.

Hallem, U. (1974). 'Translocation and complex formation of picloram and 2,4-D in rape and sunflower', *Physiol. Plant.* **32**, 78–83.

Hallem, U. (1975). 'Translocation and complex formation of root-applied 2,4-D and picloram in susceptible and tolerant species', *Physiol. Plant.* **34**, 266–272.

Hallem, U., and Eliasson, L. (1972). 'Translocation and complex formation of picloram and 2,4-D in wheat seedlings', *Physiol. Plant.* **27**, 143–149.

Hamilton, R. H., Hurter, J., Hall, J. K., and Ercegovich, C. D. (1971). 'Metabolism of phenoxyacetic acids: metabolism of 2,4-dichlorophenoxyacetic acid and 2,4,5-trichlorophenoxyacetic acid by bean plants', *J. Agr. Food Chem.* **19**, 480–483.

Haque, A., Schupan, I., and Ebing, W. (1978). 'On the metabolism of phenylurea herbicides. X. Movement and behaviour of a glucoside conjugate in plant and soil', *Chemosphere* **7**, 675–680.

Haque, A., Weisgerber, I., and Klein, W. (1976). 'Buturon-^{14}C-bound residue complex in wheat plants', *Chemosphere* **3**, 167–172.

Harborne, J. B. (1973). *Phytochemical Methods*, pp. 255–266, Chapman and Hall, London.

Hardin, J. W., Cherry, J. H., Morre, D. J., and Lembi, C. A. (1972). 'Enhancement of RNA polymerase activity by a factor released by auxin from plasmamembrane', *Proc. Natl. Acad. Sci. USA* **69**, 3146–3150.

Harvey, J., Han, J. C. Y., and Reiser, R. W. (1978). 'Metabolism of oxamyl in plants', *J. Agr. Food Chem.* **26**, 529–536.

Hawf, L. R., and Behrens, R. (1974). 'Selectivity factors in the response of plants to 2,4-DB', *Weed Sci.* **22**, 245–249.

Helgerson, S. L., Cramer, W. A., and Morre, D. J. (1976). 'Evidence for an increase in microviscosity of plasmamembranes from soybean hypocotyls induced by the plant hormone indole-3-acetic acid', *Pl. Physiol.* **58**, 548–551.

Hertel, R., and Flory, R. (1968). 'Auxin movement in corn coleoptiles', *Planta* **82**, 123–144.

Hertel, R., and Leopold, A. C. (1963). 'Versuche zur Analyse de Auxin Transports en der Koleoptile von *Zea mays*', *Planta* **59**, 535–562.

Hill, B. D., and Stobbe, E. H. (1978). 'Effects of light and nutrient levels of ^{14}C-benzoylprop-ethyl metabolism and growth inhibition in wild oat (*Avena fatua* L.)', *Weed Res.* **18**, 223–229.

Hiraga, K., Yamane, H., and Takahoshi, N. (1974). 'Biological activity of some synthetic gibberellin glucosyl esters', *Phytochem.* **13**, 2371–2376.

Hoad, G. V., Hillman, S. K., and Wareing, P. F. (1971). 'Studies on the movement of indole auxins in willow (*Salix viminalis* L.)', *Planta* **99**, 73–88.

Hodgson, R. H., Dusbabek, K. R., and Hoffer, B. L. (1974). 'Diphenamid metabolism in tomato: time course of an ozone fumigation effect', *Weed Sci.* **22**, 205–210.

Hope, A. B. (1963). 'Ionic relations of cells of *Chara australis*. VI. Fluxes of potassium', *Aust. J. Biol. Sci.* **16**, 429–441.

Hutber, G. N., Lord, E. I., and Loughman, B. C. (1978). 'The metabolic fate of phenoxyacetic acids in higher plants', *J. Exp. Bot.* **29**, 619–629.
Ingle, J., and Key, J. L. (1965). 'A comparative evaluation of the synthesis of DNA-like RNA in excised and intact plant tissues', *Pl. Physiol.* **40**, 1212–1219.
Jacobs, M., and Hertel, R. (1978). 'Auxin binding to sub-cellular fractions from *Cucurbita* hypocotyls: *in vitro* evidence for an auxin transport carrier', *Planta* **142**, 1–10.
Jacobs, W. P. (1967). 'Comparison of the movement and vascular differentiation effects of the endogenous auxin and of phenoxyacetic weedkillers on stems and petioles of *Coleus* and *Phaseolus*', *Ann. N.Y. Acad. Sci.* **144**, 102–117.
Jeffcoat, B., and Sampson, A. J. (1973). 'Mode of action of benzoylprop ethyl with reference to its field performance', in *Int. Symp. Crop Protection, Gent,* pp. 941–951.
Jenner, C. F., Saunders, P. F., and Blackman, G. E. (1968a). 'The uptake of growth substances. X. The accumulation of phenoxyacetic acid and 2,4-dichlorophenoxyacetic acid by segments of *Avena* mesocotyl', *J. Exp. Bot.* **19**, 333–352.
Jenner, C. F., Saunders, P. F., and Blackman, G. E. (1968b). 'The uptake of growth substances. XI. Variations in the accumulation of substituted phenoxyacetic acids of different physiological activity by segments of *Avena* mesocotyl', *J. Exp. Bot.* **19**, 353–369.
Johnson, K. D., Daniels, D., Dowler, M. J., and Rayle, D. L. (1974). 'Activation of *Avena* coleoptile cell wall glycosidases by hydrogen ions and auxin', *Pl. Physiol.* **53**, 224–228.
Johnson, M. P., and Bonner, J. (1956). 'The uptake of auxin by plant tissue', *Physiol. Plant.* **9**, 102–118.
Kasamo, K., and Yamaki, T. (1974). 'Effect of auxin on Mg^{2+}-activated and -inhibited ATPases from mung bean hypocotyls', *Plant Cell Physiol.* **15**, 965–970.
Kennedy, C. D., and Harvey, J. M. (1972). 'Plant growth substance action on lecithin and lecithin/cholesterol vesicles', *Pestic. Sci.* **2**, 715.
Key, J. L., Hanson, J. B., and Bils, R. F. (1960). 'Effect of 2,4-dichlorophenoxyacetic acid application on activity and composition of mitochondria from soybeans', *Pl. Physiol.* **35**, 177–183.
Key, J. L., Lin, C. Y., Gifford, E. M., and Dengler, R. (1966). 'Relation of 2,4-D induced growth aberrations to changes in nucleic acid metabolism in soybean seedlings', *Bot. Gaz.* **127**, 87–94.
Kirkwood, R. C. (1972). 'Leaf surface factors', *Proc. 11th British Weed Control Conf.* **3**, 1117–1128.
Kirkwood, R. C., Robertson, M. M., and Smith, J. E. (1968). 'Differential uptake and movement as factors influencing the activity of selected phenoxyacetic acid and phenoxybutyric acid herbicides', *Monogr. Soc. Chem. Ind.* **29**, 287–302.
Klambt, H. D. (1961). 'Wachstumsinduktion und Wuchsstoffmetabolismus in Weizenkoleoptilzylinder. III. Mitteilung-stoffwechselprodukte der Napthyl-1-essigsaure und 2,4-Dichlorophenoxyessigsaure und der Vergleich mit Jenen der Indol-3-essigsaure und Benzoesaure', *Planta* **57**, 339–353.
Kleinhofs, A., Haskins, F. A., and Gorz, H. J. (1967). 'Trans-o-hydroxycinnamic acid glucosylation in cell-free extracts of *Melilotus alba*', *Phytochem.* **6**, 1313–1318.
Kluge, M., and Heininger, B. (1973). 'Untersuchungen Uber den Efflux von Malat aus den Vacuolen der Assimilierenden Zellen von *Bryophyllum* und mogliche Einflusse dieses Vorganges auf den CAM', *Planta* **113**, 333–343.
Kopcewicz, J., Ehmann, A., and Bandurski, R. S. (1974). 'Enzymic esterification of indole-3-acetic acid to myo-inositol and glucose', *Pl. Physiol.* **54**, 846–851.
Kulandaivelu, G., and Gnanam, A. (1975). 'Effect of growth regulators and herbicides on photosynthetic partial reactions in isolated leaf cells', *Physiol. Plant.* **33**, 234–240.

Kurkdjian, A., Leguay, J. J., and Guern, J. (1979). 'Influence of fusicoccin on the control of cell division by auxins', *Pl. Physiol.* **64**, 1053–1057.
Lamb, C. J. (1978). 'Hormone binding in plants', *Nature (London)* **274**, 312–314.
Leafe, E. L. (1962). 'Metabolism and selectivity of plant growth regulator herbicides', *Nature (London)* **193**, 485–486.
Leguay, J. J., and Guern, J. (1977). 'Quantitative effects of 2,4-dichlorophenoxyacetic acid on growth of suspension-cultured *Acer pseudoplatanus* cells II. Influence of 2,4-D metabolism and intracellular pH on the control of cell division by intracellular 2,4-D concentration', *Pl. Physiol.* **60**, 265–270.
Lembi, C. A., Morre, D. J., Thompson, K. S., and Hertel, R. (1971). '*N*-1-naphthylphthalamic-acid-binding activity of a plasmamembrane-rich fraction from maize coleoptiles', *Planta* **99**, 37–45.
Leonard, O. A., Bayer, D. E., and Glenn, R. K. (1966). 'Translocation of herbicides and assimilates in red maple and white ash', *Bot. Gaz.* **127**, 193–201.
Leonard, O. A., Weaver, R. J., and Glenn, R. K. (1967). 'Effects of 2,4-D and picloram on translocation of ^{14}C-assimilates in *Vitis vinifera* L', *Weed Res.* **7**, 208–219.
Lepp, N. W., and Peel, A. J. (1971). 'Patterns of translocation and metabolism of ^{14}C-labelled IAA in phloem of willow', *Planta* **96**, 62–73.
Liao, S. H., and Hamilton, R. H. (1966). 'Intracellular localisation of growth hormones in plants', *Science* **151**, 822–824.
Likholat, T. V., and Druzhinina, T. N. (1977). 'Effects of auxin and gibberellin on activity of UDPG-4-epimerase in wheat coleoptiles', *Sov. Pl. Physiol.* **24**, 961–965.
Likholat, T. V., and Pospelov, V. A. (1974). 'The influence of β-indoleacetic acid and gibberellin on the template activity of the chromatin of wheat coleoptiles of different ages', *FEBS Lett.* **40**, 77–79.
Linscott, D. L., Hagin, R. D., and Dawson, J. E. (1968). 'Conversion of 4-(2,4-dichlorophenoxy)butyric acid to homologues by alfalfa—mechanism of resistance to this herbicide', *J. Agr. Food Chem.* **16**, 844–848.
Loffelhardt, W., Kopp, B., and Kubelka, W. (1979). 'Intracellular distribution of cardiac glycosides in leaves of *Convallaria majolis*', *Phytochem.* **18**, 1289–1291.
Long, J., and Basler, E. (1973). 'Some factors regulating auxin translocation in intact bean seedlings', *Pl. Physiol.* **51**, 128–135.
Loos, M. A. (1975). 'The phenoxyalkanoic acids', *Herbicides—Chemistry, Degradation and Mode of Action* (Eds. P. L. Kearney and D. P. Kaufman), Vol. I, 2nd edition, pp. 1–128, Marcel Dekker, New York.
Lotlikar, P. D., Remmert, L. F., and Freed, V. H. (1968). 'Effect of 2,4-D and other herbicides on oxidative phosphorylation in mitochondria from cabbage', *Weed Sci.* **16**, 161–165.
Luckwill, L. C., and Lloyd-Jones, C. P. (1960a). 'Metabolism of plant-growth regulators. I. 2,4-dichlorophenoxyacetic acid in leaves of red and black currant', *Ann. Appl. Biol.* **48**, 613–625.
Luckwill, L. C., and Lloyd-Jones, C. P. (1960b). 'Metabolism of plant-growth regulators. II. Decarboxylation of 2,4-dichlorophenoxyacetic acid in leaves of apple and strawberry', *Ann. Appl. Biol.* **48**, 626–636.
MacRobbie, E. A. C. (1971). 'Fluxes and compartmentation in plant cells', *Ann. Rev. Pl. Physiol.* **22**, 75–96.
MacRobbie, E. A. C., and Dainty, J. (1958a). 'Ion transport in *Nitellopsis obtusa*', *J. Gen. Physiol.* **42**, 335–353.
MacRobbie, E. A. C., and Dainty, J. (1958b). 'Sodium and potassium distribution and transport in the seaweed *Rhodymenia palmata* (L.) Grev., *Physiol. Plant.* **11**, 782–801.

Makeev, A. M., Makoveichuk, A. Y., and Chkanikov, D. I. (1977). 'Microsomal hydroxylation of 2,4-D in plants', *Doklady Bot. Sci.* **233**, 36–38.

Makoveichuk, A. Y., Makeev, A. M., and Chkanikov, D. I. (1978). 'Compartmentation of certain metabolites of 2,4-D in cells of cucumber leaves', *Sov. Pl. Physiol.* **25**, 320–322.

Manella, C. A., and Bonner, W. D. (1978). '2,4-D inhibits the outer membrane NADH dehydrogenase of plant mitochondria', *Pl. Physiol.* **62**, 468–469.

Marre, E. (1977). 'Effects of fusicoccin and hormones on plant cell membrane activities: observations and hypotheses', in *Regulation of Membrane Activities in Plants* (Eds. E. Marre and O. Ciferri, pp. 185–202, North Holland Publ. Co., Amsterdam.

Martin, J. T., and Juniper, B. E. (1970). *The Cuticles of Plants*, Arnold, London.

Masuda, H., and Sugawora, S. (1978). 'Studies of saccharases in the roots of sugar beet (*Beta vulgaris* L.). Part VIII. The complex formation of bound saccharase of sugar beet seedlings with cell wall structural polysaccharides', *Agr. Biol. Chem.* **42**, 1485–1490.

Masuda, Y., and Kamisaka, S. (1969). 'Rapid stimulation of RNA biosynthesis by auxin', *Plant Cell Physiol.* **10**, 79–86.

Masuda, Y., and Yamamoto, R. (1970). 'Effect of auxin on β-1,3-glucanase activity in *Avena* coleoptiles', *Devel. Growth Diff.* **11**, 287–296.

Matlib, M. A., Kirkwood, R. C., and Patterson, J. D. E. (1971). 'Binding of certain substituted phenoxy-acids by bovine serum albumin', *Weed Res.* **11**, 190–193.

Matthysse, A. G., and Phillips, C. (1969). 'A protein intermediary in the interaction of a hormone with the genome', *Proc. Natl. Acad. Sci. USA* **63**, 897–903.

Maxie, E. C., and Crane, J. C. (1967). '2,4,5-T: effect on ethylene production by fruits and leaves of the fig tree', *Science* **155**, 1548–1550.

McCready, C. C. (1963). 'Movement of growth regulators in plants. I. Polar transport of 2,4-dichlorophenoxyacetic acid in segments from the petioles of *Phaseolus vulgaris*', *New Phytol.* **62**, 3–18.

McCready, C. C., and Jacobs, W. P. (1963). 'Movement of growth regulators in plants. II. Polar transport of radioactivity from indoleacetic acid-^{14}C and 2,4-dichlorophenoxyacetic acid-^{14}C in petioles of *Phaseolus vulgaris*', *New Phytol.* **62**, 19–34.

McDougall, J., and Hillman, J. R. (1978). 'Purification of IAA from shoot tissues of *Phaseolus vulgaris* and its analysis by GC-MS', *J. Exp. Bot.* **29**, 375–386.

McIntyre, G. I., Fleming, W. W., and Hunter, J. H. (1978). 'Effect of shoot decapitation on the translocation of 2,4-D in *Cirsium arvense*', *Can. J. Bot.* **56**, 715–720.

Meagher, W. R. (1966). 'A heat-labile, insoluble, conjugated form of 2,4-dichlorophenoxyacetic acid and 2-(2,4,5-trichlorophenoxy)propionic acid in citrus peel', *J. Agr. Food Chem.* **14**, 599–601.

Milborrow, B. V., and Mallaby, R. (1975). 'Occurrence of methyl (+)-abscisate as an artefact of extraction'. *J. Exp. Bot.* **26**, 741–748.

Miller, S. D., Hudson, S. K., and Nalewaja, J. D. (1978). 'Wheat (*Triticum aestivum*) and barley (*Hordeum vulgare*) response to barban', *Weed Sci.* **26**, 226–229.

Moffit, S., and Blackman, G. E. (1972). 'The uptake of growth substances. XV. The differential effects of progressive chlorination of phenoxyacetic acid on entry into the epidermal and cut surfaces of stem segments', *J. Exp. Bot.* **23**, 128–140.

Mondal, H., Mandel, R. K., and Biswas, B. B. (1972). 'The effect of indoleacetic acid on RNA polymerase *in vitro*', *Biochem. Biophys. Res. Commun.* **49**, 306–311.

Montgomery, M. L., Chang, Y. L., and Freed, V. H. (1971). 'Comparative metabolism of 2,4-D by bean and corn plants', *J. Agr. Food Chem.* **19**, 1219–1221.

Moorby, J. (1964). 'The foliar uptake and translocation of caesium', *J. Exp. Bot.* **15**, 457–469.

More, J. E., Roberts, T. R., and Wright, A. N. (1978). 'Studies on the metabolism of 3-phenoxy benzoic acid in plants', *Pestic. Biochem. Physiol.* **9**, 268–280.

Moreland, D. E., and Hill, K. L. (1962). 'Interference of herbicides with the Hill reaction of isolated chloroplasts', *Weeds,* **10**, 229–236.

Morgan, P. W., and Hall, W. C. (1962). 'Effects of 2,4-D on the production of ethylene by cotton and grain sorghum', *Physiol. Plant.* **15**, 420–427.

Morgan, P. W., and Wayne, C. H. (1963). 'Metabolism of 2,4-D by cotton and grain sorghum', *Weeds* **11**, 130–135.

Morre, D. J., and Bracker, C. E. (1976). 'Ultrastructural alteration of plant plasmamembranes induced by auxin and calcium ions', *Pl. Physiol.* **58**, 544–547.

Morris, D. A., and Kadir, G. O. (1972). 'Pathways of auxin transport in the intact pea seedling (*Pisum sativum* L.)', *Planta* **107**, 171–182.

Morris, D. A., and Thomas, A. G. (1978). 'A microautoradiographic study of auxin transport in the stem of intact pea seedlings (*Pisum sativum* L.)', *J. Exp. Bot.* **29**, 147–157.

Mumma, R. O., and Hamilton, R. H. (1975). 'Amino acid conjugates', in *Bound and Conjugated Pesticide Residues* (Eds. D. D. Kaufman, G. G. Still, G. D. Paulson, and S. K. Bandal), pp. 68–85, Amer. Chem. Soc. Symposium Series no. 29.

Murphy, G. T. P. (1979). 'Plant hormone receptors: comparison of NAA binding by maize extracts and by a non-plant protein', *Pl. Sci. Lett.* **15**, 183–191.

Murray, M. G., and Key, J. L. (1978). '2,4-D enhanced phosphorylation of soybean nuclear proteins', *Pl. Physiol.* **61**, 190–198.

Nadakavukaren, M. J., and McCracken, D. A. (1977). 'Effects of 2,4-D on the structure and function of developing chloroplasts', *Planta* **137**, 65–69.

Neidermyer, R. W., and Nalewaja, J. D. (1969). 'Uptake, translocation and fate of 2,4-D in nightflowering catchfly and common lambsquarters', *Weed Sci.* **17**, 528–532.

Norris, L. A., and Freed, V. H. (1966). 'The absorption, translocation and metabolism characteristics of 4-(2,4-dichlorophenoxy)butyric acid in big leaf maple', *Weed Res.* **6**, 283–291.

Norris, R. F. (1974). 'Penetration of 2,4-D in relation to cuticle thickness', *Amer. J. Bot.* **61**, 74–79.

Olunuga, B. A., Lovell, P. H., and Sagar, G. R. (1977). 'The influence of plant age on the movement of 2,4-D and assimilates in wheat', *Weed Res.* **17**, 213–217.

Osmond, C. B., and Laties, G. G. (1969). 'Compartmentation of malate in relationship to ion absorption in beet', *Pl. Physiol.* **44**, 7–14.

Parups, E. V., and Miller, R. W. (1978). 'Investigation of effects of plant growth regulators on liposome fluidity and permeability', *Physiol. Plant.* **42**, 415–419.

Paterson, J. G. (1977). 'Interaction between herbicides, time of application and genotype of wild oats (*Avena fatua* L.)', *Aust. J. Agr. Res.* **28**, 671–680.

Paul, J. S., Krohne, S. D., and Bassham, J. A. (1979). 'Stimulation of CO_2 incorporation and glutamine synthesis by 2,4-D in photosynthesising leaf-free mesophyll cells', *Pl. Sci. Lett.* **15**, 17–24.

Pavlenko, A. D., Zholkevich, V. N., Butenko, R. G., and Sytnik, K. M. (1978). 'Auxin-induced entry of water into isolated protoplasts', *Doklady Bot. Sci.* **240**, 27–30.

Pemadasa, M. A. (1979). 'Stomatal responses to two herbicidal auxins', *J. Exp. Bot.* **30**, 267–274.

Pemadasa, M. A., and Jeyaseelan, K. (1976). 'Some effects of three herbicide auxins on stomatal movements', *New Phytol.* **77**, 569–573.

Penner, D., and Ashton, F. M. (1966). 'Biochemical and metabolic changes in plants induced by chlorophenoxy herbicides', *Residue Rev.* **14**, 39–113.

Pfister, K., Radosevich, S. R., and Arntzen, C. J. (1979). 'Modification of herbicide binding to photosystem II in two biotypes of *Senecio vulgaris* L.', *Pl. Physiol.* **64**, 995–999.
Pfruner, H., and Bentrup, F. W. (1978). 'Fluxes and compartmentation of K^+, Na^+ and Cl^- and action of auxins in suspension cultured *Petroselinum* cells', *Planta* **143**, 213–223.
Phung-Hong-Thai, and Field, R. J. (1979). 'The uptake and translocation of 2,4,5-T in gorse (*Ulex europaeus* L.)', *Weed Res.* **19**, 51–57.
Pickering, E. R. (1965). 'Foliar penetration pathways of 2,4-D, monuron and dalapon as revealed by microautoradiography', Ph.D. Thesis, Univ. California.
Pike, C. S., and Richardson, A. E. (1979). 'Red light and auxin effects on ^{86}rubidium uptake by oat coleoptile and pea epicotyl segments', *Pl. Physiol.* **63**, 139–141.
Pitman, M. G. (1963). 'Determination of the salt relations of the cytoplasmic phase in cells of beetroot tissue', *Aust. J. Biol. Sci.* **16**, 647–668.
Poole, R. J., and Thimann, K. V. (1964). 'Uptake of indole-3-acetic acid and indole-3-acetonitrile by *Avena* coleoptile sections', *Pl. Physiol.* **39**, 98–103.
Poovaiah, B. W., and Leopold, A. C. (1976). 'Effects of inorganic solutes on the binding of auxin', *Pl. Physiol.* **58**, 783–785.
Porter, E. M., and Bartels, P. G. (1977). 'Use of single leaf cells to study mode of action of SAN 6706 on soybean and cotton', *Weed Sci.* **25**, 60–65.
Price, C. E. (1973). 'Uptake and redistribution of *N*-methyl pyridium chloride, a model systemic compound, by wheat plants', *Proc. 7th British Insecticide and Fungicide Conf.*, pp. 161–169.
Primer, P. E. (1965). 'Investigations into the fate of some ^{14}C-labelled growth regulators of the phenoxy and naphthalenic types in apple tissue', Ph.D. Thesis, Univ Cornell.
Que Hee, S. S., and Sutherland, R. G. (1973). 'Penetration of amine salt formulations of 2,4-D in sunflower', *Weed Sci.* **21**, 115–118.
Radosevich, S. R., and Bayer, D. E. (1979). 'Effect of temperature and photoperiod on triclopyr, picloram and 2,4,5-T translocation', *Weed Sci.* **27**, 22–27.
Radwan, M. A., Stocking, C. R., and Currier, H. B. (1960). 'Histoautoradiographic studies of herbicidal translocation', *Weeds* **8**, 657–665.
Rakitin, Y. V., Zemskaya, V. A., Voronina, E. I., and Chernikova, L. M. (1966). 'Peculiarities of the detoxification of 2,4-D in plants sensitive and resistant to this herbicide', *Sov. Pl. Physiol.* **13**, 30–38.
Rao, I. M., Swamy, P. M., and Das, V. S. A. (1977). 'The reversal of scotoactive stomatal behaviour in some woody weeds by paraquat and 2,4,5-T', *Weed Sci.* **25**, 469–472.
Raven, J. A. (1975). 'Transport of indoleacetic acid in plant cells in relation to pH and electrical potential gradients, and its significance for polar IAA transport', *New Phytol.* **74**, 163–172.
Ray, P. M. (1973). 'Regulation of β-glucan synthetase activity by auxin in pea stem tissue. I. Kinetic aspects', *Pl. Physiol.* **51**, 601–608.
Ray, P. M. (1977). 'Auxin-binding sites of maize coleoptiles are localised on membranes of the endoplasmic reticulum', *Pl. Physiol.* **59**, 594–599.
Ray, P. M., Dohrmann, U., and Hertel, R. (1977a). 'Characterisation of naphthaleneacetic acid binding to receptor sites on cellular membranes of maize coleoptile tissue', *Pl. Physiol.* **59**, 357–364.
Ray, P. M., Dohrmann, U., and Hertel, R. (1977b) 'Specificity of auxin-binding sites on maize coleoptile membranes as possible receptor sites for auxin action', *Pl. Physiol.* **60**, 585–591.

Rayle, D. L. (1973). 'Auxin-induced hydrogen-ion secretion in *Avena* coleoptiles and its implications', *Planta* **114**, 63–73.

Rehfeld, D. W., and Jensen, R. G. (1973). 'Metabolism of separated leaf cells. III. Effects of calcium and ammonium on product distribution during photosynthesis with cotton cells', *Pl. Physiol.* **52**, 17–22.

Reinhold, L. (1954). 'Uptake of indole-3-acetic acid by pea epicotyl segments and carrot discs', *New Phytol.* **53**, 217–239.

Rekoslavskaya, N. I., and Gamburg, K. Z. (1977). 'Auxin induction of indoylacetyl aspartate synthesis in a culture of auxin-independent blackberry tissue', *Sov. Pl. Physiol.* **24**, 970–973.

Revel, M., and Groner, Y. (1978). 'Post-transcriptional and translational controls of gene expression in eukaryotes', *Ann. Rev. Biochem.* **47**, 1079–1126.

Richardson, R. G. (1976). 'Changes in the translocation and distribution of 2,4,5-T in blackberry (*Rubus procerus* P. J. Muell) with time', *Weed Res.* **16**, 375–378.

Richardson, R. G. (1977). 'A review of foliar absorption and translocation of 2,4-D and 2,4,5-T', *Weed Sci.* **17**, 259–272.

Roberts, T. R. (1977a). 'The metabolism of the herbicide flamprop-isopropyl in barley', *Pestic. Biochem. Physiol.* **7**, 378–390.

Roberts, T. R. (1977b). 'The metabolism of the herbicide flamprop-methyl in wheat', *Pestic. Sci.* **8**, 463–472.

Robertson, M. M., and Kirkwood, R. C. (1969). 'The mode of action of the foliage-applied translocated herbicides with particular reference to the phenoxy-acid compounds. I. The mechanism and factors influencing herbicide absorption', *Weed Res.* **9**, 224–240.

Robertson, M. M., and Kirkwood, R. C. (1970). 'The mode of action of foliage-applied translocated herbicides with particular reference to the phenoxy-acid compounds. II. Mechanism and factors influencing translocation, metabolism and biochemical inhibition', *Weed Res.* **10**, 94–120.

Robinson, S. P., Wiskich, J. T., and Paleg, L. G. (1978). 'Effects of IAA on CO_2 fixation, electron transport and phosphorylation in isolated chloroplasts', *Aust. J. Pl. Physiol.* **5**, 425–431.

Rohrbaugh, C. M., and Rice, E. L. (1949). 'Effect of application of sugar on the translocation of sodium 2,4-dichlorophenoxyacetate by bean plants in the dark', *Bot. Gaz.* **111**, 85–89.

Rubery, P. H. (1977). 'The specificity of carrier-mediated auxin transport by suspension-cultured crown gall cells', *Planta* **135**, 275–283.

Rubery, P. H. (1978). 'Hydrogen ion dependence of carrier-mediated auxin uptake by suspension-cultured crown gall cells', *Planta* **142**, 203–206.

Rubery, P. H. (1979). 'The effects of 2,4-dinitrophenol and chemical modifying reagents on auxin transport by suspension-cultured crown gall cells', *Planta* **144**, 173–178.

Rubery, P. H., and Sheldrake, A. R. (1973). 'Effect of pH and surface charge on cell uptake of auxin', *Nature, New Biol.* **244**, 285–288.

Rubery, P. H., and Sheldrake, A. R. (1974). 'Carrier-mediated auxin transport', *Planta* **118**, 101–121.

Russell, D. W. (1971). 'The metabolism of aromatic compounds in higher plants. X. Properties of the cinnamic acid 4-hydroxylase of pea seedlings and some aspects of its metabolic and developmental control', *J. Biol. Chem.* **246**, 3870–3878.

Rutherford, P. P., and Deacon, A. C. (1973). 'Inhibition by actinomycin D of the increase in hydrolase activity induced in dandelion root by treatment with 2,4-D', *Can. J. Bot.* **51**, 2516–2519.

Sakurai, N., and Masuda, Y. (1978a). 'Auxin induced changes in barley coleoptile cell-wall composition', *Pl. Cell Physiol. Tokyo* **19**, 1217–1223.

Sakurai, N., and Masuda, Y. (1978b). 'Auxin-induced extension, cell-wall loosening and changes in the wall polysaccharide content of barley coleoptile segments', *Pl. Cell Physiol. Tokyo* **19**, 1225–1233.

Sanad, A. J. (1971). 'Studies of the uptake and translocation of ^{14}C-labelled herbicides in *Agrostemma githago*, L. and *Tussilago farfara* L.', *Weed Res.* **11**, 215–223.

Sanderson, J. (1972). 'Micro-autoradiography of diffusible ions in plant tissues: problems and methods', *J. Microscopy* **96**, 245–254.

Sargent, J. A., and Blackman, G. E. (1962). 'Studies on foliar penetration. I. Factors controlling the entry of 2,4-dichlorophenoxyacetic acid', *J. Exp. Bot.* **13**, 348–368.

Sargent, J. A., and Blackman, G. E. (1965). 'Studies on foliar penetration. II. The role of light in determining the penetration of 2,4-dichlorophenoxyacetic acid', *J. Exp. Bot.* **16**, 24–47.

Sargent, J. A., and Blackman, G. E. (1969). 'Studies on foliar penetration. IV. Mechanisms controlling the rate of penetration of 2,4-dichlorophenoxyacetic acid into leaves of *Phaseolus vulgaris*', *J. Exp. Bot.* **20**, 542–555.

Sarkissian, I. V. (1970). 'Hormonal modifications of plant citrate synthase *in vitro*', *Biochem. Biophys. Res. Commun.* **40**, 1385–1390.

Saunders, J. A., and Conn, E. E. (1978). 'Presence of the cyanogenic glucoside dhurrin in isolated vacuoles from sorghum', *Pl. Physiol.* **61**, 154–157.

Saunders, P. F., Jenner, C. F., and Blackman, G. E. (1965a). 'The uptake of growth substances. IV. Influence of species and chemical structure on the pattern of uptake of substituted phenoxyacetic acids by stem segments', *J. Exp. Bot.* **16**, 683–696.

Saunders, P. F., Jenner, C. F., and Blackman, G. E. (1965b). 'The uptake of growth substances. V. Variation in the uptake of a series of chlorinated phenoxyacetic acids by stem tissue of *Gossypium hirsutum* and its relationship to differences in auxin activity', *J. Exp. Bot.* **16**, 697–713.

Schultz, D. P., and Tweedy, B. G. (1971). 'Uptake and metabolism of N,N-dimethyl-2,2-diphenylacetamide in resistant and susceptible plants', *J. Agr. Food Chem.* **19**, 36–40.

Sharman, B. C. (1978). 'Morphogenesis of 2,4-D induced abnormalities of the inflorescence of bread wheat', *Ann. Bot.* **42**, 145–153.

Sheldrake, A. R. (1973). 'Auxin transport in secondary tissues', *J. Exp. Bot.* **24**, 87–96.

Shone, M. G. T., Bartlett, B. D., and Wood, A. V. (1974). 'A comparison of the uptake and translocation of some organic herbicides and a systemic fungicide by barley. II. Relationship between uptake by roots and translocation to shoots', *J. Exp. Bot.* **25**, 401–409.

Slife, F. W., Key, J. L., Yamaguchi, S., and Crafts, A. S. (1962). 'Penetration, translocation and metabolism of 2,4-D and 2,4,5-T in wild and cultivated cucumber plants', *Weeds* **10**, 29–35.

Smith, A. E. (1979). 'Metabolism of 2,4-DB by white clover (*Trifolium repens*) cell suspension cultures', *Weed Sci.* **27**, 392–396.

Smith, C., Doo, A., and Bown, A. W. (1979). 'The influence of pH on kinetic parameters of coleoptile phosphoenolpyruvate carboxylase. Relationship to auxin stimulated dark fixation', *Can. J. Bot.* **57**, 543–547.

Smith, F. A., and Raven, J. A. (1979). 'Intracellular pH and its regulation', *Ann. Rev. Pl. Physiol.* **30**, 289–311.

Smith, R. C., and Epstein, E. (1964). 'Ion absorption by shoot tissue: technique and first findings with excised leaf tissue of corn', *Pl. Physiol.* **39**, 338–341.

Steen, R. C. (1972). 'A study of 2,4-D side-chain degradation by plants', Ph.D. Thesis, State Univ. N. Dakota.

Stenlid, G., and Saddik, K. (1962). 'The effects of some growth regulators and uncoupling agents upon oxidative phosphorylation in mitochondria of cucumber hypocotyls', *Physiol. Plant.* **15**, 369–379.

Still, G. G., Norris, F. A., and Iwan, J. (1975). 'Solubilisation of bound residues from 3,4-dichloroaniline-^{14}C and propanil-phenyl-^{14}C treated rice root tissues', in *Bound and Conjugated Pesticide Residues* (Eds. D. D. Kaufman, G. G. Still, G. D. Paulson, and S. K. Bandal), pp. 156–165, Amer. Chem. Soc. Symposium series no. 29.

Stout, R. G., Johnson, K. D., and Rayle, D. L. (1978). 'Rapid auxin and fusicoccin enhanced Rb$^+$ uptake and malate synthesis in *Avena* coleoptile sections', *Planta* **139**, 35–41.

Stowe, B. B., Epstein, E., and Vendrell, M. (1968). 'Indoles of maize and the source of woad indigo', in *Plant Growth Regulators*, pp. 102–110, S.C.I. Monographs no. 31, Staples Printers Ltd., Rochester, Kent.

Sudi, J. (1964). 'Induction of the formation of complexes between aspartic acid and indolyl-3-acetic acid or 1-naphthaleneacetic acid by other carboxylic acids', *Nature (London)* **201**, 1009–1010.

Sudi, J. (1966). 'Increases in the capacity of pea tissue to form acyl-aspartic acids specifically induced by auxins', *New Phytol.* **65**, 9–21.

Sun, C. N. (1955). 'Anomalous structure in the hypocotyl of soybean following treatment with 2,4-D', *Science* **121**, 641.

Szabo, S. S. (1963). 'The hydrolysis of 2,4-D esters by bean and corn plants', *Weeds*, **11**, 292–294.

Szabo, S. S., and Buckholtz, K. P. (1961). 'Penetration of living and non-living surfaces by 2,4-D as influenced by ionic additives', *Weeds*, **9**, 177–184.

Taylor, T. D., and Warren, G. F. (1970). 'Movement of several herbicides through excised plant tissue', *Weed Sci.* **18**, 64–74.

Teissere, M., Penon, P., and Ricard, J. (1973). 'Hormonal control of chromatin availability and of the activity of purified RNA polymerases in higher plants', *FEBS Lett.* **30**, 65–70.

Thomas, E. W., Loughman, B. C., and Powell, R. G. (1963). 'Hydroxylation of phenoxyacetic acids by stem tissue of *Avena sativa*', *Nature (London)* **199**, 73–74.

Thomas, E. W., Loughman, B. C., and Powell, R. G. (1964). 'Metabolic fate of some chlorinated phenoxyacetic acids in stem tissue of *Avena sativa*', *Nature (London)* **204**, 286.

Thomson, K. S., Hertel, R., and Muler, S. (1973). '1-*N*-naphthylphthalamic acid and 2,3,5-triiodobenzoic acid. *In vitro* binding to particulate cell fractions and action on auxin transport in corn coleoptiles', *Planta* **109**, 337–352.

Trewavas, A. (1968). 'The effect of 3-indoleacetic acid on the level of polysomes in etiolated pea tissue', *Phytochem.* **7**, 673–681.

Tukey, H. B., Hamner, C. L., and Imhofe, B. (1945). 'Histological changes in bindweed and sow thistle following applications of 2,4-D in herbicidal concentrations', *Bot. Gaz.* **107**, 62–73.

Van der Woude, W. J., Lembi, C. A., and Morre, D. J. (1972). 'Auxin (2,4-D) stimulation (*in vivo* and *in vitro*) of polysaccharide synthesis in plasmamembrane fragments isolated from onion stems', *Biochem. Biophys. Res. Commun.* **46**, 245–253.

Van Overbeek, (1956). 'Absorption and translocation of plant regulators', *Ann. Rev. Plant Physiol.* **7**, 355–372.

Veen, H. (1966). 'Transport, immobilisation and localisation of naphthylacetic acid-1-^{14}C in *Coleus* explants', *Acta Bot. Neerl.* **15**, 419–433.

Veen, H. (1974). 'Specificity of phospholipid binding to indoleacetic acid and other auxins', *Z. Naturforsch.* **29**, 39–41.

Venis, M. A. (1964). 'Induction of enzymatic activity by indolyl-3-acetic acid and its dependence on synthesis of ribonucleic acid', *Nature (London)* **202**, 900–901.

Venis, M. A. (1972). 'Auxin-induced conjugation systems in peas', *Pl. Physiol.* **49**, 24–27.

Venis, M. A. (1977). 'Receptors for plant hormones', *Adv. Bot. Res.* **5**, 53–88.

Venis, M. A., and Watson, P. J. (1978). 'Naturally occurring modifiers of auxin-receptor interaction in corn: identification as benzoxazolinones', *Planta* **142**, 103–107.

Verloop, A. (1975). 'Use of radiotracer studies in the estimation of conjugated and bound metabolites of dichlobenil in field crops', in *Bound and Conjugated Pesticide Residues* (Eds. D. D. Kaufman, G. G. Still, G. D. Paulson, and S. K. Bandal), pp. 173–177, Amer. Chem. Soc. Symposium series no. 29.

Volynets, A. P., and Pal'chenko, L. A. (1977). 'Enzymic oxidation of IAA in lupine plants treated with herbicides', *Sov. Pl. Physiol.* **24**, 446–449.

Wain, R. L. (1955a). 'A new approach to selective weed control', *Ann. App. Biol.* **42**, 151–157.

Wain, R. L. (1955b). 'Herbicidal selectivity through specific action of plants on the compound applied', *J. Agr. Food Chem.* **3**, 128–130.

Wain, R. L. (1955c). 'A new principle of weed control', *Agr. Rev. Lond.* **1**, 25–36.

Walker, N. A., and Pitman, M. G. (1976). 'Measurement of fluxes across membranes', in *Encyclopedia of Plant Physiology* (Eds. U. Luttge and M. G. Pitman), new series, Vol. II, Part A, pp. 93–126, Springer-Verlag, Berlin.

Wangermann, E. (1970). 'Autoradiographic localisation of soluble and insoluble ^{14}C from ^{14}C-indolylacetic acid supplied to isolated *Coleus* internodes', *New Phytol.* **69**, 919–927.

Wangermann, E. (1974). 'The pathway of transport of applied indolylacetic acid through internode segments', *New Phytol.* **73**, 623–636.

Wardrop, A. J., and Polya, G. M. (1977). 'Properties of a soluble auxin-binding protein from dwarf bean seedlings', *Pl. Sci. Lett.* **8**, 155–163.

Wathana, S., Corbin, F. T., and Waldrep, T. W. (1972). 'Absorption and translocation of 2,4-DB in soybean and cocklebur', *Weed Sci.* **20**, 120–123.

Whitehead, C. W., and Switzer, C. M. (1963). 'The differential response of strains of wild carrot to 2,4-D and related herbicides', *Can. J. Plant Sci.* **43**, 255–262.

Whitworth, J. W., and Muzik, T. J. (1967). 'Differential response of selected clones of bindweed to 2,4-D', *Weeds* **15**, 275–280.

Wiedman, S. J., and Appleby, A. P. (1972). 'Plant growth stimulation by sub-lethal concentrations of herbicides', *Weed Res.* **12**, 65–74.

Wieneke, J. (1975). 'Classification and analysis of pesticides bound to plant material', in *Bound and Conjugated Pesticide Residues* (Eds. D. D. Kaufman, G. G. Still, G. D. Paulson, and S. K. Bandal), pp. 166–169, Amer. Chem. Soc. Symposium series no. 29.

Wilkins, H., and Wilkins, M. B. (1975). 'The movement of 2,4-dichlorophenoxyacetic acid in root segments of *Pisum sativum* L.', *Planta* **124**, 177–189.

Williams, J. H. B. (1976). 'Studies on the metabolism of MCPA in plants', Ph.D. Thesis, Univ. Wales.

Winter, A., and Thimann, K. V. (1966). 'Bound indoleacetic acid in *Avena* coleoptiles', *Pl. Physiol.* **41**, 335–342.

Wyrill, J. B., and Burnside, O. C. (1976). 'Absorption, translocation and metabolism of 2,4-D and glyphosate in common milkweed and hemp dogbane', *Weed Sci.* **24**, 557–566.

Yamaguchi S. (1965). 'Analysis of 2,4-D transport', *Hilgardia* **36**, 349–378.

Yamaha, T., and Cardini, C. E. (1960). 'The biosynthesis of glycosides. II. Gentiobiosides', *Arch. Biochem. Biophys.* **86**, 133–137.

Yu, Y. B., Adams, D. O., and Fang, S. F. (1979). 'Regulation of auxin-induced ethylene production in mung-bean hypocotyls', *Pl. Physiol.* **63**, 589–590.

Zemskaya, V. A., Bokarev, K. S., Chernikova, L. M., and Kalibernaya, Z. V. (1973). 'Ability of corn leaf proteins to bind certain derivatives of phenoxyacetic acid', *Sov. Pl. Physiol.* **20**, 327–329.

Zemskaya, V. A., and Rakitin, Y. V. (1967). 'Localisation of 2,4-D in the cells of corn and sunflower leaves', *Sov. Pl. Physiol.* **14**, 848–852.

Zemskaya, V. A., Rakitin, Y. V., Chernikova, L. M., and Kalibernaya, Z. V. (1971). 'Kinetics of the process of 2,4-D bonding by proteins in maize leaf tissues', *Sov. Pl. Physiol.* **18**, 626–631.

Zenk, M. H. (1961). '1-(indole-3-acetyl)-β-D-glucose, a new compound in the metabolism of indole-3-acetic acid in plants', *Nature (London)* **191**, 493–494.

Zenk, M. H. (1964).'Isolation, biosynthesis and function of indoleacetic acid conjugates', in *Regulateurs Naturels de la Croissance, Vegetale* (Ed. J. P. Nitsch), pp. 241–250, CNRS, Gif-sur-Yvette, France.

Zenk, M. H., and Nissl, D. (1968). 'Evidence against an allosteric effect of indole-3-acetic acid on citrate synthetase', *Naturwiss.* **55**, 84–85.

Zimmermann, M. H., and Milburn, J. A. (1975). 'Transport in plants. I. Phloem transport', *Encyclopedia of Plant Physiology*, new series, Vol. 1, Springer-Verlag, Berlin.

Zsoldos, F., Karvaly, B., Toth, I., and Erdei, L. (1978). '2,4-D induced changes in the K^+ uptake of wheat roots at different pH values', *Physiol. Plant.* **44**, 395–399.

CHAPTER 5

Mechanisms of teratogenesis induced by organophosphorus and methylcarbamate insecticides

J. Seifert and J. E. Casida

INTRODUCTION	219
AVIAN TERATOGENESIS	219
General Features	219
Teratogenic signs	219
Test variables	222
Structure–Activity Relationships	223
Organophosphorus compounds	224
Methylcarbamates	225
Lowered NAD Level in Type *I* Teratogenesis	225
Kynurenine Formamidase Inhibition in Type *I* Teratogenesis	230
Properties and distribution of kynurenine formamidase	230
Inhibition of kynurenine formamidase	232
Cholinergic System in Type *II* Teratogenesis	236
MAMMALIAN TERATOGENESIS	240
CONCLUSIONS	240
ACKNOWLEDGEMENTS	241
REFERENCES	241

INTRODUCTION

Several organophosphorus (OP) and methylcarbamate (MC) insecticides and related compounds (Figure 5.1) are potent avian teratogens. This review evaluates two teratogenic mechanisms, one involving disruption of NAD biosynthesis and the other of the cholinergic system. It also considers the status of OP- and MC-induced teratogenesis in mammals.

AVIAN TERATOGENESIS

General Features

Teratogenic signs

Some OP and MC compounds induce mainly abnormal feathering and malformations of the lower extremities while others give primarily vertebral defects, indicating the probable involvement of more than one biochemical lesion or mode of action. Recognition of these two series of teratogenic signs combined with the associated progress on alleviating agents (Meiniel, 1976a, b) and biochemical mechanisms (Proctor and Casida, 1975; Proctor *et al.*, 1976) have led to a classification of the OP- and MC-induced avian teratogenesis as Type *I* or Type *II* (Moscioni *et al.*, 1977). The Type *I* syndrome involves micromelia and abnormal feathering (for illustrations, see Agarwal, 1956; Khera and Bedok, 1967; Greenberg and LaHam, 1969, 1970; Walker, 1971; Meiniel, 1976a) (see also Marliac and Mutchler, 1963; McLaughlin *et al.*, 1963; Roger *et al.*, 1964, 1969; Ho and Gibson, 1972; Arsenault *et al.*, 1975; Proctor and Casida, 1975;

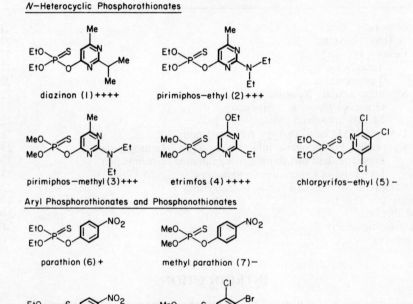

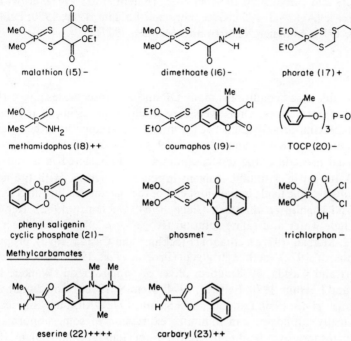

Figure 5.1 Some organophosphorus compounds and methylcarbamates tested as possible chicken teratogens. The severity of teratogenic signs at day 19 of incubation, resulting from yolk injection of 1 mg compound/egg at day 4, is indicated as −, +, + +, + + +, and + + + + for increasing magnitude of effects as reported by Proctor et al. (1976) and Moscioni et al. (1977). The 23 compounds designated by number were used in mode of action studies illustrated in Figures 5.5 and 5.9. Results for assays with chlorfenvinphos in mammals are given in Table 5.4

Proctor et al., 1976; Laley and Gibson, 1977; Moscioni et al., 1977). Type II teratogenesis is evidenced primarily by vertebral malformations such as brevicollis (for photographs, see Khera and Bedok, 1967; Lutz-Ostertag et al., 1969; Meiniel et al., 1970; Meiniel, 1973, 1974, 1976a) (see also Roger et al., 1969; Yamada, 1972; Meiniel, 1975, 1976b, c, 1977a, b; Moscioni et al., 1977). Most teratogens at high doses produce various combinations of Types I and II teratogenic signs (Moscioni et al., 1977; Eto et al., 1980) resulting in differing patterns of abnormalities.

Additional effects of OP and MC insecticides, usually associated with decreased hatchability, have been noted as follows: beak defects (Greenberg and LaHam, 1969, 1970; Roger et al., 1969; Meiniel, 1975, 1976a, b); hypoglycemia (Arsenault et al., 1975; Laley and Gibson, 1977); thyroid dysfunction (Richert and Prahlad, 1972); amuscularity (Schom et al., 1979); foot deformities (Khera et *al.,* 1966); asthenia and lethargy in some cases and induced hyperexcitability, body tremors and convulsions in other cases (Khera *et al.,* 1965); growth and weight decrease (Agarwal, 1956; Greenberg and LaHam, 1969, 1970; Paul and Vadlamudi, 1976; Autissier-Navarro and Meiniel, 1977).

Test variables

Under suitable assay conditions, many OP and MC insecticides give a 100% incidence of abnormalities, a relatively rare situation in testing of teratogens. Several variables importantly influence the intensity and sometimes the nature of the teratogenic signs (Khera and Clegg, 1969; Gebhardt, 1972).

In the usual procedure, the test compound is administered in a non-toxic volume (10–100 μl) of a suitable solvent injected via syringe with the needle passing through a hole made in the blunt end of the egg and leading directly into the yolk (McLaughlin *et al.,* 1963; Clegg, 1964; Gebhardt, 1972). Reported carriers include water and saline solution (Roger *et al.,* 1969; Landauer and Salam, 1972; Meiniel, 1976a), ethanol (Flockhart and Casida, 1972), propylene glycol (Gebhardt, 1972), methoxytriglycol (Proctor *et al.,* 1976; Moscioni *et al.,* 1977; Seifert and Casida, 1978), corn, olive, or sunflower oil (Walker, 1968; Greenberg and LaHam, 1970; Belyaev, 1974; Meiniel, 1976a), and dimethylsulphoxide (Roger *et al.,* 1969; Landauer and Salam, 1972). Although an excellent solvent, dimethylsulphoxide can minimize expression of some abnormalities evident when the teratogen is administered in water (Landauer and Salam, 1972). Following injection, the hole is sealed with paraffin wax prior to incubating the eggs.

Other treatment procedures used less frequently are injection into the allantois (Kuul, 1975) and air space (Agarwal, 1956), yolk replacement (Walker, 1968), and egg immersion into a toxicant solution (Lutz-Ostertag *et al.,* 1969). Attempts to obtain teratogenic effects with monocrotophos by feeding the hens or spraying the eggs proved unsuccessful (Schom *et al.,* 1979) whereas at high dietary levels malathion and carbaryl gave chicks with congenital malformations (Ghadiri *et al.,* 1967).

Most studies utilized eggs from chickens (*Gallus domesticus*) usually white leghorn) while others used eggs from Peking (mallard) ducks (*Anas platyrhynchos*) (e.g. Khera and Bedok, 1967), Japanese quail (*Coturnix coturnix japonica*) (e.g. Meiniel *et al.,* 1970), Bobwhite quail (*Colinus virginianus*) (Schom *et al.,* 1979), and Chukar partridge (*Alectoris chukar*) (Schom *et al.,* 1979). The severity of teratogenic signs produced depends not only on the species but also on the line or genotype (Schom and Abbott, 1977; Schom *et al.,* 1979).

The sensitivity of the embryo changes greatly during development. Treatments of hen eggs are normally made at days 4–6 of incubation (corresponding approximately to stages 24–29 of embryonic development; Hamburger and Hamilton, 1951) because at this time the embryo is most sensitive to OP and MC compounds injected into the yolk (Khera and Bedok, 1967; Greenberg and LaHam, 1969; Roger *et al.*, 1969; Ho and Gibson, 1972; Proctor *et al.*, 1976; Laley and Gibson, 1977; Moscioni *et al.*, 1977) (Figure 5.2). For example, dicrotophos is distinctly more potent when injected at day 4 than at days 2 or 6 of incubation and many compounds teratogenic at day 4 are not when injected prior to incubation (Roger *et al.*, 1969). From day 4 until day 11 or 12 there is a progressive reduction in sensitivity (Roger *et al.*, 1969) (Figure 5.2). The series of malformations is also dependent on the time of treatment. Thus, axial deformities are induced by parathion when administered at the beginning of embryonic development while only cervical abnormalities are evident with treatments later in embryonic life (Meiniel, 1977b).

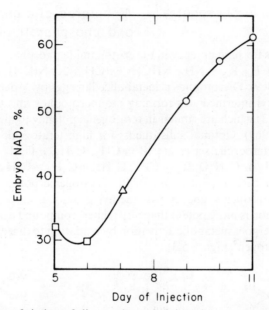

Figure 5.2 Effect of timing of dicrotophos administration on chicken embryo NAD levels and severity of teratogenic signs. Dicrotophos was injected at 1 mg/egg between days 5 and 11, embryo NAD levels determined on day 12, and teratogenic signs observed on day 19 of incubation. Increasing severity of teratogenic signs is indicated as ○, △, and □. (Data from Moscioni *et al.* (1977), reproduced by permission of Pergamon Press Ltd)

Structure–Activity Relationships

Relatively high structural specificities are encountered for both Types *I* and *II* teratogens. Teratogenic potencies for 25 OP and MC insecticides and related materials are given in Figure 5.1. More than 130 different OP and MC compounds have been tested (Roger *et al.*, 1969; Proctor *et al.*, 1976; Eto *et al.*, 1980) in deriving the structure–activity relationships discussed below.

Organophosphorus compounds

Many pyrimidyl and pyridyl phosphates and phosphorothionates (e.g. diazinon) and crotonamide phosphates (e.g. dicrotophos) are potent Type *I* teratogens.

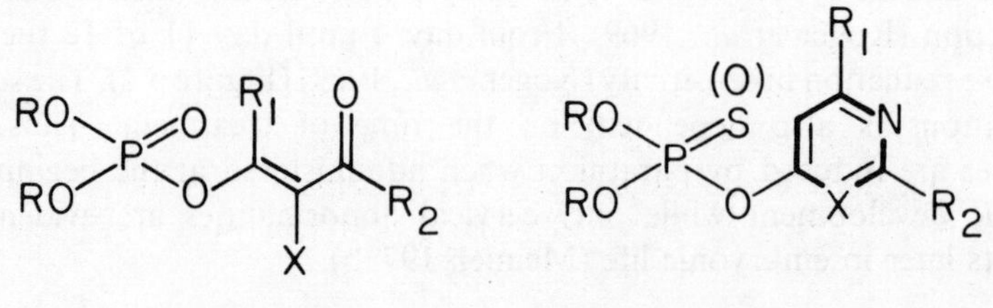

crotonamide phosphates   *N*-heterocyclic phosphates and phosphorothionates

Requirements for high potency in the crotonamide phosphates are: $R = CH_3$ or C_2H_5; $R_1 = CH_3$; $R_2 = NH_2$, $NHCH_3$, $N(CH_3)_2$, or $N(C_2H_5)_2$; $X = H$ or Cl (Roger *et al.*, 1969). Dicrotophos is metabolized in eggs by *N*-demethylation via *N*-hydroxymethyl intermediates forming monocrotophos and the amide analogue (Figure 5.3) which are similar in teratogenic potency to dicrotophos itself (Roger *et al.*, 1969). Optimal substituents for high teratogenic activity in the illustrated *N*-heterocyclic series are: $R = CH_3$, C_2H_5, n-C_3H_7, i-C_3H_7, or n-C_4H_9; $R_1 = CH_3$ or C_2H_5O; $R_2 = CH_3$, C_2H_5, n-C_3H_7, i-C_3H_7, or $(C_2H_5)_2N$; X = nitrogen or carbon (Eto *et al.*, 1980). Teratogenic potency is sometimes enhanced by introducing one or two methyl groups at carbons adjacent to nitrogen. Diazinon is more potent than any related compound examined to date. It probably undergoes metabolic activation by oxidation to diazoxon prior to its action as a teratogen (Figure 5.3).

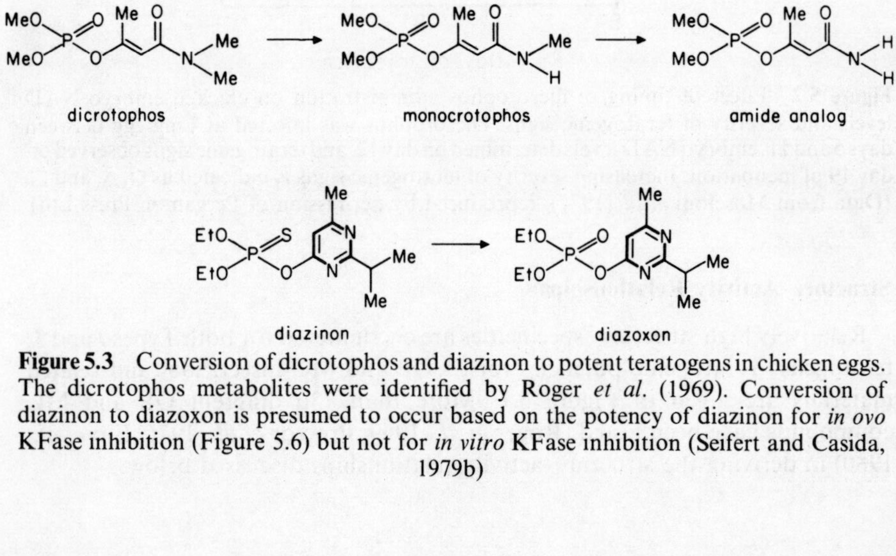

dicrotophos   monocrotophos   amide analog

diazinon   diazoxon

Figure 5.3 Conversion of dicrotophos and diazinon to potent teratogens in chicken eggs. The dicrotophos metabolites were identified by Roger *et al.* (1969). Conversion of diazinon to diazoxon is presumed to occur based on the potency of diazinon for *in ovo* KFase inhibition (Figure 5.6) but not for *in vitro* KFase inhibition (Seifert and Casida, 1979b)

Parathion is the classical Type *II* teratogen (Meiniel, 1976a, b; 1977a, b; Moscioni *et al.*, 1977). Other compounds giving predominantly or only Type *II* signs are *N*-heterocyclic diethyl phosphorothionates without ring alkyl substituents, e.g. pyridyl, pyrimidyl, and pyrazinyl derivatives (Eto *et al.*, 1980). Compounds giving a mixture of Types *I* and *II* signs include dicrotophos (Roger *et al.*, 1969; Meiniel, 1975, 1976a, b, 1977a; Moscioni *et al.*, 1977) and some pyrimidyl and/or pyridyl phosphorothionates (Moscioni *et al.*, 1977; Eto *et al.*, 1980).

Figure 5.4 illustrates small structural modifications that greatly diminish or destroy the teratogenic activities of Types *I* and *II* OP teratogens, i.e. replacing an *O*-ethyl substituent with a phenyl group in diazinon and parathion, substituting a *S*-ethyl for an *O*-ethyl group in diazoxon, removing or shifting the position of pyridyl nitrogen in two derivatives, and isomerization at the dicrotophos double bond (Eto *et al.*, 1980).

Methylcarbamates

Eserine, the first carbamate observed to have teratogenic activity (Ancel, 1945), remains the most potent compound of this type. Additional active compounds are carbaryl and its dimethylcarbamate analogue (Walker, 1968; Proctor *et al.*, 1976; Moscioni *et al.*, 1977; Eto *et al.*, 1980). The *N*-ethyl analogue of carbaryl (Figure 5.4), the dimethylcarbamate analogue of diazinon, and many other carbamate derivatives are ineffective as chicken teratogens (Eto *et al.*, 1980). Eserine gives a mixture of Types *I* and *II* teratogenic signs (Agarwal, 1956; Moscioni *et al.*, 1977) whereas carbaryl at doses up to 1 mg/egg gives essentially only Type *I* signs (Moscioni *et al.*, 1977).

Lowered NAD Level in Type *I* Teratogenesis

Several lines of evidence implicate the NAD system in Type *I* OP- and MC-induced teratogenesis.

Two teratogenic nicotinamide (NAm) antivitamins, 3-acetylpyridine (3-AP) and 6-aminonicotinamide (6-AN), give some teratogenic signs (Landauer, 1957; Landauer and Clark, 1962) in common with the OP and MC teratogens (Proctor *et al.*, 1976). NAm alleviates the 3-AP and 6-AN teratogenesis (Landauer, 1957). These compounds have been used to examine possible mechanisms by which a lowered NAD level is associated with embryonic abnormalities. 6-AN impairs chondroitin sulphate synthesis, possibly by interfering with a NAD-dependent process, which may then lead to micromelia (Overman *et al.*, 1972; Seegmiller and Runner, 1974). Malathion also interferes with chondroitin sulphate synthesis (Gill and LaHam, 1972). Pyridine nucleotide levels may control the differentiation of a limb mesodermal cell into a cell which expresses a myogenic or chondrogenic phenotype (Rosenberg and Caplan, 1974). Alternatively, the

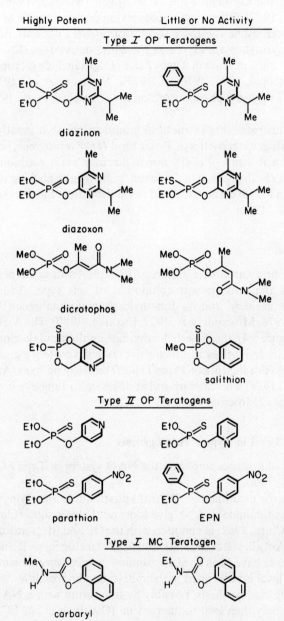

Figure 5.4 Some structural modifications with large effects on activity as chicken teratogens. Data from Eto *et al.* (1980)

action of 3-AP may involve destruction of peripheral nerves leading to serious defects in later development of the muscle (McLachlan *et al.*, 1976).

NAm alleviates the Type *I* teratogenic signs plus the beak abnormalities caused by diazinon, dicrotophos, malathion, carbaryl, and eserine (Landauer, 1949, 1957; Roger *et al.*, 1964, 1969; Upshall *et al.*, 1968; Greenberg and LaHam, 1970; Walker, 1971; Landauer and Salam, 1972; Proctor and Casida, 1975; Meiniel, 1976a, b; Proctor *et al.*, 1976). Many NAm precursors and analogues are also active (Roger *et al.*, 1964, 1969; Greenberg and LaHam, 1970; Moscioni *et al.*, 1977). The alleviating potency of compounds in the tryptophan to NAD biosynthetic pathway (i.e. tryptophan, *N*-formylkynurenine, kynurenine, 3-hydroxyanthranilic acid, quinolinic acid, nicotinic acid, nicotinamide mononucleotide, and NAD) generally increases near the end of the pathway (Roger *et al.*, 1969; Moscioni *et al.*, 1977).

Analyses of embryo NAD levels provide direct evidence for a teratogenic mechanism involving suboptimal NAD concentrations (Proctor *et al.*, 1976; Moscioni *et al.*, 1977). Three types of studies establish a close relationship between embryo NAD level and the intensity of teratogenic effects. First, the lowering of NAD level is correlated with the reduction in leg length and overall severity of teratogenesis in studies with 18 OP and MC compounds of widely varying potency (Figure 5.5A). Second, the lowered embryo NAD level also parallels the severity of the teratogenic signs in dose dependency studies with two OP teratogens (diazinon and dicrotophos) and two MC teratogens (carbaryl and eserine) (Proctor *et al.*, 1976) (Figure 5.5C). Finally, three agents (nicotinic acid, NAm, and NAD) that return the embryo NAD levels to normal are alleviating agents for the teratogenic signs whereas four related compounds (isonicotinic acid, β-picoline, picolinic acid, and 3-pyridylacetic acid) that do not restore normal NAD levels are not alleviating agents (Proctor *et al.*, 1976).

Diazinon injected at day 4 gives a progressive lowering of the embryo NAD level from days 6 to 10 (Figure 5.6). The NAD content is lowered not only in the embryo at day 12 but also in its brain, liver, and muscle at day 19 of incubation; in each case NAm administration returns the NAD level to a near normal value (Table 5.1).

Several mechanisms have been considered for the lowering of the embryo NAD level. Labelled metabolic pool studies revealed no block in conversion of NAm to NAm mononucleotide and NAD/NADP, of nicotinic acid to NAD/NADP, or of nicotinic acid to desamido-NAD (Roger *et al.*, 1969; Moscioni *et al.*, 1977). Diazinon depresses the embryo NADH level in parallel with the NAD level so the teratogen does not alter the ratio of oxidized and reduced cofactors (Moscioni *et al.*, 1977). This OP teratogen does not inhibit NAD formation from dipeptides with N- and C-terminal tryptophan residues (Moscioni *et al.*, 1977). It has been suggested that OP teratogenesis may result from alkylation of NAD by the teratogen (Schoental, 1976, 1977), an hypothesis inconsistent with the high structural specificity of the OP compounds and the

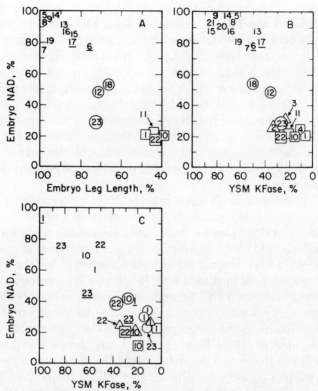

Figure 5.5 Effects of organophosphorus compounds and methylcarbamates on chicken embryo NAD level, leg length, yolk sac membrane kynurenine formamidase activity, and severity of teratogenic signs. The relative potencies of 23 OP and MC compounds injected at 1 mg/egg for lowering embryo NAD levels are compared with their activities for reducing the embryo leg length (A), inhibiting the YSM KFase activity (B), and inducing teratogenesis (A,B). Similar comparisons of NAD level, KFase inhibition, and teratogenic signs are made for two OP and two MC compounds administered at various doses (C). Numbers refer to compound designations in Figure 5.1. Severity of teratogenic signs is indicated as, for example, 1, 1, ⓘ, △, and ☐ for increasing magnitude of effects with emphasis on Type I abnormalities. Eggs were injected at day 4, YSM KFase activities assayed at days 8 and 10 (average results plotted), NAD levels determined at day 12, and leg length and teratogenic signs evaluated at day 19 of incubation. The leg length with malathion (compound 15, Figure 5.1) is for 5 mg/egg. The YSM KFase activity values at two doses of diazinon are extrapolated from a dose–response curve based on five experimental points. All values are relative to appropriate controls. Absolute values for the controls are: NAD, 932 nmol/embryo; leg length, 64 mm; YSM KFase, 20 nmol kynurenine equivalents liberated from N^1,N^α-diformylkynurenine/min per YSM. Values greater than 90% are shown in the region of 90–100%. Data from Proctor et al. (1976), Moscioni et al. (1977), and Seifert and Casida (1978)

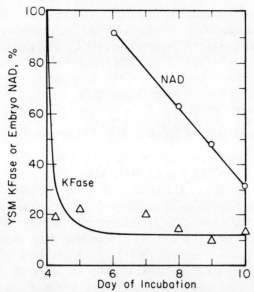

Figure 5.6 Changes of chicken embryo NAD level and yolk sac membrane kynurenine formamidase activity with time after diazinon administration. Diazinon was injected at 50–60 μg/egg at day 4 of incubation. All values are relative to appropriate controls at the same day of incubation. Data from Moscioni et al. (1977) and Seifert and Casida (1978). (Reproduced by permission of Pergamon Press Ltd)

Table 5.1 Effects of diazinon and carbaryl alone or with nicotinamide on NAD levels in whole embryos at day 12 and their tissues at day 19 of incubation

Sample	Day	NAD (%) OP or MC	OP or MC + NAm
Diazinon (30 μg/egg)			
Embryo	12	29	83
Brain	19	60	95
Liver	19	87	95
Muscle	19	65	98
Carbaryl (3 mg/egg)			
Embryo	12	23	97
Brain	19	55	96
Liver	19	73	85
Muscle	19	44	98

Teratogens were injected on day 4 of incubation with or without simultaneous administration of nicotinamide at 0.8 μmol/egg. All values are relative to appropriate controls at the same day of incubation. Data from Proctor et al. (1976) and Moscioni et al. (1977), reproduced from Moscioni et al. (1977) by permission of Pergamon Press Ltd.

identical effects of OP and MC teratogens. It appears more likely that NAD levels are lowered by a block in its biosynthesis. Dicrotophos produces a drastic decrease in conversion of [^{14}C]tryptophan to [^{14}C]NAD paralleling the lowering of the endogenous NAD levels (Moscioni et al., 1977). Thus, it appears that the metabolic block induced by the OP and MC insecticides occurs after tryptophan and before nicotinic acid or NAm (Figure 5.7).

Kynurenine Formamidase Inhibition in Type *I* Teratogenesis

OP and MC compounds are classically inhibitors of serine hydrolases (Aldridge and Reiner, 1972; Kraut, 1977). Many teratogens inhibit esterases of the yolk sac membrane (YSM) and other tissues assayed directly or after electrophoretic separation using α-naphthyl acetate as the substrate (Walker, 1971; Flockhart and Casida, 1972). Insufficient data are available to correlate inhibition of these esterases with Type *I* teratogenesis or to assign them a physiological function related to NAD formation. Thus, the hydrolases of greatest interest are those involved directly or indirectly in the metabolic pathway between tryptophan and nicotinic acid. Attention was therefore focused on enzymes that might perform an important function in this pathway and undergo the relevant OP and MC inhibition. These specifications are met by kynurenine formamidase (KFase; EC. 3.5.1.9; aryl-formylamine amidohydrolase) catalysing the conversion of *N*-formylkynurenine to kynurenine (Figure 5.7).

Properties and distribution of kynurenine formamidase

KFase is widely distributed among fungi, insects, birds, and mammals (Jakoby, 1954; Santti and Hopsu-Havu, 1968; Arndt et al., 1973; Bailey and Wagner, 1974; Tsuda et al., 1974; Moore and Sullivan, 1975; Shinohara and Ishiguro, 1977; Cumming et al., 1979) where it plays a critical role in the formation of NAD derivatives from tryptophan (Dagley and Nicholson, 1970). This enzyme is reported to exist in multiple forms, varying with the tissue and species (Krisch et al., 1975; Moore and Sullivan, 1975; Shinohara and Ishiguro, 1977; Cumming et al., 1979; Seifert and Casida, 1979a).

Studies on avian KFase during embryonic development focused on the YSM enzyme (Seifert and Casida, 1978) since at the teratogen-sensitive stage this tissue substitutes metabolically for the liver and many other non-developed organs (Romanoff, 1960). The YSM KFase activity is very low early in development but it increases markedly on growth of the YSM (Figure 5.8) (Seifert and Casida, 1978). The liver KFase activity becomes important during late embryogenesis and increases greatly on hatching (Bailey and Wagner, 1974; Seifert and Casida, 1978). These studies used (Seifert and Casida, 1978) or probably used (Bailey and Wagner, 1974) N^1,N^2-diformylkynurenine as the substrate, which resulted in

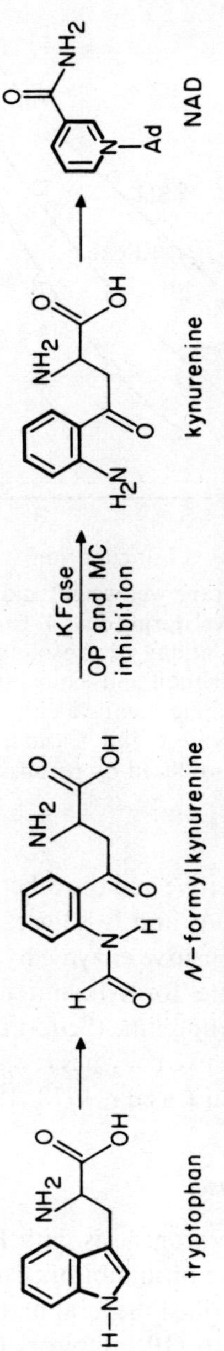

Figure 5.7 Kynurenine formamidase inhibition leads to lowered chicken embryo NAD levels and Type *I* teratogenic signs

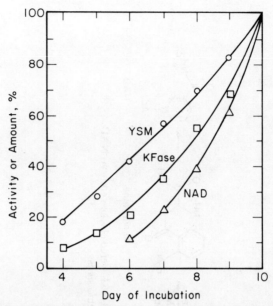

Figure 5.8 Increase of yolk sac membrane wet weight and its kynurenine formamidase activity and of chicken embryo NAD level during days 4–10 of incubation. The activity or amount is arbitrarily adjusted to 100% at day 10. Absolute values at day 10 are: NAD, 425 nmol/embryo (calculated from Romanoff and Romanoff, 1967 and Moscioni et al., 1977); KFase, 29 nmoles kynurenine equivalents liberated from N^1,N^α-diformylkynurenine/min per YSM (Seifert and Casida, 1978); YSM, 1600 mg/egg (Romanoff and Romanoff, 1967). (Reproduced by permission of Pergamon Press Ltd)

hydrolysis by two enzymes, one sensitive and the other insensitive to OP and MC compounds (referred to initially as A- and B-KFases, respectively) (Seifert and Casida, 1978). The OP- and MC-sensitive enzyme hydrolyses both diformylkynurenine and the normal substrate formylkynurenine while the insensitive enzyme hydrolyses only diformylkynurenine (Seifert and Casida, 1978). Critical evaluation of the properties of these two enzymes revealed that the OP- and MC-sensitive form is KFase (Seifert and Casida, 1978, 1979a).

Inhibition of kynurenine formamidase

Any attempted correlation of teratogenesis with KFase inhibition must be made under *in vivo* conditions where proinhibitors can be activated (e.g. Figure 5.3) and unstable compounds detoxified. Several mammalian liver KFases and chicken liver KFase are sensitive to OP inhibitors (Santti and Hopsu-Havu, 1968; Arndt et al., 1973; Bailey and Wagner, 1974; Krisch et al., 1975; Moscioni et al., 1977; Shinohara and Ishiguro, 1977) and carbaryl (Moscioni et al., 1977).

For convenience, enzyme of mouse liver was used to compare the extent of *in vivo* KFase inhibition by compounds of widely varying potency as teratogens. The results showed a surprisingly good correlation between *in vivo* mouse liver KFase inhibition, *in ovo* lowering of the chicken embryo NAD level, and the severity of Type *I* teratogenic signs produced (Figure 5.9). These structure–activity relationships, both *in vitro* and *in vivo*, have now been extended to include more than 90 OP and MC compounds with mouse liver KFase (Moscioni *et al.*, 1977; Eto *et al.*, 1980).

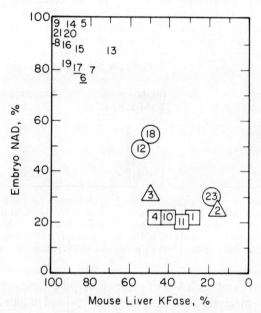

Figure 5.9 Effects of organophosphorus compounds and carbaryl on mouse liver kynurenine formamidase activity in relation to those on chicken embryo NAD level and severity of teratogenic signs. The comparisons are similar to those in Figure 5.5B but replacing YSM KFase with mouse liver KFase. The mouse liver KFase activity values are for 1 and 24 h (average result plotted) after intraperitoneal administration at $1\,\text{mg}\,\text{kg}^{-1}$. Absolute value for the control was 46 μmol kynurenine liberated from N^1,N^α-diformylkynurenine/min per mg protein. Experimental conditions described in Figure 5.5. Data from Moscioni *et al.* (1977)

Chicken YSM KFase is sensitive to *in vitro* inhibition by three potent teratogens, diazoxon, the amide analogue of dicrotophos, and eserine (Seifert and Casida, 1978). *In ovo* studies, although tedious due to the low KFase activity, proved highly informative (Seifert and Casida, 1978). In an investigation with 23 OP and MC compounds, the inhibition of YSM KFase activity was directly correlated with the lowering of the embryo NAD level and the teratogenic signs (Figure 5.5B). A similar correlation was obtained on comparing a series of teratogenic and non-teratogenic doses of four of the most potent teratogens (Figure 5.5C) (Table 5.2). In these studies, the teratogen was injected at the time of maximum sensitivity (day 4), and the YSM KFase activity and embryo NAD level determined at times of critical physiological significance (days 9 and 12, respectively). A dose of compound inhibiting YSM KFase activity by at least 50% was effective in lowering the embryo NAD level by at least 50% and inducing micromelia and abnormal feathering (Table 5.2).

Table 5.2 Correlation of potency of two organophosphorus compounds and two methylcarbamates for inhibiting yolk sac membrane kynurenine formamidase and producing lowered NAD levels and teratogenesis in chicken embryos

Compound	Dose for 50% effect (µg/egg)	
	Inhibition of YSM KFase	Lowering of embryo NAD
Diazinon	4	5
Dicrotophos	34	66
Eserine	31	60
Carbaryl	410	240

The dose of each compound required for 50% effect on YSM KFase activity and embryo NAD level also approximates the threshold level for distinct teratogenic signs. Experimental conditions described in Figure 5.5. Data from Proctor et al. (1976) and Seifert and Casida (1978), reproduced from Siefert and Casida (1978) by permission of Pergamon Press Ltd

The KFase activity of the YSM or other embryonic tissues must be inhibited for several days to maintain a low embryo NAD level and induce teratogenesis. A potent teratogen even at low dose can inhibit KFase and lower the NAD level from shortly after injection until at least day 10 of incubation (Seifert and Casida, 1978) (Figure 5.6). This persistence of inhibition is particularly remarkable when it is considered that during the period between days 4 and 10 of incubation the normal enzyme activity and NAD level increase about 10-fold on YSM and embryo growth (Figure 5.8) (Seifert and Casida, 1978). Two features are probably important for prolonged KFase inhibition and teratogenesis: maintenance of an inhibitor pool to block newly synthesized enzyme; formation of an inhibited enzyme that is not readily reactivated (Seifert and Casida, 1979b).

The inhibition and reactivation characteristics of avian KFase were evaluated with chicken liver enzyme rather than the teratogenically relevant YSM KFase. The YSM enzyme is of relatively low activity and difficult to isolate in large amounts. Chicken liver KFase appears to be identical to YSM KFase by many criteria, including comparative inhibition by OP and MC compounds (Seifert and Casida, 1979a). The kinetics of KFase inhibition by OP and MC compounds (designated by I for the phosphorylating or carbamoylating moiety and X for the leaving group, i.e. IX), examined with purified chicken liver enzyme (Seifert and Casida, 1979b), generally follow those of acetylcholinesterase (AChE) and other esterases (Aldridge and Reiner, 1972).

$$\text{KFase} + \text{IX} \underset{k_{-1}}{\overset{k_1}{\rightleftharpoons}} [\text{KFase} \ldots \text{IX}] \xrightarrow{k_2} \text{KFase-I} + \text{X}$$

$$\underbrace{\phantom{\text{KFase} + \text{IX} \rightleftharpoons [\text{KFase} \ldots \text{IX}] \to}}_{k_i}$$

The *in vitro* inhibition parameters of five OP and one MC compounds do not correlate with their teratogenic potencies, e.g. the most potent *in vitro* KFase inhibitor phenyl saligenin cyclic phosphate (Compound 21, Figure 5.1) and the best phosphorylating agent tetraethyl pyrophosphate (TEPP) are not teratogens (Table 5.3) perhaps due to *in ovo* instability. It is evident, however, that all compounds which produce Type *I* embryonic abnormalities must have a high affinity for KFase and be good phosphorylating or carbamoylating agents for the active site of this enzyme (Seifert and Casida, 1979b). Thus, the dialkoxyphosphorylated and methylcarbamoylated derivatives of chicken liver KFase are relatively resistant to reactivation of enzyme activity, either spontaneously (Figure 5.10) or by pralidoxime (structure given later) or hydroxylamine (Seifert and Casida, 1979b). It is not known to what extent dialkoxyphosphorylated KFase undergoes ageing to a form refractory to reactivation.

Table 5.3 Inhibition parameters for chicken liver kynurenine formamidase incubated with various teratogenic and non-teratogenic organophosphorus compounds and carbaryl at pH 7.4 and 25 °C

Compound	Inhibition parameters		
	k_{-1}/k_1 (M)	k_2 (min^{-1})	k_i (M^{-1} min^{-1})
Teratogens in order of decreasing potency			
Diazoxon	1.5×10^{-4}	21	1.4×10^5
Monocrotophos	5.4×10^{-3}	14	3.3×10^3
Dicrotophos	3.2×10^{-3}	36	1.2×10^3
Carbaryl	8.5×10^{-5}	15	2.6×10^5
Non-teratogenic			
TEPP	6.7×10^{-2}	131	1.9×10^3
Phenyl saligenin cyclic phosphate	8.3×10^{-7}	8	9.0×10^6

Data from Seifert and Casida (1979b), reproduced by permission of Academic Press Inc.

Teratogenic potency cannot be predicted solely on the basis of *in vitro* KFase inhibition and reactivation experiments. *In vivo* studies are required to evaluate

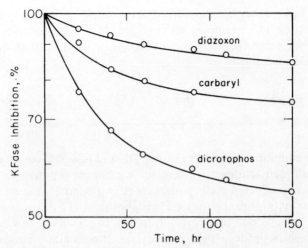

Figure 5.10 Spontaneous reactivation of chicken liver kynurenine formamidase inhibited by three avian teratogens. Recovery of enzyme at pH 7.4 was followed at 25 °C. (Data from Seifert and Casida (1979b), reproduced by permission of Academic Press, Inc.)

the balance of rates of metabolic activation and detoxification which in turn determine the ability to maintain a critical level of active teratogen for a high degree of KFase inhibition throughout the sensitive stage of embryonic development.

Cholinergic System in Type *II* Teratogenesis

The acute toxicity of OP and MC insecticides is usually attributable to AChE inhibition resulting in acetylcholine (ACh) accumulation and associated disruptions in the cholinergic nervous system (Koelle, 1963; Eto, 1974; Kuhr and Dorough, 1976). Three types of evidence indicate possible cholinergic involvement in Type *II* teratogenesis.

First, embryo ACh levels and AChE activity are altered by some teratogenic OP and MC compounds. Teratogenic doses of dicrotophos elevate embryo ACh levels (Upshall *et al.*, 1968). Dicrotophos and eserine minimize or prevent transport of unhydrolysed ACh from the yolk to the embryo (Upshall *et al.*, 1968). Embryo AChE is strongly inhibited by dicrotophos, parathion, and many other OP teratogens (Khera *et al.*, 1966; Upshall *et al.*, 1968; Greenberg and LaHam, 1970; Walker, 1971; Meiniel, 1977a) and by eserine (Upshall *et al.*, 1968). There is a relatively good correlation between *in ovo* AChE inhibition and vertebral length with the marked exception that although EPN (compound 8, Figure 5.1) is a strong inhibitor it does not alter the crown–rump length (Upshall *et al.*, 1968) (Figure 5.11). It appears possible that, as a phenylphosphonate

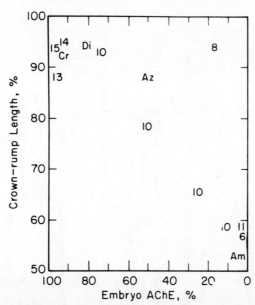

Figure 5.11 Effects of organophosphorus compounds on chicken embryo vertebral length and acetylcholinesterase activity. The relative potencies of 11 OP compounds injected at 0.03–1 mg/egg are compared for lowering embryo AChE activity and reducing the crown–rump length. Eggs were injected on day 4, AChE assayed on day 12, and crown–rump length determined on day 21 of incubation. Numbers refer to compound designations in Figure 5.1. Additional compounds are: Am, amide analogue of dicrotophos; Az, azinophosmethyl or O,O-dimethyl S-(4-oxo-1,2,3-benzotriazin-3(4H)-ylmethyl) phosphorodithioate; Cr, a crotonate analogue of mevinphos with each of the methoxy groups replaced by an ethoxy group; Di, dioxathion or S,S'-p-dioxane-2,3-diyl O,O-diethyl phosphorodithioate. Absolute values for controls are: crown–rump length, 85 mm, AChE, 10 µmol ACh hydrolysed per min per embryo. Values of greater than 90% are shown in the region of 90–100%. Data from Upshall et al. (1968)

derivative, EPN may not yield the prolonged *in ovo* AChE inhibition probably required for developmental anomalies.

Second, certain pyridinium aldoximes shown below alleviate the Type *II* teratogenic signs due to parathion, dicrotophos, and eserine without altering the Type *I* teratogenic signs in the latter two cases (Meiniel, 1974, 1975, 1976a, b, c; Landauer, 1977; Moscioni et al., 1977). Diacetylmonooxime and monoisonitrosoacetone are ineffective (Meiniel, 1976a). The vertebral defects produced by

Table 5.4 Mammalian teratogenesis induced by organophosphorus compounds and carbaryl

Compound	Species, dose and route of administration	Teratogenic signs[a,b]	References
Diazinon	Miniature swine, 5 mg kg^{-1} per day, oral	Skull and limb malformations (6%)	Earl et al. (1973)
Dimethoate	Cat, 12 mg kg^{-1} per day	Polydactyly	Khera (1979)
Chlorfenvinphos	Rat, oral	Disturbances in ossification	Tos-Luty et al. (1972)
Phosmet	Rat, 30 mg kg^{-1}, oral	Hypognathia, hydrocephaly (60%)	Kagan (1977); Martson and Voronina (1976)
Trichlorfon	Rat, 80 mg kg^{-1}, oral	Oedema, hydrocephaly	Martson and Voronina (1976); Kagan (1977)
	Rat, 432 mg kg^{-1} per day, oral	Malformations of the skull and other organs (30%)	Staples et al. (1976)
	Rat, 0.005–9 mg m^{-3}, inhalation	Skeletal defects	Gofmekler and Tabakova (1970)
Carbaryl	Mouse, 100 mg kg^{-1} per day, subcutaneous	Microphthalmia	Courtney et al. (1970); Durham and Williams (1972)
	Guinea pig, 300 mg kg^{-1}, oral	Vertebral and other malformations (10%)	Robens (1969)
	Rabbit, 200 mg kg^{-1} per day, oral	Omphalocele (7%)	Murray et al. (1979)
	Dog, 13 mg kg^{-1} per day, oral	Abdominal-thoracic fissures, brachygnathia and other skeletal malformations (18%)	Smalley et al. (1968)
	Miniature swine, 16 mg kg^{-1} per day oral	Limb malformations (4%)	Earl et al. (1973)

[a] Mild to severe embryonic abnormalities have been noted on single dose, intraperitoneal administration in the following cases: methyl parathion at 60 mg kg^{-1} in mice (Tanimura et al., 1967); diazinon at 100–200 mg kg^{-1}, parathion at 3.5 mg kg^{-1} and dichlorvos at 15 mg kg^{-1} in rats (Kimbrough and Gaines, 1968); demeton [mixture of S- (and O-)ethylthioethyl O,O-diethyl phosphorothioates] at 7–10 mg kg^{-1} and fenthion [O-(4-methylthio-m-tolyl) O,O-dimethyl phosphorothioate] at 40–80 mg kg^{-1} in mice (Budreau and Singh, 1973).

[b] No embryonic abnormalities were observed under the test conditions with: diazinon in hamsters and rabbits (Robens, 1969) and dogs (Earl et al., 1973); parathion in rats (Fish, 1966); methyl parathion in rats (Fish, 1966; Tanimura et al., 1967); Fuchs et al., 1976); dichlorvos in rats and rabbits (Thorpe et al., 1972); malathion in rats (Kalow and Marton, 1961; Kimbrough and Gaines, 1968); coumaphos in cattle (Bellows et al., 1975); phosmet in rats (Staples et al., 1976); trichlorphon in rats (Tsaregorodtseva and Talanov, 1973); DFP (O,O-di-isopropyl fluorophosphonate) in rats (Fish, 1966); amiphos [S-(2-acetylaminoethyl) O,O-dimethyl phosphorodithioate] in mice (Hashimoto et al., 1972); formothion [S-(N-formyl-N-methylcarbamoylmethyl) O,O-dimethyl phosphorodithioate] and thiometon (S-ethylthioethyl O,O-dimethyl phosphorodithioate) in rabbits (Klotzsche, 1970); crufomate (4-tert-butyl-2-chlorophenyl methyl methylphosphoramidate) in rats (Rumsey et al., 1969) and cattle (Rumsey et al., 1974; Bellows et al., 1975); carbaryl in mice (Murray et al., 1979), rats (Tsaregorodtseva and Talanov, 1973; Weil et al., 1972, 1973), hamsters (Robens, 1969), guinea pigs (Weil et al., 1973), rabbits (Robens, 1969), and monkeys (Dougherty et al., 1971).

two carbamate cholinesterase inhibitors, carbachol and neostigmine, are also alleviated by pralidoxime (Landauer, 1977). The active alleviating agents are antidotes for acute OP poisoning by virtue of their ability to dephosphorylate inhibited AChE (Aldridge and Reiner, 1972). A similar action of these oximes in the avian embryo would implicate disruption of the cholinergic system in Type *II* teratogenesis.

Finally, several cholinergic effectors (decamethonium, hexamethonium, gallamine, succinylcholine, tetramethylammonium, and trimethylphenylammonium) produce vertebral defects (Upshall *et al.*, 1968; Roger *et al.*, 1969; Landauer, 1975a, b; Meiniel, 1978) similar to those of Type *II* OP teratogens.

MAMMALIAN TERATOGENESIS

All commercial OP and MC insecticides have been examined to various degrees for possible teratogenic effects in mammals. Some of these data appear in journal articles while the remainder are in the files of industrial laboratories or governmental regulatory agencies. Table 5.4 presents a sample of the published information. OP and MC teratogenesis is rarely encountered in studies with mammals. Variations or discrepancies in the results may be due in part to differences in chemical purity, species, and test methods.

No information is available on the mechanisms of OP- and MC-induced teratogenesis in mammals. The low incidence of abnormalities is a deterrent to studies of this type. Findings from avian systems have therefore not been fully evaluated in mammals. It is of interest, however, that the NAm antagonist 6-AN is a potent teratogen in mammals (Chamberlain and Goldyne, 1970) and that carbaryl lowers the oxidized pyridine nucleotide levels of tissues in orally treated rats (Kuzminskaya and Pavlova, 1970).

In contrast to the avian embryo, the mammalian embryo is placental and therefore has the protection afforded by maternal absorption, detoxification, and excretion of foreign compounds. In early pregnancy embryo NAD originates predominantly from maternal sources of nicotinic acid and other precursors. Teratological manifestations of KFase inhibition are therefore restricted to non-placental species or unusual circumstances in mammals where reduced enzyme activity significantly impairs maintenance of normal levels of NAD or other essential biochemicals derived from kynurenine. Cholinergic block is unlikely in the embryo at OP and MC doses without serious effect on the mother.

CONCLUSIONS

Avian teratogenesis is evident with many OP and MC insecticides administered by yolk injection. Potent OP teratogens include diazinon, etrimfos, pirimiphos-ethyl, dicrotophos, monocrotophos, and parathion. Carbaryl is a potent MC teratogen. OP- and MC-induced avian teratogenesis includes at least two types of developmental anomalies. Micromelia and abnormal feathering are associated with a lowered embryo NAD level due to a biosynthetic block resulting from KFase inhibition. These embryonic abnormalities, designated as Type *I* teratogenic signs, are alleviated by NAm and other NAD precursors circumventing the primary metabolic block. Vertebral defects are attributable to AChE inhibition and associated disruptions of the cholinergic system. These malformations, the Type *II* teratogenic signs, are alleviated by pralidoxime. Many OP and MC compounds inhibit both AChE and KFase, giving various combinations of Types *I* and *II* teratogenic signs.

Mammalian teratogenesis is rare with OP and MC insecticides, at least in part because the placental character of the mammalian embryo allows protection by maternal defence systems. There is no evidence for any relationship between OP and MC teratogenesis in mammals and disruption of the NAD or cholinergic system.

ACKNOWLEDGEMENTS

The portion of research considered from the Pesticide Chemistry and Toxicology Laboratory at Berkeley was supported by the National Institute of Environmental Health Sciences Program Project Grant PO1 ES00049.

REFERENCES

Agarwal, I. P. (1956). 'Morphogenetic effects of eserine sulphate. 1. The skeletal abnormalities', *J. Anim. Morph. Physiol.* **3**, 63–74.

Aldridge, W. N., and Reiner, E. (1972). *Enzyme Inhibitors as Substrates,* American Elsevier Publishing Company, Inc., New York.

Ancel, P. (1945). 'L'achondroplasie. Sa réalisation Expérimentale—Sa pathogénie', *Annales d'Endocrinologie* **6**, 1–24.

Arndt, R., Junge, W., Michelssen, K., and Krisch, K. (1973). 'Isolation and molecular properties of formamidase from rat liver cytoplasm', *Hoppe-Seyler's Z. Physiol. Chem.* **354**, 1583–1590.

Arsenault, A. L., Gibson, M. A., and Mader, M. E. (1975). 'Hypoglycemia in malathion-treated chick embryos', *Can. J. Zool.* **53**, 1055–1057.

Autissier-Navarro, C., and Meiniel, R. (1977). 'Action du dicrotophos sur le développement *in vitro* du tibia de l'embryon de Poulet', *C. R. Soc. Biol.* **171**, 1235–1239.

Bailey, C. B. and Wagner, C. (1974). 'Kynurenine formamidase. Purification and characterization of the adult chicken liver enzyme and immunochemical analyses of the enzyme of developing chicks', *J. Biol. Chem.* **249**, 4439–4444.

Bellows, R. A., Rumsey, T. S., Kasson, C. W., Bond, J., Warwick, E. J., and Pahnish, O. F. (1975). 'Effects of organic phosphate systemic insecticides on bovine embryonic survival and development', *Amer. J. Vet. Res.* **36**, 1133–1140.

Belyaev, V. I. (1974). 'Effect of organophosphorus compounds introduced into the yolk sac on the development of chick embryos', *Khim. Sel'sk. Khoz.* **12**, 623–626; from *Chem. Abstr.* **82**: 52380m (1975).

Budreau, C. H., and Singh, R. P. (1973). 'Teratogenicity and embryotoxicity of demeton and fenthion in CF#1 mouse embryos', *Toxicol. Appl. Pharmacol.* **24**, 324–332.

Chamberlain, J. G., and Goldyne, M. E. (1970). 'Intra-amniotic injection of pyridine nucleotides or adenosine triphosphate as countertherapy for 6-aminonicotinamide (6-AN) teratogenesis', *Teratology* **3**, 11–16.

Clegg, D. J. (1964). 'The hen egg in toxicity and teratogenicity studies', *Food Cosmet. Toxicol.* **2**, 717–727.

Courtney, K. D., Gaylor, D. W., Hogan, M. D., Falk, H. L., Bates, R. R., and Mitchell, I. A. (1970). 'Teratogenic evaluation of pesticides: a large-scale screening study', *Teratology* **3**, 199.

Cumming, R. B., Walton, M. F., Fuscoe, J. C., Taylor, B. A., Womack, J. E., and Gaertner, F. H. (1979). 'Genetics of formamidase-5 (brain formamidase) in the mouse: localization of the structural gene on chromosome 14', *Biochem. Genet.* **17**, 415–431.

Dagley, S., and Nicholson, D. E. (1970). *An Introduction to Metabolic Pathways*, Wiley, New York.

Dougherty, W. J., Golberg, L., and Coulston, F. (1971). 'The effect of carbaryl on reproduction in the monkey (*Macacca mulatta*)', *Toxicol. Appl. Pharmacol.* **19**, 365.

Durham, W. F., and Williams, C. H. (1972). 'Mutagenic, teratogenic, and carcinogenic properties of pesticides', *Ann. Rev. Entomol.* **17**, 123–148.

Earl, F. L., Miller, E., and Van Loon, E. J. (1973). 'Reproductive, teratogenic, and neonatal effects of some pesticides and related compounds in beagle dogs and miniature swine', in *Pesticides and The Environment: A Continuing Controversy* (Ed. W. B. Deichmann), pp. 253–266, Intercontinental Medical Book Corp., New York.

Eto, M. (1974). *Organophosphorus Pesticides: Organic and Biological Chemistry*, CRC Press, Inc., Cleveland, Ohio.

Eto, M., Seifert, J., Engel, J. L., and Casida, J. E. (1980). 'Organophosphorus and methylcarbamate teratogens: structural requirements for inducing embryonic abnormalities in chickens and kynurenine formamidase inhibition in mouse liver', *Toxicol. Appl. Pharmacol.* **54**, 20–30.

Fish, S. A. (1966). 'Organophosphorus cholinesterase inhibitors and fetal development', *Amer. J. Obst. Gynec.* **96**, 1148–1154.

Flockhart, I. R., and Casida, J. E. (1972). 'Relationship of the acylation of membrane esterases and proteins to the teratogenic action of organophosphorus insecticides and eserine in developing hen eggs', *Biochem. Pharmacol.* **21**, 2591–2603.

Fuchs, V., Golbs, S., Kuehnert, M., and Osswald, F. (1976). 'Studies on the prenatal toxic action of parathion-methyl on Wistar rats in comparison to cyclophosphamide and trypan blue', *Arch. Exp. Veterinaermed.* **30**, 343–350, from *Chem. Abstr.* **86**: 819z (1977).

Gebhardt, D. O. E. (1972). 'The use of the chick embryo in applied teratology', *Adv. Teratol.* **5**, 97–111.

Ghadiri, M., Greenwood, D. A., and Binns, W. (1967). 'Feeding of malathion and carbaryl to laying hens and roosters', *Toxicol. Appl. Pharmacol.* **10**, 392.

Gill, G. R., and LaHam, Q. N. (1972). 'Histochemical and radiographic investigations of malathion-induced malformations in embryonic chick limbs', *Can. J. Zool.* **50**, 349–351.

Gofmekler, V. A., and Tabakova, S. A. (1970). 'Action of chlorophos on the embryogenesis of rats', *Farmakol. Toksikol.* **33**, 735–737; from *Chem. Abstr.* **74**: 75563x (1971).

Greenberg, J., and LaHam, Q. N. (1969). 'Malathion-induced teratisms in the developing chick', *Can. J. Zool.* **47**, 539–542.

Greenberg, J., and LaHam, Q. N. (1970). 'Reversal of malathion-induced teratisms and its biochemical implications in the developing chick', *Can. J. Zool.* **48**, 1047–1053.

Hamburger, V., and Hamilton, H. L. (1951). 'A series of normal stages in the development of the chick embryo', *J. Morphol.* **88**, 49–92.

Hashimoto, Y., Makita, T., and Noguchi, T. (1972). 'Teratogenic studies of *O,O*-dimethyl S-(2-acetylaminoethyl) dithiophosphate (DAEP) in ICR-strain mice', *Oyo Yakuri* **6**, 621–626, from *Chem. Abstr.* **78**: 38962p (1973).
Ho, M., and Gibson, M. A. (1972). 'A histochemical study of the developing tibiotarsus in malathion-treated chick embryos', *Can. J. Zool.* **50**, 1293–1298.
Jakoby, W. B. (1954). 'Kynurenine formamidase from *Neurospora*', *J. Biol. Chem.* **207**, 657–663.
Kagan, Y. S. (1977). *Toksikologia Fosfororganitscheskikh Pestitsidov*, Izdavatelstvo Meditsina, Moscow, USSR.
Kalow, W., and Marton, A. (1961). 'Second-generation toxicity of malathion in rats', *Nature* **192**, 464–465.
Khera, K. S., (1979). 'Evaluation of dimethoate (Cygon 4E) for teratogenic activity in the cat', *J. Environ. Pathol. Toxicol.* **2**, 1283–1288.
Khera, K. S., and Bedok, S. (1967). 'Effects of thiol phosphates on notochordal and vertebral morphogenesis in chick and duck embryos', *Food Cosmet. Toxicol* **5**, 359–365.
Khera, K. S., and Clegg, D. J. (1969). 'Perinatal toxicity of pesticides', *Can. Med. Ass. J.* **100**, 167–172.
Khera, K. S., LaHam, Q. N., Ellis, C. F. G., Zawidzka, Z. Z., and Grice, H. C. (1966). 'Foot deformity in ducks from injection of EPN during embryogenesis', *Toxicol. Appl. Pharmacol.* **8**, 540–549.
Khera, K. S., LaHam, Q. N., and Grice, H. C. (1965). 'Toxic effects in ducklings hatched from embryos inoculated with EPN or Systox., *Food Cosmet. Toxicol.* **3**, 581–586.
Kimbrough, R. D., and Gaines, T. B. (1968). 'Effect of organic phosphorus compounds and alkylating agents on the rat fetus', *Arch. Environ. Health* **16**, 805–808.
Klotzsche, C. (1970). 'Teratologic and embryotoxic investigations with formothion and thiometon, *Pharm. Acta Helv.* **45**, 434–440; from *Chem. Abstr.* **73**: 65402j (1970).
Koelle, G. B. (Ed.) (1963). *Cholinesterases and Anticholinesterase Agents, Handbuch der Experimentellen Pharmakologie*, Vol. 15, Springer-Verlag, Berlin.
Kraut, J. (1977). 'Serine proteases: structure and mechanism of catalysis', *Ann. Rev. Biochem.* **46**, 331–358.
Krisch, K., Arndt, R., Junge, W., Menge, U., and Michelssen, K. (1975). 'Purification and some properties of formamidase from rat and pig liver', *Acta Vitamin. Enzymol.* **29**, 302–306.
Kuhr, R. J., and Dorough, H. W. (1976). *Carbamate Insecticides*, CRC Press, Inc., Cleveland, Ohio.
Kuul, A. K. (1975). 'Effect of phthalophos on chicken embryos', *Veterinariya (Moscow)*, 104–106; from *Chem. Abstr.* **84**: 1000w (1976).
Kuzminskaya, U. A., and Pavlova, I. I. (1970). 'Effect of DDT and Sevin on the pyridine nucleotide level in animal tissues', *Gig. Primen. Toksikol. Pestits. Klin. Otravlenii* **8**, 110–115; from *Chem. Abstr.* **77**: 135966e (1972).
Laley, B. O., and Gibson, M. A. (1977). 'Association of hypoglycemia and pancreatic islet tissue with micromelia in malathion-treated chick embryos', *Can. J. Zool.* **55**, 261–264.
Landauer, W. (1949). 'Le problème de l'électivité dans les expériences de tératogenèse biochimique', *Arch. Anat. Microsc. Morphol. Exp.* **38**, 184–189.
Landauer, W. (1957). 'Niacin antagonists and chick development', *J. Exp. Zool.* **136**, 509–530.
Landauer, W. (1975a). 'Cholinomimetic teratogens: studies with chicken embryos', *Teratology* **12**, 125–140.
Landauer, W. (1975b). 'Cholinomimetic teratogens. II. Interaction with inorganic ions', *Teratology* **12**, 271–276.
Landauer, W. (1977). 'Cholinomimetic teratogens. V. The effect of oximes and related cholinesterase reactivators', *Teratology* **15**, 33–42.

Landauer, W., and Clark, E. M. (1962). 'The interaction in teratogenic activity of the two niacin analogs 3-acetylpyridine and 6-aminonicotinamide', *J. Exp. Zool.* **151**, 253–258.

Landauer, W., and Salam, N. (1972). 'Aspects of dimethyl sulfoxide as solvent for teratogens', *Develop. Biol.* **28**, 35–46.

Lutz-Ostertag, Y., Meiniel, R., and Lutz, H. (1969). 'Action du Parathion sur le développement de l'embryon de Caille', *C. R. Acad. Sci. Paris, Ser. D* **286**, 2911–2913.

Marliac, J.-P., and Mutchler, M. K. (1963). 'Use of the chick embryo technique for detecting potentiating effects of chemicals', *Fed. Proc.* **22**, 188.

Martson, L. V., and Voronina, V. M. (1976). 'Experimental study of the effect of a series of phosphoroorganic pesticides (Dipterex and Imidan) on embryogenesis', *Environ. Health Perspect.* **13**, 121–125.

McLachlan, J., Bateman, M., and Wolpert, L. (1976). 'Effect of 3-acetylpyridine on tissue differentiation of the embryonic chick limb', *Nature* **264**, 267–269.

McLaughlin, J., Marliac, J.-P., Verrett, M. J., Mutchler, M. K., and Fitzhugh, O. G. (1963). 'The injection of chemicals into the yolk sac of fertile eggs prior to incubation as a toxicity test', *Toxicol. Appl. Pharmacol.* **5**, 760–771.

Meiniel, R. (1973). 'Malformations squelettiques axiales chez les embryons de Poulet et de Caille issus d'oeufs traités par le parathion à différents stades de l'incubation', *C. R. Soc. Biol.* **167**, 459–462.

Meiniel, R. (1974). 'Action protectrice de la pralidoxime vis-à-vis des effets tératogènes du parathion sur le squelette axial de l'embryon de Caille', *C. R. Acad. Sci. Paris, Ser. D* **279**, 603–606.

Meiniel, R. (1975). Pralidoxime, specific antiteratogen compound for bidrin-induced axial deformities in quail embryos', *C. R. Hebd. Séances Acad. Sci. Ser. D* **280**, 1019–1022; from *Chem. Abstr.* **82**: 165611f (1975).

Meiniel, R. (1976a). 'Prévention des anomalies induites par deux insecticides organophosphorés (parathion et bidrin) chez l'embryon de Caille', *Arch. Anat. Mircosc. Morphol. Exp.* **65**, 1–15.

Meiniel, R. (1976b). 'Plurality in the determinism of organophosphorus teratogenic effects', *Experientia* **32**, 920–922.

Meiniel, R. (1976c). 'Expression of parathion axial teratogenesis after giving various compounds known to have antiteratogenic or antitoxic action in the adult or embryo of vertebrates after exposure to organic phosphates. Study on quail embryo (*Coturnix coturnix Japonica*)', *C. R. Hebd. Séances Acad. Sci. Ser. D* **283**, 1085–1087; from *Chem. Abstr.* **86**: 38339f (1977).

Meiniel, R. (1977a). 'Cholinesterase activities and expression of axial teratogenesis in the quail embryo exposed to organophosphates', *C. R. Hebd. Séances Acad. Sci. Ser. D* **285**, 401–404; from *Chem. Abstr.* **87**: 178684k (1977).

Meiniel, R. (1977b). 'Teratogenesis of axial abnormalities induced by an organic phosphorus insecticide (parathion) in the bird embryo', *Wilhelm Roux's Arch. Dev. Biol.* **181**, 41–63.

Meiniel, R. (1978). 'Neuroactive compounds and vertebral teratogenesis in the bird embryo', *Experientia* **34**, 394–396.

Meiniel, R., Lutz-Ostertag, Y., and Lutz, H. (1970). 'Effets tératogènes du parathion (insecticide organo-phosphoré) sur le squelette embryonnaire de la Caille Japonaise (*Coturnix coturnix Japonica*)', *Arch. Anat. Microsc.* **59**, 167–183.

Moore, G. P., and Sullivan, D. T. (1975). 'The characterization of multiple forms of kynurenine formamidase in *Drosophila melanogaster*', *Biochim. Biophys. Acta* **397**, 468–477.

Moscioni, A. D., Engel, J. L., and Casida, J. E. (1977). 'Kynurenine formamidase inhibition as a possible mechanism for certain teratogenic effects of organophosphorus and methylcarbamate insecticides in chicken embryos', *Biochem. Pharmacol.* **26**, 2251–2258.

Murray, F. J., Staples, R. E., and Schwetz, B. A. (1979). 'Teratogenic potential of carbaryl given to rabbits and mice by gavage or by dietary inclusion', *Toxicol. Appl. Pharmacol.* **51**, 81–89.

Overman, D. O., Seegmiller, R. E., and Runner, M. N. (1972). 'Coenzyme competition and precursor specificity during teratogenesis induced by 6-aminonicotinamide', *Develop. Biol.* **28**, 573–582.

Paul, B. S., and Vadlamudi, V. P. (1976). 'Teratogenic studies of fenitrothion on white leghorn chick embryos', *Bull. Environ. Cont. Toxicol.* **15**, 223–229.

Proctor, N. H., and Casida, J. E. (1975). 'Organophosphorus and methylcarbamate insecticide teratogenesis: diminished NAD in chicken embryos', *Science* **190**, 580–582.

Proctor, N. H., Moscioni, A. D., and Casida, J. E. (1976). 'Chicken embryo NAD levels lowered by teratogenic organophosphorus and methylcarbamate insecticides', *Biochem. Pharmacol.* **25**, 757–762.

Richert, E. P., and Prahlad, K. V. (1972). 'Effect of the organophosphate O,O-diethyl S-[(ethylthio)methyl] phosphorodithioate on the chick', *Poult. Sci.* **51**, 613–619.

Robens, J. F. (1969). 'Teratologic studies of carbaryl, diazinon, norea, disulfiram, and thiram in small laboratory animals', *Toxicol. Appl. Pharmacol.* **15**, 152–163.

Roger, J.-C., Chambers, H., and Casida, J. E. (1964). 'Nicotinic acid analogs: effects on response of chick embryos and hens to organophosphate toxicants', *Science* **144**, 539–540.

Roger, J.-C., Upshall, D. G., and Casida, J. E. (1969). 'Structure-activity and metabolism studies on organophosphate teratogens and their alleviating agents in developing hen eggs with special emphasis on bidrin', *Biochem. Pharmacol.* **18**, 373–392.

Romanoff, A. L. (1960). *The Avian Embryo*, Macmillan, New York.

Romanoff, A. L., and Romanoff, A. J. (1967). *Biochemistry of the Avian Embryo. A Quantitative Analysis of Prenatal Development*, Interscience Publ., Wiley, New York.

Rosenberg, M. J., and Caplan, A. I. (1974). 'Nicotinamide adenine dinucleotide levels in cells of developing chick limbs: possible control of muscle and cartilage development', *Develop. Biol.* **38**, 157–164.

Rumsey, T. S., Cabell, C. A., and Bond, J. (1969). 'Effect of an organic phosphorus systemic insecticide on reproductive performance in rats', *Amer. J. Vet. Res.* **30**, 2209–2214.

Rumsey, T. S., Samuelson, G., Bond, J., and Daniels, F. L. (1974). 'Teratogenicity to 35-day fetuses, excretion patterns and placental transfer in beef heifers administered 4-*tert*-butyl-2-chlorophenyl methyl methylphosphoroamidate (Ruelene®)', *J. Animal Sci.* **39**, 386–391.

Santti, R. S., and Hopsu-Havu, V. K. (1968). 'Formamidase in guinea pig liver, I: purification and characterization', *Hoppe-Seyler's Z. Physiol. Chem.* **349**, 753–766.

Schoental, R. (1976). 'Alkylation of coenzymes and the acute effects of alkylating hepatotoxins', *FEBS Lett.* **61**, 111–114.

Schoental, R. (1977). 'Depletion of coenzymes at the site of rapidly growing tissues due to alkylation: the biochemical basis of the teratogenic effects of alkylating agents, including organophosphorus and certain other compounds', *Biochem. Soc. Trans.* **5**, 1016–1017.

Schom, C. B., and Abbott, U. K. (1977). 'Temporal, morphological, and genetic responses of avian embryos to azodrin, an organophosphate insecticide', *Teratology* **15**, 81–87.

Schom, C. B., Abbott, U. K., and Walker, N. E. (1979). 'Adult and embryo responses to organophosphate pesticide: Azodrin', *Poultry Sci.* **58**, 60–66.

Seegmiller, R. E., and Runner, M. N. (1974). 'Normal incorporation rates for precursors of collagen and mucopolysaccharide during expression of micromelia induced by 6-aminonicotinamide', *J. Embryol. Exp. Morph.* **31**, 305–312.

Seifert, J., and Casida, J. E. (1978). 'Relation of yolk sac membrane kynurenine formamidase inhibition to certain teratogenic effects of organophosphorus insecticides and of carbaryl and eserine in chicken embryos', *Biochem. Pharmacol.* **27**, 2611–2615.

Seifert, J., and Casida, J. E. (1979a). 'Multiple forms of chicken kynurenine formamidase', *Comp. Biochem. Physiol.* **63C**, 123–127.

Seifert, J., and Casida, J. E. (1979b). 'Inhibition and reactivation of chicken kynurenine formamidase: *in vitro* studies with organophosphates, *N*-alkyl carbamates, and phenylmethanesulfonyl fluoride', *Pestic. Biochem. Physiol.* **12**, 273–279.

Shinohara, R., and Ishiguro, I. (1977). 'New formamidase having substrate specificity for *o*-formylaminoacetophenone in pig liver', *Biochim. biophys. Acta* **483**, 409–415.

Smalley, H. E., Curtis, J. M., and Earl, F. L. (1968). 'Teratogenic action of carbaryl in beagle dogs', *Toxicol. Appl. Pharmacol.* **13**, 392–403.

Staples, R. E., Kellam, R. G., and Haseman, J. K. (1976). 'Developmental toxicity in the rat after ingestion or gavage of organophosphate pesticides (Dipterex, Imidan) during pregnancy', *Environ. Health Perspect.* **13**, 133–140.

Tanimura, T., Katsuya, T., and Nishimura, H. (1967). 'Embryotoxicity of acute exposure to methyl parathion in rats and mice', *Arch. Environ. Health* **15**, 609–613.

Thorpe, E., Wilson, A. B., Dix, K. M., and Blair, D. (1972). 'Teratological studies with dichlorvos vapour in rabbits and rats', *Arch. Toxicol.* **30**, 29–38.

Tos-Luty, S., Latuszynska, J., Przylepa, E., and Szukiewicz, Z. (1972). 'Effect of chlorfenvinphos on embryonal development of Wistar rats', *Bromatol. Chem. Toksykol.* **5**, 331–338; from *Chem. Abstr.* **78**: 24959f (1973).

Tsaregorodtseva, G. N., and Talanov, G. A. (1973). 'Embryotoxic and teratogenic effect of chlorophos, TCM-3, Sevin and dicresyl on white rats', *Tr. Vses. Nauchno-Issled. Inst. Vet. Sanit.* **47**, 150–155; from *Chem. Abstr.* **84**: 26595g (1976).

Tsuda. H., Noguchi, T., and Kido, R. (1974). 'Formamidase in rat brain', *J. Neurochem.* **22**, 679–683.

Upshall, D. G., Roger, J.-C., and Casida, J. E. (1968). 'Biochemical studies on the teratogenic action of bidrin and other neuroactive agents in developing hen eggs', *Biochem. Pharmacol.* **17**, 1529–1542.

Walker, N. E. (1968). Use of yolk-chemical mixtures to replace hen egg yolk in toxicity and teratogenicity studies', *Toxicol. Appl. Pharmacol.* **12**, 94–104.

Walker, N. E. (1971). 'The effect of malathion and malaoxon on esterases and gross development of the chick embryo', *Toxicol. Appl. Pharmacol.* **19**, 590–601.

Weil, C. S., Woodside, M. D., Bernard, J. B., Condra, N. I., King, J. M., and Carpenter, C. P. (1973). 'Comparative effect of carbaryl on rat reproduction and guinea pig teratology when fed either in the diet or by stomach intubation', *Toxicol. Appl. Pharmacol.* **26**, 621–638.

Weil, C. S., Woodside, M. D., Carpenter, C. P., and Smyth, H. F., Jr. (1972). 'Current status of tests of carbaryl for reproductive and teratogenic effect', *Toxicol. Appl. Pharmacol.* **21**, 390–404.

Yamada, A. (1972). 'Teratogenic effects of organophosphorus insecticides in the chick embryo', *Osaka Shiritsu Daigaku Igaku Zasshi,* **21**, 345–355; from *Chem. Abstr.* **79**: 62350d (1973).

CHAPTER 6

The correlation between *in vivo* and *in vitro* metabolism of pesticides in vertebrates

C. H. Walker

INTRODUCTION	247
METHODS OF STUDYING METABOLISM *IN VIVO*	248
The Direct Assessment of Excreted Metabolites	249
Excretory routes	249
Excretion in urine	250
Collection of faeces or droppings	250
Excretion of metabolites in bile	251
Excretion by other routes	251
The Indirect Assessment of *in vivo* Metabolism	252
The measurement of biological half-life and clearance	252
LD_{50}	253
Differences in pharmacological action	254
METHODS OF STUDYING METABOLISM *IN VITRO*	254
Organ perfusion	254
Tissue Explants	255
Tissue Slices	255
Cell Suspensions and Cultures	255
Subcellular Fractions	256
Enzyme Preparations	257
CORRELATIONS BETWEEN *IN VIVO* AND *IN VITRO* METABOLISM	258
Qualitative Comparisons	258
Quantitative Comparisons	265
Comparisons involving the direct determination of metabolites	265
Comparisons involving pharmacokinetics of substrate	268
Correlations of *in vitro* activity with toxicity	276
The Use of *in vitro* Data to Determine the Phylogenetic Distribution of Enzymes	277
'A' esterases	277
Hepatic microsomal monooxygenase	279
CONCLUSIONS	280
REFERENCES	282

INTRODUCTION

As with many other fields of biochemical research, the metabolism of xenobiotics has been studied both in the living animal and *in vitro*. In this way it has sometimes been possible to build up a more complete picture of metabolic processes than could have been achieved by either of these approaches pursued alone. Gross metabolic changes, and the factors controlling them, may be studied in the whole animal, but individual processes and the operation of particular enzymes are better investigated under the closely controlled conditions which can be maintained in test tube experiments.

Whilst a combination of *in vivo* and *in vitro* techniques represents an ideal approach to the study of xenobiotic metabolism, the extent to which it can be employed is limited. Because of the cost, and of the difficulty and/or undesirability of working with living animals, *in vivo* studies are necessarily limited in number and in scope. *In vitro* studies, on the other hand, tend to be inexpensive, relatively rapid, and easy to use for a wide range of animals, including many species which it is not feasible to investigate *in vivo*. Not surprisingly, interest has grown in recent years with regard to the possibility of using *in vitro* systems as models for the *in vivo* metabolism of xenobiotics. This interest has extended from the comparative biochemistry of drug-metabolizing enzymes to the biochemical toxicology of pesticides, drugs, and other foreign compounds. From the practical point of view, industrialists have been interested in the possibility of using such *in vitro* tests in connection with screening, and obtaining registration for new pesticides and drugs.

The purpose of the present review is to examine the relationship between the *in vivo* and the *in vitro* metabolism of pesticides (and of certain other xenobiotics, where appropriate) and to comment on the possibility of using *in vitro* techniques as models for metabolism in the living animal. The methods employed for studying *in vivo* and *in vitro* metabolism will be discussed before considering the question of comparison. Attention will be paid to both qualitative and quantitative comparisons.

METHODS OF STUDYING METABOLISM *IN VIVO*

Enzymes concerned with the metabolism of foreign compounds are present in most tissues of the body (Oesch *et al.*, 1977) although much of the activity of many of the important drug-metabolizing enzymes is concentrated in the liver. The final products of metabolism are usually water-soluble conjugates, which are efficiently excreted in urine or in bile. Consequently the levels of metabolites and conjugates tend to be very low in tissues and in organs, and it is not usually practicable to measure these concentrations *in situ*. Usually studies of *in vivo* metabolism depend upon an assessment of excreted metabolites and conjugates, which, in the absence of other information, give no indication of the tissues in which the metabolites were formed. Such work is usually performed with radiolabelled substrates. These arguments do not, of course, apply to persistent liposoluble metabolites such as pp'DDE, or to active metabolites which bind to cellular components (e.g. certain epoxides of polycyclic aromatic hydrocarbons).

In vivo metabolism may also be assessed by indirect means such as the rate of disappearance of the substrate, or the nature and duration of the toxicological or pharmacological effect produced by the substrate or by its active metabolite(s).

Of general importance in the design of *in vivo* experiments is the selection of the route of administration. This issue will be discussed at appropriate points in the text.

The Direct Assessment of Excreted Metabolites

Excretory routes

Alone amongst the vertebrates, mammals void urine independently of faeces. Birds, amphibians, reptiles, and fish combine urine and faeces within the cloaca. The extent to which metabolites and conjugates are excreted via the urine as against the bile depends upon both the structure of the xenobiotic, and the species in question. Organic anions are excreted almost entirely in the urine if they are of low molecular weight (<300) but almost entirely in the bile if they are of high molecular weight (>500). This conclusion applies to all species so far studied, but there are striking species differences regarding the preferred excretory route for anions of intermediate molecular weight (300–500) (Hirom *et al.*, 1972a). It is possible to define an approximate threshold molecular weight for individual species for anions, above which significant (i.e. more than 10% of dose) excretion occurs in the bile. The following thresholds are suggested: the female rat (325 ± 50), the guinea pig (400 ± 50), and the rabbit (475 ± 50). The dog apparently has a threshold similar to the rat, and the rhesus monkey a threshold similar to the guinea pig and the rabbit. For compounds of similar molecular weight, stereochemical factors can influence the preferred excretory route (Hirom *et al.*, 1972b). It has been suggested that these trends are due to a process of selective reabsorption of compounds (usually anionic conjugates) from the bile (Clark *et al.*, 1971). If water-soluble compounds are introduced into the biliary system of the rat by retrograde biliary infusion, compounds of low molecular weight appear to be taken into the blood stream more readily than are compounds of relatively high molecular weight (Clark *et al.*, 1971). This observation supports the idea of selective reabsorption and could explain species variations in terms of differential absorption. These results do, however, require careful interpretation (Millburn, 1976). Retrograde infusion was performed at pressures which do not normally exist, and which may promote absorption Also, the levels of compounds were not measured in the blood, so it is not clear whether materials found in the biliary system had first been taken into the blood and then returned to the bile.

Organic cations do not behave in the same way as organic anions (Millburn, 1976). In all species so far tested, organic cations have a threshold molecular weight for biliary excretion of about 200; there are no significant species differences.

Differences between species with regard to the preferred route of excretion of foreign compounds have important toxicological and pharmacological implications and need to be taken into account when designing experiments on *in vivo* metabolism.

Excretion in urine

The collection of urine is relatively simple in mammals, where the only requirement is for a cage which allows the separation of urine from faeces. It is not possible, however, to measure rates of excretion by this means, because urine accumulates in the bladder. It is necessary to use some other procedure (e.g. the insertion of a catheter into the ureter) if the measurement of excretion rates is to be attempted.

With birds, reptiles, and amphibians, urine may be separated from faeces by establishing a catheter in the ureter, or by colostomy.

Collection of faeces or droppings

Many studies of *in vivo* metabolism have relied upon the determination of metabolites and conjugates, often in radiolabelled forms, in faeces or droppings. There are a number of pitfalls associated with this approach. Extracts of faeces or droppings require rigorous clean-up before metabolites can be determined, with the consequence that there can be substantial losses of material. There is also the problem of microbial action within the gut, or within the faeces or droppings after they have been voided. In the special case of ruminant animals, metabolism may be carried out by the rumen microflora when dosing is oral. Thus, there is the difficulty of distinguishing between microbial metabolites and metabolites which are produced within the vertebrate animal. A further complication arises because of the enterohepatic circulation of metabolites. Conjugates excreted into bile are often hydrolysed within the gut and the metabolites so released are then reabsorbed into the blood and returned to the bile after reconjugation. This causes a delay in the removal of metabolites in faeces or droppings, with the consequence that the actual rate of excretion into the bile cannot be determined. Metabolites returned to the liver may undergo further transformation there, and they may have an inhibitory effect upon the rate of metabolism of the original foreign compound. To summarize, *in vivo* experiments involving the analysis of droppings or faeces have a number of limitations; recoveries of metabolites may be low, bacterial metabolites confuse the picture, and rates of excretion into bile cannot be measured.

Excretion of metabolites in bile

In view of the problems associated with the determination of metabolites in faeces and droppings, there are obvious advantages in studying metabolites in bile, especially if the intention is to estimate rates of formation *in vivo*. Although there is a great deal of information upon metabolites in bile (see Smith, 1973) not much of it is concerned with the estimation of rates of excretion. Most of the work has been performed upon animals undergoing terminal cannulation, frequently under anaesthesia. Under these conditions, it is not possible to obtain reliable estimates of metabolic rates under physiological conditions, especially where anaesthetics have been used which can inhibit enzymic reactions. These difficulties can be overcome by using rats fitted with re-entrant bile duct cannulae (Chipman and Cropper, 1977). With this technique, animals can be experimented upon weeks after the operation, without the use of anaesthetics. Intermittent bile collections can be made, so that there is a normal flow of bile into the duodenum for most of the time. Furthermore, it is possible to carry out long term studies (e.g. on persistent insecticides) to investigate changes in the metabolic pattern with time.

Excretion by other routes

Foreign compounds and their metabolites and conjugates are sometimes excreted by routes other than urine or bile. For the sake of completeness, these will be briefly mentioned although they have not been important in making *in vivo/in vitro* comparisons.

Volatile compounds may be 'excreted' into the alveolar space of the lungs, and expelled during exhalation. Certain compounds are excreted in saliva and sweat (see Parke, 1968).

Highly liposoluble substances are not usually excreted to any important extent in their original forms. However, certain persistent liposoluble insecticides and their stable metabolites (e.g. dieldrin, DDT, and DDE) undergo only very slow biotransformation, so that direct excretion, although slow, is of some significance in their removal. DDT and DDE are removed in the droppings of quail over a period of weeks after intraperitoneal injection (Ahmed and Walker, 1979), perhaps by slow diffusion from the lymph into the gut as has been suggested in the case of hexachlorobenzene excretion (Müller *et al.,* 1978). Dieldrin, DDT, and DDE are excreted in the eggs of birds and in the milk of cows and humans (Still *et al.,* 1973; Kan, 1978). In both cases, the compounds are found largely in the lipid-rich fractions, i.e. in the yolk of eggs and in the fat of milk.

The Indirect Assessment of *in vivo* Metabolism

The measurement of biological half-life and clearance

Where a foreign compound is not excreted to any significant extent in its original form, the biological half-life may give a useful indication of the overall rate of metabolism *in vivo*. Important questions in experimental design are:

(1) by which route is the compound to be administered?
(2) which tissues are to be used for the determination of half-life?

When compounds are given orally or intraperitoneally to mammals, much of the dose is taken directly to the liver in portal blood. (In birds some material may go directly to the kidneys via the renal portal system.) The situation may be more complicated if administration is intramuscular, intravenous, or topical. In some cases oral or intraperitoneal administration is followed by efficient removal by the liver, with little tendency for the compound to be stored in the tissues. Here interpretation is relatively simple and the half-life in blood may give a good indication of the rate at which metabolism occurs. Compounds which have pronounced liposolubility tend to be effectively distributed around the body with the complication of storage in fat depots. With such compounds, it is desirable to measure half-lives in several tissues, e.g. blood, depot fat, muscle, and brain (see for example Robinson *et al.*, 1969). If animals are dosed for some time before measuring tissue concentrations, equilibrium may be reached between the different compartments of the body, and the half-lives are often similar in different tissues.

In most cases more than one enzyme system is involved in the initial enzymic transformation of a foreign compound and it is not possible to estimate the rate of any one *in vivo* process from half-lives alone. In other cases nearly all of the initial transformation may be accounted for by a single enzyme or even by a single metabolic conversion, e.g. over 90% of the *in vivo* conversion of dieldrin is to 9-hydroxy dieldrin in the rat (Hutson, 1976). Here it may be possible to make estimations of metabolic rate *in vivo*. The efficiency of elimination of a drug by a particular organ is measured by its clearance (Cl). Clearance may be obtained by dividing the velocity of drug elimination (v) by the incoming drug concentration (C_{in}). If direct measurements of blood composition can be made across the liver, then hepatic drug clearance is the product of liver blood flow (Q) and the extraction ratio (E) of the drug across the organ.

$$Cl = \frac{v}{C_{in}} = QE$$

Determination of these and other kinetic parameters *in vivo* is discussed by Pang *et al.* (1978).

LD_{50}

The toxicity of a compound is sometimes related to the rate at which it is metabolized. Although metabolism commonly leads to a loss of toxicity, there are some important exceptions to this rule where metabolism results in activation. Phosphorothionate insecticides such as parathion, malathion, diazinon, pirimiphos-methyl, and dimethoate are converted into active oxons by microsomal mono-oxygenase attack (see Eto, 1974). Dieldrin is apparently activated when it is converted to a *trans*-diol in or near nervous tissue (Wang *et al.*, 1971). Thus LD_{50}s may be positively or negatively correlated with *in vivo* metabolic rates. Other factors such as distribution, storage, susceptibility of site of action, and rates of intake and excretion also influence the susceptibility of individuals, species, and strains to insecticides. Although the rate of metabolism of a compound is sometimes important in determining toxicity, in other instances it is unimportant or insignificant. Two main aspects are of interest in the present context:

(1) the general question of selective toxicity between groups, species, sexes, and age groups; and
(2) the specific question of resistance.

Sometimes very large differences exist between groups which are mainly attributable to differences in enzyme activity *in vivo*. A well-known example is the much greater toxicity of malathion to mammals than to insects; this is related to the higher activity of carboxyesterase in the former group than in the latter (see O'Brien, 1967). Another example which will be discussed in more detail later, is the greater toxicity of organophosphates such as diazinon and pirimiphos-methyl to birds than to mammals (Machin *et al.*, 1975; Brealey *et al.*, 1980).

The basis for selective toxicity between contrasting groups is often complex and usually involves a number of factors other than metabolism. On the other hand, differences between resistant and susceptible strains of the same species are sometimes relatively simple. Many cases of resistance of insects to insecticides are mainly due to differences in metabolic detoxification. Thus enhanced DDT-dehydrochlorinase or microsomal mono-oxygenase are associated with DDT resistance in house flies; enhanced microsomal mono-oxygenase and glutathione transferase, with diazinon resistance in house flies (Lewis and Lord, 1969); and enhanced carboxyesterase, with malathion resistance in *Tribolium* (see Brown, 1971). Such resistance is also found in vertebrate species, e.g. a strain of pine mouse (*Microtus pinetorum*) has developed resistance to endrin which is connected with an increased rate of excretion of metabolites (Petrella and Webb, 1973). The dependence of selective toxicity upon metabolic differences may be studied by the use of synergists which inhibit the enzyme in question and can thereby reduce or eliminate the difference between susceptible and resistant organisms.

Differences in pharmacological action

The duration of a pharmacological effect may provide an indirect measurement of the rate of metabolism of a foreign compound. One example is the measurement of the sleeping time of an animal after it has been treated with a barbiturate. The dependence of barbiturate sleeping times upon metabolic activity is indicated by the large reduction in duration of action in laboratory animals such as rats and mice following the induction of liver enzymes (Remmer, 1970).

METHODS OF STUDYING METABOLISM *IN VITRO*

There are two main categories of choice in *in vitro* work, first the organ or tissue to be used, and second, the level of organization to be studied. Although many tissues have measurable metabolic activity towards foreign compounds, the liver is of particular interest by virtue of its size and the high specific activity of certain of its enzymes. This should not, of course, distract attention from the toxicological importance of biotransformations in particular tissues which represent only a very small proportion of the total turnover. The activation/deactivation of insecticides and carcinogens in target tissues may be absolutely critical in determining toxic action although the amount of biotransformation usually only represents a small proportion of that which occurs in the whole animal. However, as stated earlier, the study of metabolic changes in individual tissues *in vivo* is subject to very serious technical problems, and does not fall within the scope of the present review. Regarding the second category of choice, it is possible to perform *in vitro* studies at varying levels of complexity. At the simplest level, when working with isolated enzymes or subcellular fractions, the experimenter has the maximum degree of control over experimental conditions, e.g. by the exclusion of competing processes and reactions, and the precise control of pH and of concentrations of substrates, cofactors and inhibitors. It is possible to characterize the enzymes involved with regard to structure, mechanism of action, substrate specificity, and susceptibility to inhibitors. Moving from whole homogenates, to tissue slices, cell suspensions, and finally to perfused organs, the degree of experimental control becomes less as the situation becomes more complex and approximates more closely to that which exists *in vivo*. If the metabolism of a compound is studied using enzymes or subcellular fractions there are advantages in performing parallel experiments with cell suspensions or perfused organs to facilitate comparison with *in vivo* metabolism.

Organ Perfusion

Although perfused organs have been extensively used in biochemical research, their employment in the study or foreign compounds has been limited. One relevant example is the study of endrin metabolism in perfused rat liver (Altmeier et al., 1969). The disadvantages of organ perfusion are that it is a relatively difficult technique to operate, and is correspondingly expensive.

Tissue Explants

This is another technique which has only been employed to a limited extent for studying xenobiotic metabolism although it has been widely used in other areas of biochemical research. Samples of tissue are cut into small pieces prior to incorporation into a suitable incubation medium. A test chemical, usually in radiolabelled form, is then added, and after incubation metabolites are identified (Sullivan et al., 1972).

Tissue Slices

Intact cells are retained in this type of preparation. Liver and kidney slices have been used to study comparative metabolism by mono-oxygenases, glucuronyl transferases, and sulphate-conjugating enzymes (Bartlet and Kirinya, 1976).

Cell Suspensions and Cultures

This technique has been developed relatively recently (see Bridges and Fry, 1977). Of particular interest are suspensions of hepatocytes, which are released from thin slices of liver by digestion with collagenase and hyaluronidase. After suspension in a suitable incubation medium, they are usually metabolically viable for at least 1 h at 37 °C. The technique has the advantage of retaining the whole cell intact so that integrated processes can be studied. It is relatively cheap and easy to run, and holds promise for use in comparative work on different species, strains, sexes, age groups, and tissues. It provides a link between experiments upon enzymes and subcellular fractions on the one hand, and *in vivo* studies on the other.

When employing cell suspensions, it is essential to test the viability of the preparation. Both substrates and solvents may cause damage to cells, especially with regard to the permeability of the plasma membrane. Most methods of assessing viability depend upon the exclusion from the cell of materials such as dyes, cofactors, and substrates (Bridges and Fry, 1977).

The use of cell cultures as opposed to cell suspensions for studying xenobiotic metabolism raises certain problems. In the first place there are technical difficulties because sterile techniques are required and because the use of relatively small numbers of cells demands very sensitive methods of analysis for metabolites. Cultures of dividing cells contain cytochrome P448 rather than cytochrome P450 (Owens and Nebert, 1975), thus influencing the pattern of oxidative metabolism.

Subcellular Fractions

Homogenates of liver and other tissues may be separated into fractions by differential centrifugation. Typically, homogenates are spun at 9000–11,000 g for 30 min to bring down the plasma membrane, mitochondria, lysosomes, nucleus, and other cellular structures. Soluble proteins and a proportion of the endoplasmic reticulum remain in this supernatant. The remainder of the smooth and the rough endoplasmic reticulum is brought down as microsomes when this supernatant is spun at high speed (usually 100,000–105,000 g for about 1 h).

Hepatic microsomes, or low-speed supernatants, provide a convenient source of important drug-metabolizing enzymes of the endoplasmic reticulum, notably mono-oxygenase, epoxide hydrolase, non-specific esterase, and glucuronyl transferase. The high-speed supernatant (cytosol) contains glutathione-S-transferases, sulphokinases, and esterases. The low-speed precipitate is complex in its composition and contains the enzymes of the microsomal fractions as well as those associated with mitochondria, lysosomes, and other cellular structures.

The endoplasmic reticulum of the liver is of particular interest in the present context. If the 9,000–11,000 g supernatant rather than the microsomal fraction is used as a source of the enzyme, the metabolic situation is relatively complex. Soluble enzymes may become involved in the metabolic process, and cofactors are available for certain microsomal enzymes, e.g. UDPGA for glucuronyl transferase and NADPH for mono-oxygenase. More closely controlled conditions are possible with microsomes where enzymes such as mono-oxygenase and glucuronyl transferase can be operated selectively by regulating the appropriate cofactor. Apart from the advantage of being able to study certain enzymes individually by these means, there is also the possibility of investigating the kinetics of linked reactions. For instance, it is possible to study the interrelationship between mono-oxygenase and glucuronyl transferase.

With subcellular fractions such as microsomes it is possible to determine Lineweaver–Burke plots, and to calculate values for the kinetic constants app. V_{max} and app. K_m (see later for further discussion). From this information it is possible to estimate the rate of reaction *in vivo* if the concentration of substrate in the relevant tissue fraction is known.

Comparison has been made between the activity of tissue slices and hepatic microsomes with regard to microsomal mono-oxygenase (Gerayesh-Nejad *et al.*, 1975). Although no allowance was made for the loss of microsomal protein during preparation, microsomal mono-oxygenase activities after 5 min incubation were comparable to or greater than the corresponding activities in liver slices. Similarly (for certain oxidations) Billings *et al.* (1977) showed that rat hepatic microsomes were more active, per nmol P450, than were hepatic cell suspensions. It was not clear whether the substrate concentration in the vicinity of the enzyme was similar in the two situations.

The importance of the hepatic microsomal fraction in the metabolism of many xenobiotics has made this a popular preparation for studies of metabolism. It has the advantages of being relatively cheap and easy to employ.

Enzyme Preparations

Both water-soluble and membrane-bound enzymes play a role in the metabolism of insecticides and other xenobiotics, and can be investigated at various levels of purification. Water-soluble enzymes are in some respects the easiest of these two groups to investigate. Body fluids and high-speed supernatants of tissue homogenates provide sources of these. Of particular interest in the present context are the A and B esterases of plasma, and the glutathione transferases and sulphate-conjugating enzymes of the high-speed supernatants of tissue homogenates.

Membrane-bound enzymes are more difficult to study because they must be removed from their normal hydrophobic environment as a first step in purification. Treatment of membranes with detergents, phospholipases, or by sonication are usually needed to release the enzymes, and this may result in modification of properties. Sometimes this may mean the deactivation of the enzyme. Frequently, as with microsomal mono-oxygenase, activity may be restored by recombining the constituents of the enzyme system in a phospholipid environment (Lu *et al.*, 1974). In the case of glucuronyl transferases, the free enzyme can show higher activity than the membrane-bound form, but, with the loss of the constraint imposed by the membrane, nucleotides other than UDPGA can interact with the system (Zakim and Vessey, 1976).

The great advantage of purified enzymes is that, subject to the qualification given, they may be used for studies of substrate specificity, cofactor requirement, susceptibility to inhibitors, kinetic properties, and mechanism of action. Once again soluble enzymes present a relatively simple situation—substrate concentrations are easily determined and K_m and V_{max} may be determined in the conventional way. With membrane-bound enzymes the determination of substrate concentration is not straightforward (see Lenk, 1976). Where purified membrane-bound enzymes are suspended in an aqueous medium, the situation is not comparable to that which exists *in vivo*, so kinetic analysis is of doubtful value. If such enzymes are incorporated into liposomes, a situation which is comparable to that which exists in microsomes, the substrate concentration should be determined in the membrane. The concentration reached in the membrane will depend upon the partition coefficient of the substrate. With highly liposoluble substrates such as dieldrin and its analogues, practically all of the substrate added to an incubation medium will be taken up by the liposomes (or microsomes) until saturation is reached (see Chipman and Walker, 1979). With more polar substances the substrate will be divided more evenly between the aqueous medium and the lipid. Although standard kinetic studies are frequently performed upon membrane-bound enzymes, the interpretation of the results is open to discussion. First, it is not clear to what extent reaction rates may be determined by the availability of the substrate within the membrane, rather than the interaction of the substrate with the enzyme. Second, destabilization of the membrane may occur as the substrate concentration rises (Schwenk et al., 1976; Pfaff, personal communication). The kinetic constants obtained from such studies should be termed app.K_m and app.V_{max}.

CORRELATIONS BETWEEN *IN VIVO* AND *IN VITRO* METABOLISM

This subject was reviewed recently by Terriere (1979). For convenience, qualitative studies will be distinguished from quantitative studies. The classification is somewhat arbitrary since even the qualitative studies usually include some estimation of quantities of metabolites, or, at least, a distinction between major and minor metabolites. For present purposes, quantitative studies will be those which attempt an estimation of the rate of metabolism, half-life, or clearance *in vivo*.

Qualitative Comparisons

Sullivan et al. (1972) conducted a comparative study of the metabolism of [^{14}C]carbaryl (ring and *N*-methyl labelled), using small pieces of fresh tissue (typically 2 mm cubes) suspended in an incubation medium for periods of up to 18 h. Samples of the medium were run on DEAE–cellulose columns using a gradient elution programme detailed in Figure 6.1. The fractions were radiocounted. In parallel *in vivo* experiments, [^{14}C]carbaryl was administered to rats, dogs, and guinea pigs, while the unlabelled insecticide was given to humans. Urine was collected over a 24 h period, and metabolites and conjugates were extracted. As can be seen from Figure 6.1 there was a general similarity between the metabolic pattern found in the rat *in vivo* and that found *in vitro*; similar results were obtained with the dog, the guinea pig, and man. The metabolites and conjugates separated by this method were only partially characterized. The main peak with the rat (D) was found to be a mixture of conjugates, the major component being 5,6-dihydro-5,6-dihydroxycarbaryl glucuronide (Figure 6.2). Peak F was a glucuronide of hydroxycarbaryl (position of substitution unspecified), peak G was naphthyl glucuronide, and peak I was a sulphate conjugate. The relative sizes of different peaks were different between the *in vivo* and *in vitro* situations. Since the main peak D contained several components, the comparability of the two sets of data was limited. In a further study using the explant technique, Chin et al. (1974) compared the metabolism of carbaryl by different human tissues, and found that overall metabolic activity ranked in descending order was: liver > lung > kidney > placenta > vaginal mucosa > uterus > uterine leiomyoma (Chin et al., 1974). This technique, which did not

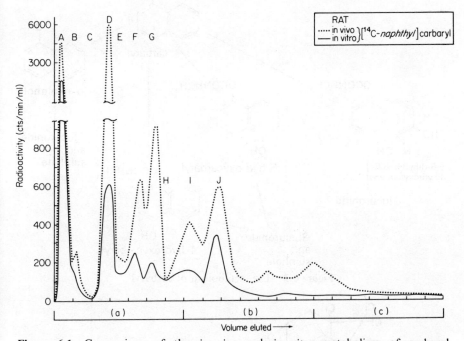

Figure 6.1 Comparison of the *in vivo* and *in vitro* metabolism of carbaryl. DEAE–cellulose chromatogram of *in vivo* and *in vitro* rat metabolites of [^{14}C-naphthyl] carbaryl. Gradient elution with the following solvent systems: (a) 0.01 M Tris-HCl buffer, pH 7.5, to 0.05 M Tris-HCl buffer, pH 7.5; (b) 0.05 M Tris-HCl buffer, pH 7.5, to 0.1 M Tris-HCl buffer, pH 7.5; (c) 0.1 M Tris-HCl buffer, pH 7.5, to 0.5 M Tris-HCl buffer, pH 7.5. (After Sullivan *et al.*, 1972)

require the use of cofactors, demonstrated a number of Phase I and Phase II transformations, notably hydroxylation, epoxide hydration, glucuronide formation, and sulphate formation.

Chipman *et al.* (1979) studied the metabolism of the dieldrin analogue HCE by male rats, rabbits, Japanese quail, and feral pigeons. *In vitro* studies were with liver microsomes fortified with NADPH. Collections of urine and bile were made for 2 h after the administration of an intraperitoneal dose of 15 mg kg^{-1} [^{14}C]HCE. Bile collections were made from animals fitted with re-entrant bile duct cannulae. Urine collections were made from colostomized pigeons; no urine collections were attempted from the Japanese quail. Bile conjugates were hydrolysed with β-glucuronidase, urine conjugates by hydrolysis with 2M HCl. The metabolites released were extracted with ether and analysed by gas chromatography. In all four species, *in vivo* and *in vitro*, the metabolites were predominantly oxidative (Figure 6.3, Table 6.1). In microsomes HCE underwent a series of transformations dependent upon NADPH and oxygen. The initial product was a hydroxy epoxide (HHC) which was converted to a dihydroxy

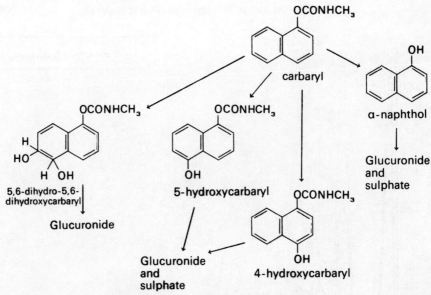

Figure 6.2 The metabolism of carbaryl

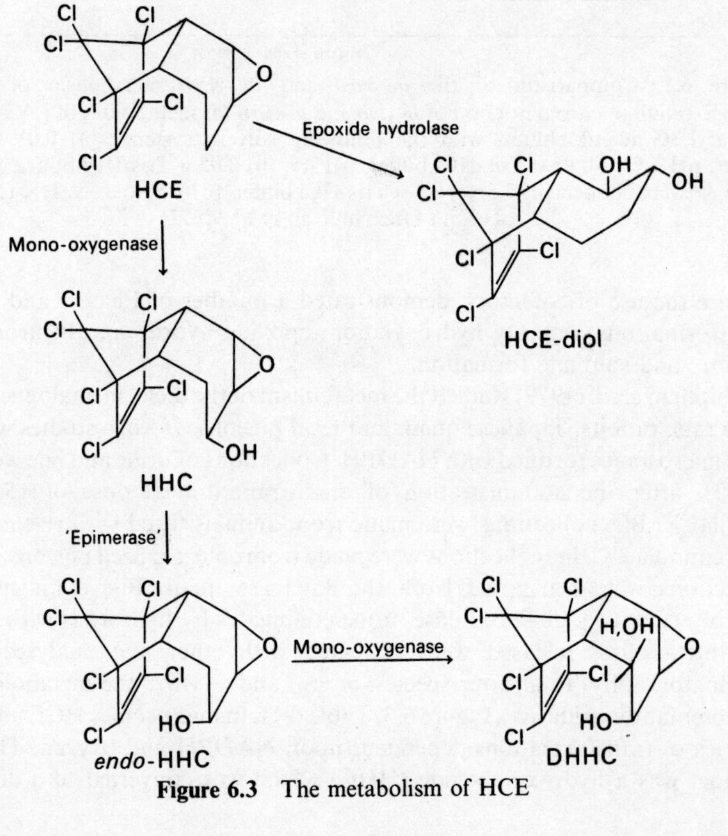

Figure 6.3 The metabolism of HCE

Table 6.1 HCE metabolism

(a) *In vivo*

Species	Number of animals	Total ^{14}C in bile or urine which was extractable	HHC	endo-HHC	DHHC	HCE trans-diol	Others
			Individual metabolites as a % total metabolites extracted				
		In bile					
Rat	7	89	61	7	14	4	14
Pigeon	4	91	50	46	4	0	0
Japanese quail	4	57	63	19	5	0	13
		In urine					
Rabbit	3	87	68	10	19	3	
Pigeon	4	95	89	6.4	4.6	0	

(b) *In vitro* (microsomes)

Species	Number of experiments	% Substrate converted	g Liver represented per incubation	HHC	endo-HHC	DHHC	HCE trans-diol	Others
				Individual				
Rat	6	60	1.5	41	33	20	5	0
Rabbit	3	75	0.8	41	4	10	30	15
Pigeon	3	55	1.0	45	31	20	0	4
Japanese quail	3	25	1.0	70	12	8	7	0

epoxide (DHHC) via its epimer (*endo*-HHC). This sequence of reactions was also found *in vivo* in all four species (Figure 6.4 gives results for rat bile). HHC dominated at first but then declined as the dihydroxy epoxide (DHHC) increased. These results emphasize the importance of considering the course of metabolism with time when making *in vivo* versus *in vitro* comparisons. The extent to which secondary oxidative attack occurred *in vivo* was dependent upon the level of original substrate present, and presumably, upon the efficiency of conjugation mechanisms (e.g. glucuronidation) in removing hydroxy metabolites before they could undergo further metabolic change.

The microsomal system predicted correctly the formation of a *trans*-diol in the rat, rabbit, and Japanese quail *in vivo*. (In the latter case this was found only in droppings.) The *trans*-diol was not produced by the pigeon either *in vitro* or *in vivo*.

Although microsomes provided a reasonable indication of major *in vivo* metabolites, there were certain discrepancies. The *trans*-diol accounted for a smaller proportion of the total metabolites *in vivo* than might have been expected. A contributory factor to this may be the fact that microsomal monooxygenase has a lower K_m for this type of substrate than does epoxide hydrolase (Chipman, 1978). Higher substrate concentrations were employed *in vitro* than existed *in vivo*, so that epoxide hydrolase should make a proportionally greater contribution to total metabolism in microsomes than in the whole animal. Certain minor metabolites were found *in vivo* but not *in vitro*. The reasons for this have yet to be established but possible explanations include extrahepatic metabolism, or the activity of gut microorganisms. Finally, the intermediate metabolite *endo*-HHC was maintained at a higher level *in vitro* than *in vivo*. One possible reason for this is that glucuronide conjugation, which did not occur under *in vitro* conditions, limited the opportunity for secondary oxidative attack.

The *in vivo* studies were complicated by the fact that there were species differences regarding the preferred route of excretion. Thus nearly all of the ^{14}C was excreted in the urine of the rabbit but in the bile of the rat, with the pigeon

Figure 6.4 HCE metabolites in rat bile and in liver microsomes. Pattern of HCE oxidative metabolism observed in the rat with respect to time. (A) *In vivo*: metabolites found in bile with intermittent collections. Each point represents a mean value from four rats. Thirty-seven per cent of the ^{14}C dose (15 mg kg^{-1}) was excreted into the bile within 10 h of administration of [^{14}C]HCE. The oxidative metabolites HHC, endo-HHC, and DHHC shown in this figure accounted for 82% of the total excreted radioactivity (see Figure 6.3 for structures). (B) *In vitro*: metabolism by hepatic microsomes reinforced with NADPH and O_2. Each point represents a mean value from duplicate incubations (8 μg[^{14}C]-HCE g liver equivalent^{-1}). The oxidative metabolites HHC, endo-HHC, and DHHC shown in this figure accounted for 95% of the total metabolites found (see Figure 6.3 for structures). (Reproduced from Chipman *et al.* 1979, by permission of Pergamon Press Ltd)

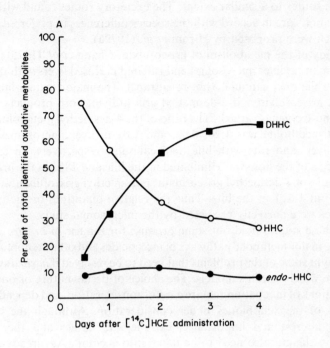

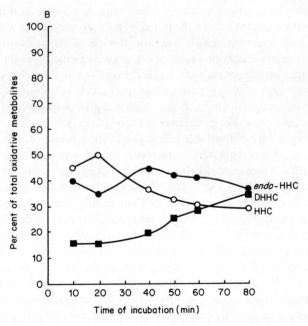

using both routes to a similar extent. The excretory routes found with HCE in rats and rabbits are in accord with the species differences in preferred excretory route which were proposed by Hirom *et al.* (1972a).

In a study of the metabolism of griseofulvin, Chang *et al.* (1975) compared metabolism in rat liver microsomes and isolated perfused livers with that in rats fitted with bile duct cannulae (not re-entrant). The major metabolites were 4-desmethyl griseofulvin and 6-desmethyl griseofulvin, both products of microsomal mono-oxygenase attack. The ratio of the 4-desmethyl metabolite to the 6-desmethyl metabolite was 1.20, 0.89, and 1.01 in liver microsomes, isolated perfused liver, and rats with bile duct cannulae respectively. In cannulated animals 65% of the dose was eliminated in the bile and 18% in the urine within 4 h. The ratio of 4-desmethyl griseofulvin : 6-desmethyl griseofulvin was 1 : 13 in the urine but 1.6 : 1 in the bile. Thus the relative quantities of the two major metabolites were correctly predicted by the microsomal study.

All of these studies hold out some promise for the use of *in vitro* models to predict the major metabolic pathways of insecticides and other xenobiotics. They also highlight some of the problems that need to be resolved if *in vitro* systems are to be more widely used as models. The choice of tissue, nature of preparation, and conditions of incubation require careful consideration, and depend upon the properties of the xenobiotics under consideration. Although the liver is of particular interest and importance, extrahepatic tissues and the action of microbes in the gut also need to be taken into account. As already indicated, there is a wide choice of preparations which may be suitable as models. When there is one major metabolic step mediated by a single enzyme, as with HCE, the choice may be a relatively simple one, and the use of the relevant subcellular fraction, or even of an enzyme preparation, may give much useful information. Where a more complex metabolic pattern exists, as with carbaryl, there are obvious advantages in using tissue explants, or cell suspensions, perhaps in combination with subcellular fractions which will give more precise information on particular processes. The properties of individual xenobiotics may give some guidance vis à vis the most suitable techniques in the absence of any information on metabolism. Carboxy esters, phosphate esters, and epoxides all possess specific groups which are likely to be attacked by particular enzymes. Compounds of very low water solubiltiy (e.g. chlorinated insecticides such as dieldrin) are likely to undergo much of their initial biotransformation in the hepatic endoplasmic reticulum. On the other hand, compounds of pronounced water solubility are likely to be metabolized to a considerable extent by soluble enzymes of the cytosol or plasma. The selection of assay conditions is especially critical with subcellular fractions and preparations of enzymes, less so with explants and cell suspensions, where the cell environment is maintained. Following changes over a period of time may be important, as in the case of HCE, where secondary oxidations occur as the substrate concentration falls.

Quantitative Comparisons

Comparisons involving the direct determination of metabolites

Chipman et al. (1979) studied the rate of excretion of ^{14}C in bile after dosing with radiolabelled dieldrin and two of its analogues (HCE and HEOM). Mature male rats fitted with re-entrant bile duct cannulae were given single doses of 15 mg kg^{-1} by intraperitoneal injection, and bile collections were made over a period of 270 min. Maximum ^{14}C excretion rates were reached between 20 and 40 min following dosing, after which these rates declined quickly in the case of HCE and HEOM but more slowly in the case of dieldrin (Figure 6.5). The maximal excretion rates were 60–90 times greater for HCE and HEOM than for dieldrin (Table 6.2). After treatment of the animals with phenobarbital, the rate of excretion of ^{14}C (by rats dosed with dieldrin) increased threefold ($P < 0.01$) but there was no significant increase in the excretion rate after dosing with HCE.

The principal metabolic changes for the three compounds are shown in Figures 6.3, 6.6, and 6.7. The metabolism of HCE and dieldrin was predominantly oxidative, most of the excreted ^{14}C being in the form of the conjugates HHC glucuronide, and 9-hydroxy dieldrin glucuronide respectively. The metabolism of HEOM was partly oxidative partly hydrative (Table 6.3). An estimation of the maximum rate of formation of HHC *in vivo* was obtained from a Lineweaver–Burke plot for microsomal metabolism, assuming the HCE concentration *in vivo* to be the maximum concentration found in hepatic microsomes from dosed animals. The estimated rate of HHC formation was within 10% of the maximal rate of excretion in bile.

With both HCE and HEOM the maximal rate of excretion of metabolites was only about 20% of the rate of overall metabolism determined in microsomes at high substrate concentration. Although induction with phenobarbital increased the microsomal mono-oxygenase activity with respect to HCE 17-fold in terms of unit liver weight, there was no significant increase in the excretion rate of the metabolites. It was concluded that the rate of excretion of HCE metabolites was not determined by the rate of metabolism in the liver, and probably approximated to the rate of uptake by the liver. The result could not be explained on the grounds of a saturation of transport processes in the liver, because increasing the dose above 15 mg kg^{-1} caused corresponding increases in the rate of excretion of HCE metabolites. Because of the increase which occurred after induction, the rate of excretion of dieldrin metabolites was evidently dependent upon the rate of metabolism in the liver. The very slow rate of excretion of dieldrin metabolites approximated to the rate of metabolism of dieldrin at high substrate concentrations in microsomes (Table 6.3).

Willson et al. (1979) compared the *in vivo* and *in vitro* kinetics of aminopyrine metabolism in male rats. *In vitro*, they measured the rate of formation of HCHO

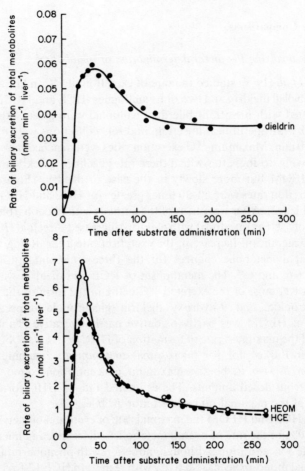

Figure 6.5 ^{14}C excretion in rat bile after dosing with dieldrin and two of its analogues. Typical biliary excretion patterns of total metabolites of dieldrin, HCE and HEOM in the rat. Substrates were administered as ethanolic solution by intraperitoneal injection at a dose rate of 15 mg kg^{-1}. (Reproduced from Chipman and Walker (1979), by permission of Pergamon Press Ltd)

in microsomes reinforced with NADPH. *In vivo* they determined the rate of elimination of $^{14}CO_2$ via the lungs after dosing animals intravenously with approximately 30 mg kg^{-1} of the radiolabelled drug. Hepatic injury was induced by dosing with CCl_4 and acetaminophen, and this caused reductions in the rates of microsomal metabolism at single saturating substrate concentration by 67% and 15% for the two treatments respectively. *In vivo* the rate of $^{14}CO_2$ elimination over 2 h was reduced by 47% ($P < 0.001$) after CCl_4 treatment and there was a small but non-significant decrease after acetaminophen treatment. The treatment with CCl_4 brought a decrease in app.V_{max} for aminopyrine

Table 6.2 The excretion of HCE and dieldrin metabolites from control and phenobarbital treated male rats

Intraperitoneal dose of organochlorine substrate ($15\,mg\,kg^{-1}$)	Treatment	Maximum values obtained for total metabolite excretion into bile			Bile flow rate ($ml\,h^{-1}\,kg^{-1}$)	Liver wt (% body wt)
		($nmol\,min^{-1}\,kg\,body\,wt^{-1}$)	($nmol\,min^{-1}\,g\,liver^{-1}$)	($nmol\,ml\,bile^{-1}$)		
^{14}C-HCE ($n=9$)	Control	203.6 ± 6.5	4.58 ± 0.13	3013 ± 263	4.23 ± 0.37	4.47 ± 0.17
^{14}C-HCE ($n=6$)	Pheno-barbital	217.4 ± 16.4 (ns)	4.34 ± 0.29	2545 ± 130	5.15 ± 0.14 (I)	5.00 ± 0.09 (II)
^{14}C-dieldrin ($n=5$)	Control	3.17 ± 0.55	0.073 ± 0.012	49.2 ± 9.8	4.0 ± 0.33	4.41 ± 0.05
^{14}C-dieldrin ($n=7$)	Pheno-barbital	9.36 ± 1.43 (IV)	0.18 ± 0.025 (III)	117.6 ± 18.5 (III)	4.91 ± 0.2 (II)	5.2 ± 0.1 (IV)
^{14}C-HEOM ($n=4$)	Control	298.0 ± 10.6	6.60 ± 0.19	—	—	—

Significantly higher than control values (Student's t-test): I = $P<0.1$; II = $P<0.05$; III = $P<0.02$; IV = $P<0.01$; ns = not significant.

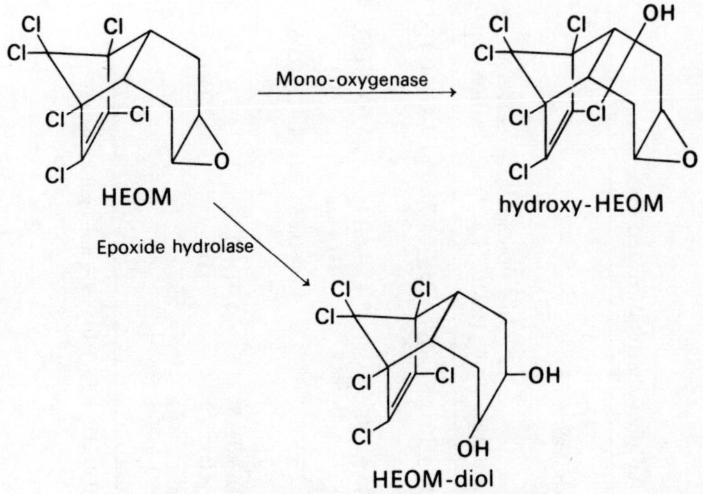

Figure 6.6 The metabolism of HEOM

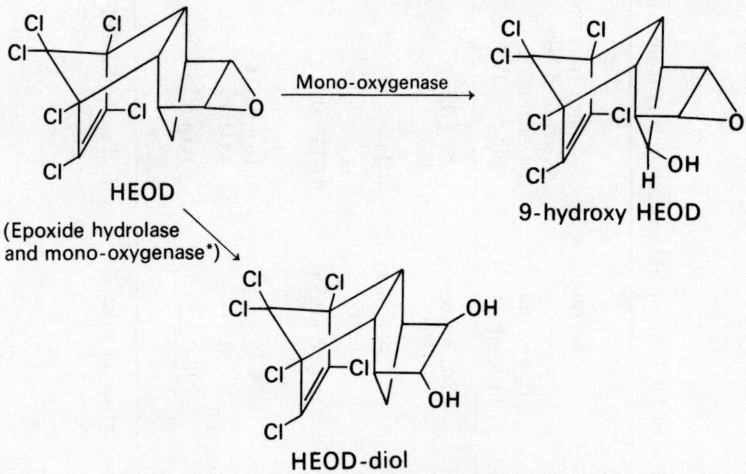

Figure 6.7 The metabolism of HEOD (dieldrin)

metabolism by microsomes, but no corresponding change in the K_m. The cytochrome P450 content per g liver was reduced by 30%. It was concluded that the maximal velocity of demethylation was the important physiological determinant of the rate of aminopyrine elimination by the liver.

Comparisons involving pharmacokinetics of substrate

A number of investigators have attempted to establish a relationship between *in vitro* kinetic parameters such as specific activity, app.V_{max} and app.K_m on the

Table 6.3 The metabolism of HCE, HEOM and dieldrin by rat liver microsomes

Substrate dose	Mode of metabolism	Control		Phenobarbital treated	
		nmol of substrate metabolized per mg microsomal protein	per g liver	nmol of substrate metabolized per mg microsomal protein	per g liver
HCE ($n = 6$) 16 μg mg^{-1} microsomal protein	Hydroxylation	0.53 ± 0.05	24.00 ± 2.3	5.54 ± 0.45‡	404.3 ± 32.8
HEOM ($n = 2$) 17 μg mg^{-1} microsomal protein	Hydroxylation	0.58 ± 0.04	26.1 ± 1.8	—	—
	Hydration	0.36 ± 0.03	16.2 ± 1.35	—	—
Dieldrin ($n = 1$) 1.3 μg mg^{-1} microsomal protein	Hydroxylation	—	—	0.0028†	0.204

All figures are means ($\pm$ S.E.M.)

† Data for dieldrin metabolism taken from Hutson (1976) where hepatic microsomes were obtained from a male CFE rat and incubation was for 30 min. Sodium phenobarbitone was administered in the diet (100 p.p.m. for 2 weeks).

Microsomal preparations from male Wistar rats were preheated for 90 s with reaction medium prior to addition of HCE and HEOM. Incubations were then carried out for 3 min and terminated by partitioning with diethyl ether.

‡ Sodium phenobarbitone was administered (75 mg kg^{-1} i.p.) once a day for three successive days prior to preparing hepatic microsomes.

one hand, and clearance and half-life on the other. Plasma antipyrine half-lives were determined in individual dogs (Vesell *et al.*, 1973) and rabbits (Statland *et al.*, 1973) and these were compared with the activities of hepatic microsomal monooxygenase. Enzyme activity was determined using as substrates aniline and ethyl morphine for dogs, and antipyrine for rabbits. Correlation coefficients of between -0.78 and -0.79 were found in the case of the three comparisons in the two separate studies. Pirttiaho *et al.* (1978) measured antipyrine half-lives and clearances in humans, and compared these to the cytochrome P450 levels in liver (Figure 6.8). In normal subjects there was a correlation between plasma

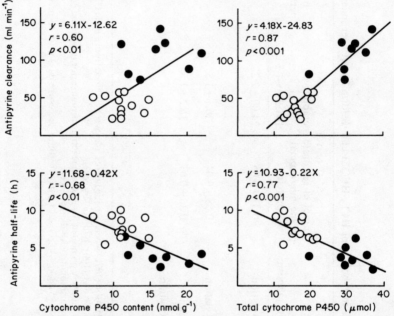

Figure 6.8 The correlation of antipyrine kinetics and cytochrome P450 in humans with normal liver parenchyma. Closed circles represent subjects who have received inducing drugs; open circles those who have no history of such treatment. (Reproduced from Pirttiaho *et al.* (1978), by permission of Medicine & Hygiene)

antipyrine half-life and cytochrome P450 concentration in microsomes. The correlation was considerably improved if cytochrome P450 was expressed as the total quantity per liver instead of the concentration per gram of liver. In 92 subjects showing histological abnormalities of the liver, a linear relationship between cytochrome P450 and plasma antipyrine half-life was not found.

In their study of aminopyrine metabolism in rats, Willson *et al.* (1979) demonstrated a correlation between a reduction in the rate of microsomal oxidation at high substrate concentration and an increase in the half-life of the drug following treatment with CCl_4. This, together with the data upon $^{14}CO_2$ elimination using this substrate, provided clear evidence for the importance of metabolic capacity in determining the rate of removal of aminopyrine *in vivo*.

Collins *et al.* (1978) studied the metabolism of phenytoin in pregnant and non-pregnant female rats. The authors regarded phenytoin in the rat as a suitable model system, because the low extraction of the drug by rat liver minimizes the effects of blood flow, which is evidently the principal non-enzymatic factor involved. The kinetic constants app.K_m and app.V_{max} for each of the two major metabolic conversions of this drug were determined by microsomal studies. [^{14}C]Phenytoin was administered intravenously and the rate of loss of the drug from plasma was determined, leading to the determination of the half-life and the clearance. Estimations of half-life and clearance (Cl) from *in vitro* data were obtained using the following equations:

$$Cl = f \times \frac{\text{app.}V_{max}}{\text{app.}K_m}$$

where

Cl = *in vivo* clearance of total drug (bound and free);

f = fraction of drug unbound;

app.V_{max} = specific enzyme activity × microsomal protein content of liver × ratio of liver wt : body wt

Total clearance was the summation of the two individual clearance values for each of the separate metabolic pathways.

$$t_{\frac{1}{2}} = 0.693 \times \frac{V_d}{Cl}$$

where V_d (apparent volume of distribution)

$$= \frac{\text{amount of phenytoin in whole body homogenate}}{\text{concentration of phenytoin in plasma}}$$

In pregnant rats, predicted half-lives were somewhat longer, and clearance rates somewhat slower than were found experimentally, but in non-pregnant rats there was no significant difference between measured and predicted values. A good estimate of phenytoin half-life was also obtained for 7 day old rats. The predicted half-life was very inaccurate for 1 day old rats, but this was based upon a K_m estimation for adult rats, which is unlikely to be valid for neonates. A promising feature of this study was the success in predicting half-lives and clearances from app.K_m and app.V_{max} determined *in vitro* together with the parameters V_d and f which are simple to determine using samples taken from dosed animals. The theoretical background of this requires further discussion.

Clearance is described by the following relationship:

$$Cl = Q \times E = \frac{Q \times (f \times Cl_1)}{Q + (f \times Cl_1)}$$

where

Cl_1 = intrinsic clearance $\left(\dfrac{\text{app.}V_{\max}}{\text{app.}K_m}\right)$;

E = fraction of drug extracted by liver;

Q = blood flow;

f = unbound fraction of drug in plasma;

The intrinsic clearance (Cl_1) provides a convenient measure of the volume of liver water which is cleared of a compound in unit time. As such it gives an assessment of the rate of hepatic elimination independently of hepatic blood flow and plasma binding. $V_{\max}/K_m$ approximates to $V/[S]$ where metabolism is first order with respect to the substrate (V = rate of metabolism, $[S]$ = substrate concentration) (Wilkinson and Shand, 1975).

Rane et al. (1977) tested the validity of this relationship in the case of various drugs added to isolated perfused rat liver. Microsomal studies were performed to determine app.K_m and app.$V_{\max}$. Reasonably good agreement was found between predicted and experimentally determined values for clearance. In their *in vivo* studies with phenytoin, Collins et al. (1978) found that the extraction of the drug by the liver was only 3% at a blood flow of 1 ml min^{-1} per g liver. Thus $E = 0.03$ and $Q \gg f \times Cl_1$. The above relationship was therefore simplified as follows:

$$Cl = f \times Cl_1 = f \times \frac{\text{app.}V_{\max}}{\text{app.}K_m}$$

It should be emphasized that this simplification would not have been valid if the extraction had been high.

Lutz et al. (1977) developed a pharmacokinetic model to predict the rate of loss of four chlorinated biphenyls and their metabolites from different tissues of the rat. The model was tested by administering 0.6 mg kg^{-1} of each compound by intravenous injection, and measuring rates of loss from blood. The compounds were 4-chloro-, 4,4'-dichloro-, 2,2'4,5,5'-pentachloro-, and 2,2',4,4',5,5'-hexachlorobiphenyl (henceforward 1-CB, 2-CB, 5-CB, and 6-CB respectively, see Figure 6.9). The following expression was used to determine the rate of loss of parent compound from blood:

$$V_B \frac{dC_B}{dt} = Q_L \frac{C_L}{R_L} + Q_M \frac{C_M}{R_M} + Q_S \frac{C_S}{R_S} + Q_F \frac{C_F}{R_F}$$

$$- (Q_L + Q_M + Q_S + Q_F) C_B + Mg(t)$$

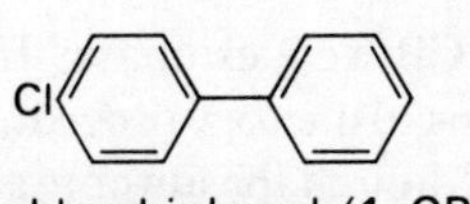

METABOLISM OF PESTICIDES IN VERTEBRATES

4-chlorobiphenyl (1-CB)

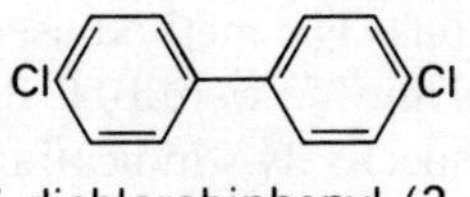

4,4'-dichlorobiphenyl (2-CB)

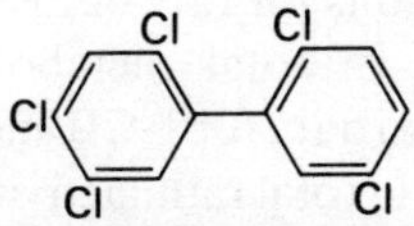

2,2',4,5,5'-pentachlorobiphenyl (5-CB)

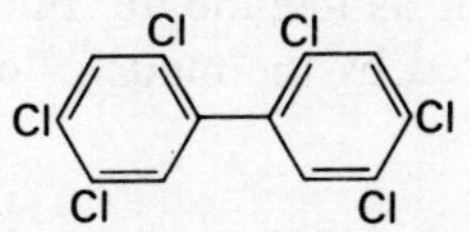

2,2',4,4',5,5'-hexachlorobiphenyl (6-CB)

Figure 6.9 The structure of some chlorinated biphenyls

where

V_B = volume of blood compartment (ml);

C_B, C_L, C_M, C_S, and C_F are the concentrations of parent compound in blood, liver, muscle, skin, and fat respectively;

R_L, R_M, R_S, and R_F are the distribution ratios tissue/blood in liver, muscle, skin, and fat respectively;

Q_L, Q_M, Q_S, and Q_F are the blood flow rates (ml min^{-1}) in liver, muscle, skin, and fat respectively;

M = dose (nmol);

$g(t)$ = injection function.

The expression used for the liver was:

$$V_L \frac{dC_L}{dt} = Q_L \left[C_B - \frac{C_L}{R_L} \right] - K_M \frac{C_L}{R_L}$$

where

V_L = volume of liver compartment (ml);

K_m = the metabolic rate constant (ml min^{-1}).

The absolute rate of disappearance of the parent compound (nmol min^{-1}) given by $K_m \times C_L$.

The K_ms for 5-CB and 6-CB were estimated from the long term excretion of radioactivity, a somewhat unsatisfactory procedure on account of the problem of enterohepatic circulation, although the authors state that reabsorption from the gut was not extensive. The rate constant for 1-CB was determined by *in vitro* studies upon liver preparations. The method used for 2-CB was not stated. The metabolic rate constants (ml min^{-1}) were 10.0, 2.0, 0.39, and 0.045 for 1-CB, 2-CB, 5-CB, and 6-CB respectively, indicating a very large variation in biodegradability.

In addition to these expressions for rate of loss of substrate, similar formulae were developed for the loss of total metabolites. The comparison of the theoretical and the experimental data for 1-CB and 6-CB is shown in Figures 6.10 and 6.11. For 1-CB the decline in total radioactivity was well predicted up to 20 h, but subsequently there was a greater persistence of ^{14}C than expected. After 10 h most of the ^{14}C was present as metabolite. For 6-CB, elimination was much slower and was well predicted by the model. For this compound, simulations

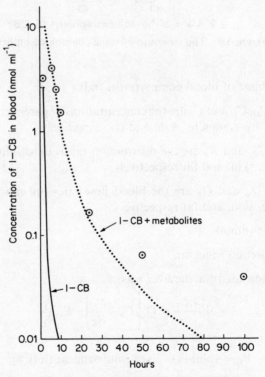

Figure 6.10 The loss of 1-CB from blood. The points represent experimental data for total 1-CB (1-CB + metabolites, the latter expressed as equivalents). After 10 h nearly all residual material was accounted for as metabolites. The lines represent simulations obtained by use of equations. (After Lutz *et al.*, 1977)

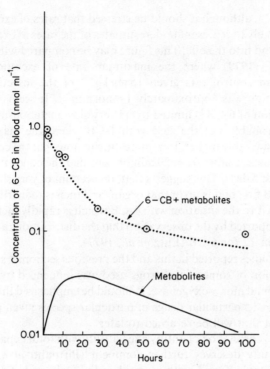

Figure 6.11 The loss of 6-CB from blood. The points represent total 6-CB equivalents (including metabolites). The lines represent simulations obtained by use of equations. (After Lutz et al., 1977)

were also successful in other tissues over a 100 h period. Long term predictions of levels in fat and blood (<42 days) were also reasonably accurate so long as account was taken of the increase in total body fat during the course of the experiment.

Increases in the level of chlorination brought decreases in the rates of metabolism of these four compounds. (With regard to PCBs in general, the authors note that position of substitution as well as number of chlorine atoms is important in determining metabolic rate.) The rates of clearance and excretion were found to be related to the metabolic capacity, the rates of elimination following the order 1-CB > 2-CB > 5-CB > 6-CB. The preferred route of excretion changes from the urine to the bile as one passes from 1-CB to 6-CB in accordance with the proposals of Hirom et al. (1972a). These results provide further evidence for the dependence of rates of excretion upon the metabolic capacity in the liver. This is particularly evident with 6-CB where only c. 3% of the dose is excreted in 20 h compared with 15–70% in the case of the other three compounds. The most rapid rate of excretion of 6-CB metabolites was

<0.1 nmol min^{-1}, although it should be stressed that rates of excretion in urine and faeces are likely to represent underestimates of the rates of excretion into the tubular lumen and into the bile. This figure may be compared with the results of Chipman *et al.* (1979) where the maximum rate of excretion of dieldrin metabolites from control rats given 15 mg kg^{-1} of the insecticide by intraperitoneal injection was approximately 1 nmol min^{-1}. It seems likely that the rate of elimination of 6-CB is limited by its very slow metabolism. On the other hand, this is probably not the case with 1-CB where the maximum rate of excretion is considerably higher. It is interesting to note that the concentration of 1-CB in fat declines steadily from 20 min on, but that the concentration of 6-CB rises over the first 5 days. This suggests that, on account of very slow metabolism in the liver, 6-CB first builds up in that organ, and then is redistributed to the fat depots in contrast to the situation with 1-CB which is rapidly metabolized. This suggestion is supported by the observation that the distribution ratio liver:blood is 1 for 1-CB but 12 for 6-CB (Lutz *et al.*, 1977).

In all of the studies reported in this and the previous section, it appears that the rate of elimination of some of the drugs used was influenced by the activity of hepatic microsomal mono-oxygenase. It should be emphasized that this tentative conclusion refers to particular drugs in particular species given particular dose regimes, a point that will be returned to later.

The successful correlation of antipyrine half-life with total hepatic cytochrome P450 in one study deserves further comment (Pirttiaho *et al.*, 1978). This correlation was only found with normal individuals and did not apply to individuals showing histological evidence of liver damage. Although such correlations have been found between individuals of the same species and age, it should not be assumed that the same will be true when comparing species, strains, sexes, or age groups. It is now clear that cytochrome P450 exists in a number of different forms with contrasting and usually overlapping substrate specificities. Since the proportion of different forms of P450 varies between species, strains, sexes, and age groups, there can be no simple relationship between cytochrome P450 content and oxidative activity when making comparisons of this kind.

Correlations of in vitro activity with toxicity

A number of investigators have attempted to correlate the toxicity of organophosphate insecticides with metabolic capability in different species.

Chlorfenvinphos shows a remarkable degree of selective toxicity amongst mammals, the acute oral LD$_{50}$s (mg kg^{-1}) in the rat, mouse, rabbit, and dog being 10, 100, 500, and >12000. The high susceptibility of the rat compared with the dog was apparently connected with a number of factors including the greater sensitivity of rat brain cholinesterase and the greater uptake of chlorfenvinphos by rat brain (Hutson and Hathway, 1967). The principal primary metabolic transformation in rats and dogs was desethylation by microsomal monooxygenase. When this reaction was studied in liver slices incubated at 37 °C, the initial reaction rates for the rat, mouse, rabbit, and dog were found to be in the ratio 1:8:24:88 (Donninger *et al.*, 1972). Thus there was a reasonable correlation between the rates of this reaction and the toxicity, although it should be borne in mind that a number of other factors were involved in the determination of selective toxicity.

O'Brien (1967) was able to relate the toxicity of dimethoate to the total degradative activity of the liver in the guinea pig, rat, sheep, mouse, cow, and hen. The hydrolysis of this compound involves the degradation of both —P—O—CH$_3$ and —CONH— bonds. The lowest activity and the greatest toxicity were found in the domestic fowl, the only bird studied.

Machin *et al.* (1975) investigated the toxicity of diazinon to the sheep, cow, rat, guinea pig, pig, turkey, chicken, and duck. They found a good inverse correlation between the activity of plasma 'A' esterase towards the active metabolite diazoxon (Figure 6.12), and the toxicity of diazinon to this range of species. It appeared that diazoxon formed in the liver was effectively degraded in the blood by the sheep and other mammals, but that this hydrolysis was slow and ineffective in birds. Brealey *et al.* (1980) reported a similar situation with the related compound pirimiphos-methyl which also produces an oxon that is a good substrate for 'A' esterase (Figure 6.12). The general question of the role of 'A' esterases in the control of the toxicity of organophosphates to mammals and birds will be discussed in the next section.

The Use of *in vitro* Data to Determine the Phylogenetic Distribution of Enzymes

'A' esterases

Aldridge (1953) proposed that enzymes which hydrolyse organophosphates be termed 'A' esterases in contrast to esterases which are inhibited by organophosphates ('B' esterases).

Since 'A' esterases act as detoxifying enzymes towards organophosphates, there is considerable interest in their role in determining patterns of selective toxicity. As noted earlier, Machin *et al.* (1975) suggested that the selective toxicity of diazinon between five mammals on the one hand and three birds on the other was due to a much higher plasma 'A' esterase activity in the former than in the latter. In a later study (Machin *et al.*, 1978) a similar situation was reported for coumaphos, and to a lesser extent chlorpyriphos (Figure 6.12), and possibly also parathion. There was, however, no significant correlation between plasma 'A' esterase activity towards the oxon and LD$_{50}$ in the case of bromophos and fenchlorphos. Brealey *et al.* (1980) conducted a survey of plasma 'A' esterase activity towards pirimiphos-methyl oxon and paraoxon. In 14 species of birds

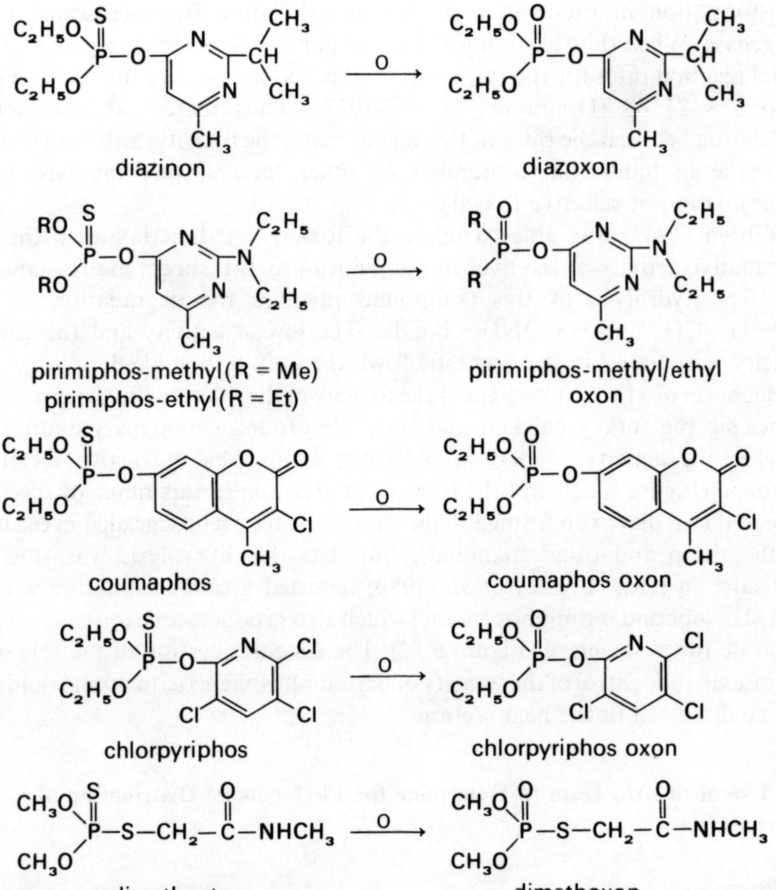

Figure 6.12 The structure of some organophosphates which are more toxic to birds than to mammals

representing six different orders, the plasma esterase activities (expressed as nmol min^{-1} per ml of plasma) were always low, ranging from 0 to 71 for pirimiphos-methyl oxon and from 0 to 0.63 for paraoxon. Mammalian activities were very much higher than these, and were in no case less than 13 times more active in the same assay procedure. These large differences in plasma 'A' esterase activity between the two groups were correlated with a considerably higher toxicity towards birds than to mammals in the case of pirimiphos-methyl, pirimiphos-ethyl, diazinon, and dimethoate. It is interesting to note that the first three compounds are all derivatives of pyrimidine, and that the oxons of pirimiphos-methyl and diazinon are hydrolysed more rapidly by 'A' esterase than is paraoxon (Brealey et al., 1980). Thus it would appear that the *in vitro* estimation of plasma 'A' esterase successfully predicts a phylogenetic difference in the activity of the enzyme *in vivo* which is reflected in the selective toxicity of certain organophosphates between birds and mammals.

Hepatic microsomal mono-oxygenase

This complex of enzymes is involved in the initial biotransformation of many liposoluble pesticides and drugs. Often the enzymatic attack involves the introduction of a hydroxyl group which is available for subsequent conjugation to form glucuronides, sulphates, or other conjugates.

In a recent survey of the literature, hepatic microsomal mono-oxygenase activities were compared in 36 vertebrate species. Twelve different assay procedures had been used to obtain the original data so, to facilitate comparison, all values were expressed relative to the male rat which was ascribed an arbitrary value of 1. A correction was made for the liver weight: body weight ratio in each individual species (Walker, 1978). If animals of similar body weight were compared, the groups studied could be arranged in the following order of decreasing enzyme activity: mammals > birds > fish. In a further review (Walker, 1980) log relative hepatic mono-oxygenase activity was plotted against log body weight for a range of vertebrates. Although there was a great deal of individual variation, mammals, birds, and fish produced different regression lines indicating that mammals had microsomal mono-oxygenase activities 10–25 times greater than those of fish, whilst birds occupied an intermediate position. The relative mono-oxygenase activities were compared with the biological half-lives of antipyrine, dieldrin, hexobarbitone, and phenylbutazone, four compounds which undergo extensive mono-oxygenase attack in the first stage of their biotransformation (Walker, 1978). Plotting nine different vertebrate species independently, there was an inverse relationship between the log relative hepatic microsomal mono-oxygenase activity and the log of the relative half-life (male rat = 1). This suggested that mono-oxygenase activity was an important factor in determining the half-lives of these four compounds in different species under the experimental conditions used, and agrees in principle with the work of Vesell *et al.* (1973) and Statland *et al.* (1973) on antipyrine half-lives in individual animals.

The role of hepatic microsomal mono-oxygenase in determining the toxicity of insecticides is sometimes complex as, for instance, in the case of certain organophosphates, where it may have both an activating and a deactivating function. Many carbamate insecticides appear to present a simpler picture, because mono-oxygenase appears to have only a deactivating function. Schafer (1972) reported LD_{50} values of 20 different carbamate pesticides for the male rat, and for one or both of two birds—the starling (*Sturnus vulgaris*) and the American blackbird (*Agelaius phoeniceus*). Eighteen of the carbamates were more toxic to either or both of the birds than to the male rat, whilst the remaining two compounds were only marginally more toxic to the rat. It would be interesting to know whether the greater toxicity of carbamates to birds than to mammals is connected with the relatively low mono-oxygenase activities found in the former group.

Within the group of birds discussed above, five species of fish-eating sea birds showed particularly low hepatic microsomal mono-oxygenase activities, comparable in magnitude to those found in fish (Knight et al., 1980; Walker, 1980). It seems likely that these low activities are an important factor in determining the capacity of species like the shag (*Phalacrocorax aristotelis*) and the cormorant (*Phalacrocorax carbo*) to bioaccumulate persistent liposoluble substances such as dieldrin, DDE, and PCBs.

CONCLUSIONS

In the foregoing sections, a number of studies were cited which demonstrated correlations of metabolism *in vitro* with metabolism *in vivo*. Although this evidence gives some encouragement for the predictive value of *in vitro* models, it must be re-emphasized that this conclusion only applies to some of the very limited selection of compounds so far tested, under particular experimental conditions. The development of this approach depends upon further experimentation with other xenobiotics which have contrasting chemical, pharmacological, and toxicological properties.

It is difficult in *in vitro* work to supply the substrate in a similar manner and at a similar concentration, to that which exists *in vivo*. In the first place, the substrate concentration in the vicinity of the enzyme *in vivo* may be very difficult to determine. Second, with most *in vitro* work, one is dealing with a 'closed' system, in contrast to the 'open' system that exists *in vivo*. It would be difficult, if not impossible, to simulate the rate of arrival of substrate, and the rate of removal of metabolites and conjugates which is found in the living animal. Where it is possible to determine substrate concentrations in the vicinity of enzyme systems *in vivo* estimates of metabolic rate may be obtained from *in vitro* data. It is important that such kinetic studies should be carried out over the whole range of concentrations that occur *in vivo*. A single substrate may be metabolized by more than one enzyme, or more than one form of the same enzyme, and these different enzymes or forms may have contrasting K_m values. Where this is so, the balance between different metabolic pathways will depend upon substrate concentrations *in vivo*.

The values of half-life and clearance can be dependent upon the dosage given (see for example Gabler and Hubbard, 1972). If any processes concerned with the removal of a foreign compound become saturated with increasing dose, then changes in half-life and clearance are to be expected. At low concentrations a relatively high proportion of a drug may be tightly bound, e.g. to plasma protein, with the consequence that it is of only limited availability. With increasing concentration a higher proportion of the drug should become available as the binding sites approach saturation. Similarly, at high dose levels, processes of transport or metabolism may approach saturation so that further increments in dose will bring corresponding changes in half-life and clearance. Thus, the question of dose rates needs to be given careful consideration before attempting to use *in vitro* data to predict half-lives and clearances.

In the living animal, xenobiotics are subject to a number of processes, including metabolism, before their metabolites/conjugates are finally eliminated from the body. In the present context attention has been focused upon cases where the rate of removal of substrate or the rate of production of metabolites have been related to app.V_{max}, app.K_m, and specific activity of key enzymes. On the other hand, there are cases where a substantial difference in metabolic capacity apparently has no effect upon the actual rate of metabolism which occurs *in vivo*. For example, the rate of excretion of HCE metabolites is not affected by induction (Chipman *et al.*, 1979). Here, the rate of metabolism is presumably determined by factors which establish the rate at which the substrate reaches the hepatic endoplasmic reticulum, e.g. polarity, tendency to bind to plasma proteins.

Pharmacokinetic models incorporating metabolic constants, such as that developed by Lutz *et al.* (1977), take into account many of the points mentioned above and provide a means of estimating metabolic rates, half-lives, and excretion rates *in vivo*. They do, however, require a considerable amount of experimental data before they can be used to make predictions. Collins *et al.* (1978) were able to predict phenytoin clearances and half-lives in rats using only K_m and V_{max} from microsomal studies together with V_d and f obtained by simple measurements upon tissues from dosed animals. Unfortunately this approach is only valid where the extraction of the compound in question is low. From the point of view of the kinetics of metabolic processes in the liver, more attention could be paid to estimation of the rates at which different types of compound can be taken up by the liver as a function of dose. Where such rates approach the app.V_{max} for the relevant metabolic process(es) the enzymatic activity of the liver should be an important determinant of metabolic rate, half-life, and excretion rate, and species, sex, and strain differences in enzyme activity should be important in determining differential half-lives, excretion rates, and toxicity (Walker *et al.*, 1979). Where, on the other hand, the metabolic capacity indicated by app.V_{max} greatly exceeds the rate of arrival of substrate, it is unlikely to have a significant influence upon metabolic rate, although species differences in the balance of different enzyme systems should still be reflected in characteristic patterns of excreted metabolites, even where the overall rate of substrate metabolism is similar for all.

Microsomal systems have been more widely used than other *in vitro* techniques, to predict metabolism *in vivo*. For the most part they have just been used to study oxidations mediated by hepatic microsomal mono-oxygenase. In view of the success of this approach, the way is open to use them for other purposes. For instance, it is possible to study the coupling of microsomal hydroxylation to glucuronidation by including UDPGA in the incubation medium. Biotransformations mediated by other microsomal enzymes such as epoxide hydrolase, and 'B' esterase can also be studied in this fraction. In addition other subcellular fractions such as high- and low-speed supernatants, 'mitochondria', and 'lysosomes' can be used to complement work on microsomes. In the end, however, this approach is limited because it does not preserve the cell environment. There is a strong case for greater use of cell suspensions and tissue explants to complement studies performed upon subcellular fractions.

In vitro metabolic studies should give a useful indication of the risks of bioaccumulation presented by particular pesticides and other pollutants. Special attention should be given to substrates such as dieldrin, DDE, and PCBs which undergo very slow biotransformation and which, therefore, are liable to have very long biological half-lives.

Apart from the use of *in vitro* studies to investigate particular metabolic problems, they have value in a wider context. They provide a means of building up a picture of the phylogenetic distribution of those enzymes which are concerned with xenobiotic metabolism and also of generating ideas about the evolutionary development of such enzymes in response to the selective pressure of chemicals in the diet and in the general environment. The successful phylogenetic classification of these enzymes coupled with a fuller understanding of their function would provide a useful theoretical background for more practically orientated studies in the field of biochemical toxicology. In particular, it should be of value in the prediction of certain patterns of selective toxicity in pesticides, and in aiding the extrapolation of pharmacological and toxicological data from laboratory animals to man.

REFERENCES

Ahmed, M. M., and Walker, C. H. (1979). 'The metabolism of DDT *in vivo* by the Japanese quail', *Pestic. Biochem. Physiol.* **10**, 40–48.

Aldridge, W. N. (1953). 'Serum esterases', *Biochem. J.* **53**, 110–124.

Altmeier, G., Klein, W., and Korte, F. (1969). 'Metabolismus endrin-^{14}C in perfundierten ratten', *Tetrahedron Letters* **49**, 4269–4271.

Bartlet, A. L., and Kirinya, L. M. (1976). 'Activities of mixed function oxidase, UDP-glucuronyl transferase and sulphate conjugation enzymes in galliformes and anseriformes', *Quart. J. Exp. Physiol.* **61**, 105–119.

Billings, R. E., McMahon, R. E., Ashmore, J., and Wagle, S. R. (1977). 'The metabolism of drugs in isolated rat hepatocytes—a comparison with *in vivo* drug metabolism and drug metabolism in subcellular liver fractions', *Drug. Metab. Disp.* **5**, 518.

Brealey, C. J., Walker, C. H., and Baldwin, B. C. (1980). 'A-esterase activities in relation to the differential toxicity of pirimiphos-methyl to birds and mammals', *Pestic. Sci.,* **11**, 546–554.

Bridges, J. W., and Fry, J. R. (1977). 'Drug metabolism in cell suspensions and cultures', in *Drug Metabolism from Microbes to Man* (Eds. D. V. Parke and R. L. Smith), pp. 43–55, Taylor and Francis, London.

Brown, A. W. A. (1971). 'Pest resistance to pesticides', in *Pesticides in Environment*, Vol. 1, pp. 457–551, Marcel Dekker, New York.

Chang, R. L., Zampagliône, N., and Lin, C. (1975). 'Correlation of ^{14}C-Griseofulvin metabolism in rat liver microsomes, isolated perfused livers, and in rats with bile duct cannulas', *Drug Metab. Disp.* **3**, 487–493.

Chin, B. H., Eldridge, J. M., and Sullivan, L. J. (1974). 'Metabolism of carbaryl by selected human tissues using an organ-maintenance technique', *Clin. Toxicol.* **7**, 37–56.

Chipman, J. K. (1978). 'The comparative metabolism and excretion of certain chlorinated cyclodiene insecticides in vertebrate species', Ph.D. Thesis, University of Reading.

Chipman, J. K., and Cropper, N. C. (1977). 'A technique for chronic intermittent bile collection from the rat', *Res. Vet. Sci.* **22**, 366–370.

Chipman, J. K., Kurukgy, M., and Walker, C. H. (1979). 'Comparative metabolism of a dieldrin analogue: hepatic microsomal systems as models for metabolism in the whole animal', *Biochem. Pharmacol.* **28**, 69–75.

Chipman, J. K., and Walker, C. H. (1979). 'The metabolism of dieldrin and two of its analogues: the relationship between rates of microsomal metabolism and rates of excretion of metabolites in the male rat', *Biochem. Pharmacol.* **28**, 1337–1345.

Clark, A. G., Hirom, P. C., Millburn, P., and Smith, R. L. (1971). 'Absorption of some organic compounds from the biliary system of the rat', *J. Pharm. Pharmacol.* **23**, 150–152.

Collins, J. M., Blake, D. A., and Egner, P. G. (1978). 'Phenytoin metabolism in the rat', *Drug Metab. Disp.* **6**, 251–257.

Donninger, C., Hutson, D. H., and Pickering, B. A. (1972). 'The oxidative dealkylation of insecticidal phosphoric acid triesters by mammalian liver enzyme', *Biochem. J.* **126**, 701–707.

Eto, M. (1974). *Organophosphorus pesticides: Organic and Inorganic Chemistry*, C.R.C. Press, Boca Raton, Florida.

Gabler, W. L., and Hubbard, G. L. (1972). 'Metabolism *in vitro* of 5,5-diphenylhydantoin', *Biochem. Pharmacol.* **21**, 3071–3073.

Gerayesh-Nejad, S., Jones, R. S., and Parke, B. V. (1975). 'Comparison of rat hepatic microsomal mixed-function oxidase activities in microsomal preparations, isolated hepatocytes, and liver slices', *Biochem. Soc. Trans.* **3**, 403–405.

Hirom, P. C., Millburn, P., Smith, R. L., and Williams, R. T. (1972a). 'Species variations in the threshold molecular weight factor for the biliary excretion of organic anions', *Biochem. J.* **129**, 1071–1077.

Hirom, P. C., Millburn, P., Smith, R. L., and Williams, R. T. (1972b). 'Molecular weight and chemical structure as factors in the biliary excretion of sulphonamides', *Xenobiotica* **2**, 205–214.

Hutson, D. H. (1976). 'Comparative metabolism of dieldrin in the rat (CFE) and in two strains of mouse (CFI and LACG)', *Food Cosmet. Toxicol.* **14**, 577–591.

Hutson, D. H., and Hathway, D. E. (1967). 'Toxic effects of chlorfenvenphos in dogs and rats', *Biochem. Pharmacol.* **16**, 949–962.

Kan, C. A. (1978). 'Accumulation of organochlorine residues in poultry—a review', *J. Agr. Food Chem.* **26**, 1051–1055.

Knight, G. C., Walker, C. H., Harris, M., and Cabot, D. C. (1980). 'The activities in sea birds of two hepatic microsomal enzymes which metabolise liposoluble pollutants', *Comp. Biochem. Physiol.* **68**, in press.

Lenk, W. (1976). 'Application and interpretation of kinetic analyses from the microsomal drug-metabolising oxygenases', *Biochem. Pharmacol.* **25**, 997–1005.

Lewis, J. B., and Lord, K. A. (1969). 'Metabolism of some organophosphorus insecticides by strains of housefly', *Proc. Br. Insectic. Fungic. Conf. 5th,* 465–471.

Lu, A. Y. H., Levin, W., and Kuntzman, R. (1974). 'Reconstituted liver microsomal enzyme system that hydroxylates drugs, other foreign compounds and endogenous substrates', *Biochem. Biophys. Res. Commun.* **60**, 266.

Lutz, R. J., Dedrick, R. L., Matthews, H. B., Eling, T. E., and Anderson, M. W. (1977). 'A preliminary pharmacokinetic model for several chlorinated biphenyls in the rat', *Drug Metab. Disp.* **5**, 386–396.

Machin, A. F., Anderson, P. H., Quick, M. P., Waddell, D. R., Skibniewska, K. A., and Howells, L. C. (1978). 'The metabolism of diazinon in the liver and blood of species of varying susceptibility to diazinon poisoning', *Xenobiotica* **7**, 104.

Machin, A. F., Rogers, H., Cross, A. J., Quick, M. P., and Howells, L. C. (1975). 'Metabolic aspects of the toxicology of diazinon, I. Hepatic metabolism in the sheep, cow, pig, guinea pig, rat, turkey, chicken and duck', *Pestic. Sci.* **6**, 459–472.

Millburn, P. (1976). 'The excretion of xenobiotic compounds in bile', in *The Hepatobiliary System* (Ed. W. Taylor), pp. 109–129. Plenum, New York.

Müller, F., Scheunert, I., Rozman, K., Kögel, W., Freitag, D., Richter, E., Coulston, F., and Korte, F. (1978). 'Comparative metabolism of hexachlorobenzene and pentachloronitrobenzene in plants, rats, and rhesus monkeys', *Ecotox. and Env. Safety* **2**, 437–445.

O'Brien, R. D. (1967). *Insecticides: Action and Metabolism,* pp. 265–267. Academic Press, New York.

Oesch, F., Glatt, H-R., and Schmassmann, H-V. (1977). 'The apparent ubiquity of epoxide hydratase in rat organs', *Biochem. Pharmacol.* **26**, 603–607.

Owens, I. S., and Nebert, D. W. (1975). AHH induction in mammalian liver derived cell cultures', *Mol. Pharmacol.* **11**, 94–105.

Pang, K. S., Rowland, M., and Tozer, T. N. (1978). '*In vivo* evaluation of Michaelis–Menten constants of hepatic drug-eliminating systems', *Drug Metab. Disp.* **6**, 197–200.

Parke, D. V. (1968). *The Biochemistry of Foreign Compounds,* Pergamon, Oxford.

Petrella, V. J., and Webb, R. E. (1973). 'Excretion of ^{14}C endrin and its metabolites in endrin—susceptible and resistant strains of pine mice (*Microtus pinetorum*).*, Fed. Proc. Fed. Amer. Soc. Exp. Biol.* **32**, Abstract No. 593, p. 320.

Pirttiaho, H. I., Sotaniemi, E. A., Pelkonen, R. o., and Pitkänen, U. (1978). 'The influence of liver size on the relationship between *in vivo* and *in vitro* studies of drug metabolism', *Eur. Journ. Drug Metab. Pharmacokinetics* **3**, 217–222.

Rane, A., Wilkinson, G. R., and Shand, D. G. (1977). 'Prediction of hepatic extraction ratio from *in vitro* measurement of intrinsic clearance', *J. Pharmacol. Exp. Ther.* **200**, 420–424.

Remmer, H. (1970). 'Induction of drug-metabolising enzymes on different species', in 'The problems of species, differences and statistics in toxicology', *Proc. Eur. Soc. Study. Drug Toxicity* **XI**, 14–18.

Robinson, J., Roberts, M., Baldwin, M., and Walker, A. I. T. (1969). 'The pharmacokinetics of HEOD (dieldrin) in the rat', *Food Cosmet. Toxicol.* **7**, 317–332.

Schafer, E. W. (1972). 'The acute oral toxicity of 369 pesticidal, pharmaceutical and other chemicals to wild birds', *Toxicol. Appl. Pharmacol.* **21**, 315–330.

Schwenk, M., Burr, R., and Pfaff, E. (1976). 'Influence of viability on bromosulphophthalein uptake by isolated hepatocytes', *Naunyn-Schmiedebergs Arch. Pharmacol.* **295**, 99–102.

Smith, R. L. (1973). *The Excretory Function of Bile,* Chapman and Hall, London.
Statland, B. E., Astrup, P., Black, C. H., and Oxholm, E. (1973). 'Plasma antipyrine half life and hepatic microsomal antipyrine hydroxylase activity in rabbits', *Pharmacology* **10**, 329–337.
Still, J. W., Brown, W. H., and Whiting, F. M. (1973). 'Some characteristics of metabolism of organochlorine pesticides in bovine', *Fed. Proc.* **32**, 1995–2000.
Sullivan, A. J., Chin, B. H., and Carpenter, C. P. (1972). '*In vitro* vs. *in vivo* chromatographic profiles of carbaryl anionic metabolites in man and lower animals', *Toxicol. Appl. Pharmacol.* **22**, 161–174.
Terriere, L. C. (1979). 'The use of *in vitro* techniques to study the comparative metabolism of xenobiotics', *ACS Symposia Series* No. 97, pp. 285–320.
Vesell, E. S., Lee, C. J., Passananti, G. T., and Shively, C. A. (1973). 'Relationship between plasma antipyrine half lives and hepatic microsomal drug metabolism in dogs', *Pharmacology* **10**, 317–328.
Walker, C. H. (1978). 'Species differences in microsomal mono-oxygenase activities and their relationships to biological half lives', *Drug Metab. Rev.* **7**, 295–323.
Walker, C. H. (1980). 'Species variations in some hepatic microsomal enzymes', in *Progress in Drug Metabolism.* (Eds. J. W. Bridges and L. F. Chasseaud), Vol. 0, pp. 113–164, Wiley, London.
Walker, C. H., Chipman, J. K., and Kurukgy, M. (1979). 'Microsomal systems as models for *in vivo* metabolism', *Ecotox. and Env. Safety* **3**, 39–46.
Wang, C. M., Narahashi, T., and Yamada, M. (1971). 'The neurotoxic action of dieldrin and its derivatives in the cockroach', *Pest. biochem. Physiol.* **1**, 84.
Wilkinson, G. R., and Shand, D. G. (1975). 'A physiological approach to hepatic drug clearance', *Clin. Pharmacol. Ther.* **18**, 377–390.
Willson, R. A., Hart, F. E., and Hew, J. T. (1979). 'Comparison of *in vivo* and *in vitro* drug metabolism in experimental hepatic injury in the rat', *Gastroenterology* **76**, 697–703.
Zakim, D., and Vessey, D. A. (1976). 'The effect of lipid–protein interactions on the kinetic parameters of microsomal UDP-glucuronyl-transferase', in *Enzymes Biol. Membranes* (Ed. A. Martonosi), Vol. 2, pp. 443–461, Plenum Press, New York.

CHAPTER 7

The metabolism of insecticides in man

D. H. Hutson

INTRODUCTION	287
ORGANOCHLORINE INSECTICIDES	289
Dieldrin	289
Endrin	292
DDT	296
ORGANOPHOSPHORUS INSECTICIDES	299
Dichlorvos	301
Chlorfenvinphos	303
Ethyl Parathion and Methyl Parathion	305
General	306
CARBAMATE INSECTICIDES	307
Carbaryl (Sevin)	308
Propoxur (Baygon)	311
Other Carbamates	311
PYRETHROID INSECTICIDES	312
ENZYMES INVOLVED IN THE BIOTRANSFORMATION OF FOREIGN COMPOUNDS	314
Oxidation	314
Reduction	317
Hydrolysis	318
Carboxylesters	318
Organophosphorus compounds	319
The hydration of epoxides	320
Conjugation Reactions	320
Glutathione conjugation	320
Glucuronide formation	322
Sulphation	323
Acetylation	324
Methylation	324
Conjugation with amino acids	324
CONCLUSIONS	325
REFERENCES	325

INTRODUCTION

The major insecticide classes act on the nervous system of the target organism and, as absolute insect/mammal selectivity is rarely achieved, the insecticides generally exhibit bioactivity towards mammals at appropriate doses. There are three major classes of insecticides in terms of amounts in use (Ridgeway et al., 1978):

(1) the organochlorines (e.g. DDT, dieldrin);
(2) the organophosphates (e.g. parathion, diazinon);
(3) the carbamates (e.g. carbaryl).

Important minor groups are:

(1) the natural toxin derivatives (e.g. rotenone, nicotine);
(2) the pyrethroids (e.g. permethrin) developed initially from natural products but which are on the verge of a dramatic increase in use;
(3) the growth regulants (e.g. the juvenile hormone analogues) which have modes of action specific to their targets.

Over the past decade, usage of the major classes has changed, from the order shown above, to organophosphates > carbamates > organochlorines.

The toxic action of an insecticide in a mammal is dependent on its intrinsic reactivity to the biochemical target (e.g. the reactivity of paraoxon towards brain acetylcholinesterase); however, toxicity *in vivo* is considerably modulated by the rates of absorption, metabolism, and elimination of the insecticide. The modulation usually leads to a decrease in toxicity but can sometimes result in an increase in toxicity. Thus, metabolic fate plays an important role in toxicity and persistence. Species differences in toxicity are often due to differential rates of routes of biotransformation. These differences can pose a problem in that they complicate the safety evaluation process. However, they are a fact of life, and it is sometimes possible to use them constructively in the study of modes of toxic action and in developing predictive measurements.

The impetus behind the early studies of species differences in metabolism, pioneered by R. T. Williams and colleagues at St. Mary's Hospital, London and continued in many other laboratories, was the search for an experimental animal model for man. It is probably true to say that the search has failed in that no model has been found which holds for all compound classes. The pig (a physiologist's model) and the non-human primates (chosen because of their close evolutionary relationship with man), for example, are not general models. Even a model species selected on the basis of studies on one compound class can usually be invalidated by further study within the same compound class. A reflection of the unsatisfactory state of the search for an animal model is the continued dominance of the rat for use in long term toxicity studies. Other species are favoured for the study of teratogenicity, mutagenicity, etc., the choice being made on the basis of technical suitability.

The extrapolation of pesticide toxicological data to man, at present, is by judgement from the results of a variety of toxicity and metabolism studies in several species. If there are no observed problems and if there is a uniformity of metabolism in several species, a compound is assessed as low risk. The development of a drug, on the other hand, requires that limited human experimentation takes place between animal experimentation and clinical trials. Pesticide development is different in that (1) human exposure is minimized at all stages of use and (2) there is nothing equivalent to a therapeutic dose at which metabolism/toxicity studies should be carried out. The main impetus for studies of the metabolism of pesticides in man stems from the desirability of biomedical monitoring. An estimation of exposure or body burden of a pesticide can be gained from the analysis of blood or urine (depending on the persistence of the compound) for the compound itself or its metabolites. Quantitative measurements can be made on manufacturing plant workers, formulators, and spraymen. The results can afford confirmation of metabolic pathways discovered in animals. In addition, where species differences exist, they allow the placement of man in the range of species studied. In some situations, *in vitro* metabolism studies using human tissues in comparison with animal tissues is often useful. The correlation between *in vitro* results and metabolism *in vivo* is treated in some detail in Chapter 6 of this volume.

ORGANOCHLORINE INSECTICIDES

Dieldrin

Dieldrin (Figure 7.1, 1.1) is the common name for a technical product which contains at least 85% of 1,8,9,10,11,11-hexachloro-4,5-*exo*-epoxy-2,3,7,6-*endo*-2,1,7,8-*exo*-tetracyclo[6.2.1.1.$^{3,6}0^{2,7}$]dodec-9-ene. It is one of a small group of chlorinated insecticides manufactured from hexachlorocyclopentadiene; others of this group include aldrin, endrin, chlordane, heptachlor, endosulphan, and mirex. The name used above is that in current usage derived from the von Bayer–IUPAC system. The compound will be referred to here as dieldrin and the current numbering system (Bedford, 1974) will be used. Thus 12-hydroxydieldrin (1.2) and 4,5-*trans*-dihydroaldrindiol (1.3) are used in place of the older names (9-hydroxydieldrin and 6,7-*trans*-dihydroaldrindiol). Dieldrin is one of the most intensively and extensively studied compounds in pesticide toxicology (Jager, 1970). Its discovery and wide use in the 1950s immediately preceded the development of a new generation of analytical methods (gas–liquid chromatography and the electron-capture detector). It was subsequently found at low concentrations in many sectors of the environment, including human diet and human fat. Its stability in biological systems is due both to the cage structure and to its high degree of chlorination. These factors conspire to present only a few bonds of low chemical reactivity to the enzymes capable of foreign compound metabolism. For example, even the epoxide group of dieldrin is so chemically stable that it survives gentle refluxing for 30 min in a 10% potassium hydroxide solution in isopropanol.

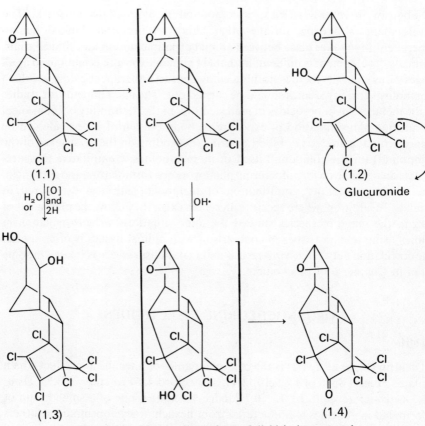

Figure 7.1 Biotransformations of dieldrin in mammals

As well as being well studied in mammals, dieldrin has been the subject of one of the more ambitious pesticide investigations in man. This research, a kinetic study, was initiated at a time when metabolism studies were non-routine. What little work had been done led researchers to the uneasy conclusion that dieldrin was not metabolized in mammals. The presence of this and other organochlorine pesticides in man's diet stimulated concern that man's body burden of organochlorines would increase continually. The kinetic study (Hunter and Robinson, 1967; Hunter et al., 1969) was designed with two objectives:

(1) to define the quantitative relationship between the oral intake of dieldrin and its concentrations in blood and fat (thereby allowing meaningful monitoring of manufacturing, formulating, and spraying operatives); and
(2) to find the uptake and elimination kinetics of dieldrin in these tissues.

The study also provided an opportunity for monitoring the health of human subjects under defined exposure conditions using full clinical, physiological, and laboratory examination. Metabolic studies could not be incorporated because neither the technology nor the knowledge of dieldrin metabolism was available at the time.

Twelve male volunteers, in groups of three, received 0, 10, 50, and 211 µg of dieldrin per day for up to 2 years. The study was continued for a further 8 months after cessation of treatment. Venous blood samples were taken monthly and samples of adipose tissue were taken by needle or open biopsy at 18 and 24 months and at 8 months post-exposure. No effects on the health of any of the volunteers were found; a further encouraging discovery was that, under continuous intake, the rate of increase of dieldrin concentration in the blood declined progressively and at about 15 months a plateau value was approached (0.014 µg ml^{-1} at an intake of 211 µg day^{-1}). The ratio of the concentration of dieldrin in adipose tissue to that in blood was 136 (95% confidence limits 109–170). Both values were dependent on the amount of dieldrin ingested daily. When dieldrin administration was terminated, its concentration in blood fell exponentially with a half-life of 141–592 days. It was of interest that both the body burden of dieldrin (blood and fat values) and its rate of elimination (post-exposure blood values) were characteristic not only of the amount ingested, but also of the *individual*. This suggested that individual variations in the rate of metabolism of dieldrin had a profound effect on its kinetics.

The biotransformation of dieldrin has now been studied extensively in rat and mouse (Richardson *et al.*, 1968; Baldwin *et al.*, 1972; Hutson, 1976a), rabbit, rhesus monkey, and chimpanzee (Müller *et al.*, 1975) and sheep (Feil *et al.*, 1970); various steps have been studied *in vitro*. The mechanisms of biotransformation of dieldrin in mammals have been reviewed in some detail (Bedford, 1975; Bedford and Hutson, 1976). The major route of metabolism involves a hepatic mono-oxygenase-catalysed hydroxylation to afford *syn*-12-hydroxydieldrin (1.2) which is excreted in the bile as a glucuronide conjugate (rats and mice). After deconjugation by intestinal bacteria, and some enterohepatic circulation, this metabolite is excreted in the faeces. Another route of metabolism is conversion into 4,5-*trans*-dihydroaldrindiol, (1.3). This metabolite is also conjugated and excreted in the bile/faeces. The reaction is not catalysed by epoxide hydrolase but probably occurs via 4-hydroxylation, rearrangement, and stereoselective reduction of the resulting keto group (Bedford, 1975). A minor metabolite, excreted in the urine of male rats and probably formed as an alternative to 12-hydroxydieldrin (Bedford and Hutson, 1976), is the caged structure known as dieldrin pentachloroketone (1.4). Unchanged dieldrin is not eliminated in the urine; indeed, it is likely that dieldrin is not eliminated as such by any route and that metabolism must precede elimination.

This background knowledge, together with human data from a variety of sources, allows us to construct a picture of the metabolism of dieldrin in man. There is considerable interindividual variation in the hepatic microsomal mono-oxygenase in humans (see below). Thus, if this enzyme is important in dieldrin metabolism, we should expect the individual variations found in the kinetic study. Neither dieldrin nor its known metabolites (1.2, 1.3, 1.4) were detected in the urine (<0.001 µg ml^{-1}) of manufacturing plant workers (Richardson and Robinson, 1971), however, *syn*-12-hydroxydieldrin (1.2) was found in the faeces and characterized unequivocally. It is possible that urinary conjugates may have been missed by the methods used for extraction. However, it seems likely from the radiochemical studies of Feldman and Maibach (1974) that dieldrin metabolites are eliminated mostly in the faeces. Six male volunteers were given intravenous injections of [^{14}C]dieldrin (several micograms, unspecified). Only 3% of the radioactivity was eliminated in the urine in 5 days. This value is very similar to that obtained with the rat (dose 3 mg kg^{-1}) (Hutson, 1976a). It should be stressed, however, that the metabolism of dieldrin in rats is more rapid than that in man. A kinetic study of dieldrin in rats (Robinson *et al.*, 1969) has shown that the body burden, as judged by the analysis of blood and adipose tissue, is released with a half-life of 10.2 days. The hydroxylation of dieldrin by rat liver microsomes is a very slow reaction (Hutson, 1976a); if the ratio of the rates of metabolism by rat and human liver preparations is of the same order as that for metabolism *in vivo*, extremely low rates of hydroxylation would be predicted with human liver *in vitro*. Such an experiment is unlikely to afford useful results.

In summary, the metabolism of dieldrin in man is slower than that in the common experimental animals but the major biotransformation pathway and elimination route is common to man and animals. The kinetic studies of Robinson and co-workers have allowed a useful assessment of the burden of dieldrin in adipose tissue from the analysis of dieldrin in blood.

Endrin

Endrin (Figure 7.2, 2.1) is the common name for the *endo, endo* stereoisomer of dieldrin; its substituents are numbered as are those of dieldrin. It was introduced as an insecticide with dieldrin about 25 years ago. It is more acutely toxic than dieldrin to mammals (having an acute oral LD$_{50}$ value of 7.5–17.5 mg kg^{-1} compared with 50–60 mg kg^{-1} for dieldrin) (Barnes and Heath, 1964; Bedford *et al.*, 1975b). The values are likely to be in a similar ratio for man. Many more cases of fatal intoxication have occurred with endrin than with dieldrin. Most of these have resulted from suicides or from one incident in which flour was contaminated with endrin (Jager, 1970, p. 86).

Endrin has received less toxicological study than has dieldrin. The reason for this is that endrin is not a persistent pesticide. It has not been detected in the general environment, e.g. in the adipose tissue of the general population (England), seals (North Sea), or penguins (Antarctica) (Hutson *et al.*, 1975; Jager, 1970, p. 86). Furthermore, it could not be found in the blood or adipose tissue of occupationally exposed workers (Hayes, 1967). The higher acute toxicity of endrin and its lower residues in body fluids have prevented the

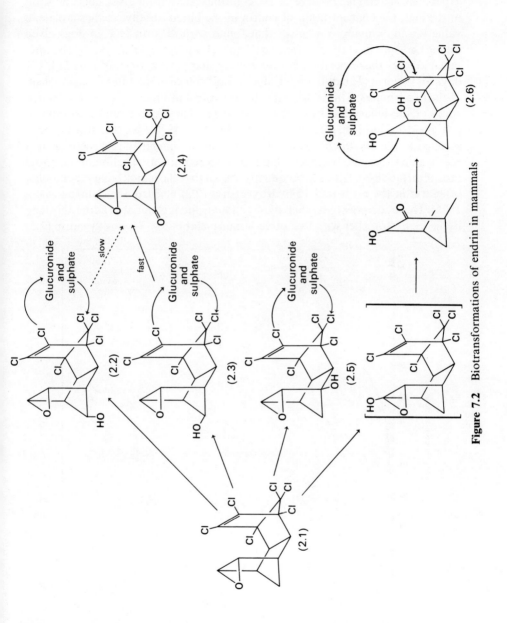

Figure 7.2 Biotransformations of endrin in mammals

undertaking of a kinetic study of the type carried out with dieldrin. Studies on experimental animals, however, have confirmed that, for a given concentration in the diet, the concentration of endrin in the blood rapidly reaches a plateau value which is much lower than that found with dieldrin (e.g. in dogs given 0.1 mg kg^{-1} per day) the difference is 33-fold (Richardson *et al.*, 1967). The time taken to reach the respective plateaus was less than a week for endrin but 114–121 days for dieldrin. The difference is due to a difference in the rates of metabolism of the two isomers. This is dramatically illustrated in Figure 7.3, derived from a study of the biliary secretion of radioactivity following the administration of [^{14}C]dieldrin and [^{14}C]endrin to rats (Cole *et al.*, 1970). Neither insecticide is secreted in the bile unchanged and, in view of the similarity in structures, it is reasonable to assume that the initial rate of secretion is a reflection of the initial rate of metabolism. It was subsequently shown that the major biliary metabolite of endrin in the rat is *anti*-12-hydroxyendrin (2.2) glucuronide (Hutson *et al.*, 1975). This undergoes enterohepatic circulation and is eliminated as the aglycone in the faeces together with two other minor metabolites, 3-hydroxyendrin (2.5)

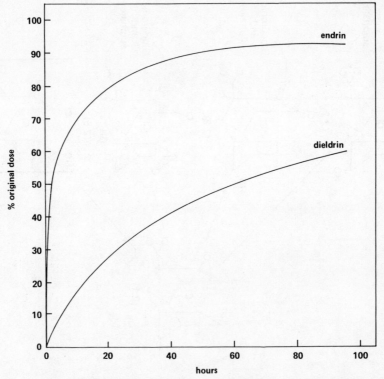

Figure 7.3 Cumulative excretion profile of radioactivity in bile following a single intravenous dose (0.25 mg kg^{-1}) of [^{14}C]dieldrin and of [^{14}C]endrin to male Holtzmann rats (Cole *et al.*, 1970)

and 4,5-*trans*-dihydroisodrindiol (2.6). The major urinary metabolite of endrin in male rats (only 1–2% of the dose) is 12-ketoendrin (2.4). This metabolite is formed by the action of microsomal mono-oxygenase on *syn*-12-hydroxyendrin (2.3) (Hutson and Hoadley, 1974), i.e. the isomer in which the hydroxyl group is orientated towards the epoxide oxygen atom. 12-Ketoendrin is an interesting metabolite in that it is 5–6 times more toxic to rats than is its parent, endrin (Bedford *et al.*, 1975b). A study of the acute toxicity of endrin and its metabolites to rats (Bedford *et al.*, 1975b) has indicated that 12-ketoendrin may be the acute toxicant in endrin poisoning in that species. The metabolism of endrin in the rabbit is superficially different to that in the rat (Bedford *et al.*, 1975a) and confuses the extrapolation to man. The major metabolite is still *anti*-12-hydroxyendrin. However, this is conjugated with sulphate and eliminated in the urine. Some *syn*-12-hydroxyendrin was also detected as its sulphate in urine (possibly conjugation and elimination prevented further oxidation to 12-ketoendrin). The respective glucuronide conjugates were also eliminated in the urine as were the glucuronides of 3-hydroxyendrin and the 4,5-*trans* diol (2.6). The molecular weight threshold for biliary secretion of anions in the rat is 325 ± 50, and in the rabbit, is 475 ± 50 (Hirom *et al.*, 1972). Therefore, it should be no surprise that conjugates of the endrin metabolites are eliminated in the bile/faeces of rats and in the urine of rabbits. The molecular weight threshold in man lies between those of the rat and rabbit but closer to the latter. These molecular weight thresholds must always be taken into account when attempting to monitor metabolites in human urine or when comparing the metabolism of a xenobiotic in man and experimental animals. The primary hydroxylation steps, however, occur at similar rates and in a similar order in both the rat and the rabbit, namely: *anti*-C-12 $\gg$ C-3 > *syn*-C-12 > C-4. The ratios are, for rat 50:7:1.5:1, and for rabbit, 40:5:4:1 (Bedford and Hutson, 1976)

The consistent failure to detect endrin in the blood of humans, including that of workers on the manufacturing plant, indicates that the insecticide is metabolized more rapidly than its isomer, dieldrin, in man. This isomer effect has been explained (Bedford and Hutson, 1976) in terms of the steric influence of the epoxide atom on C-12 hydroxylation. Hydroxylation at *syn*-C-12 is inhibited, therefore the metabolism of dieldrin is slow. Hydroxylation at *anti*-C-12 is relatively rapid and accounts for the relatively rapid metabolism of endrin. Even *syn*-12-hydroxyendrin is rapidly hydroxylated at its *anti*-C-12 position (affording 12-ketoendrin). These differences are illustrated in Figure 7.4 in which the thickness of the arrows indicates the approximate extent to which each process occurs. The bulky hexachlorinated fragment inhibits attack at C-3 and C-4 of both isomers.

Anti-12-hydroxyendrin has been detected in the faeces of occupationally exposed personnel (Baldwin, 1980). Its glucuronide has been detected in the urine of these workers (Baldwin and Hutson, 1980) by a method involving periodate cleavage of the conjugate. The toxic metabolite, 12-ketoendrin, which can be

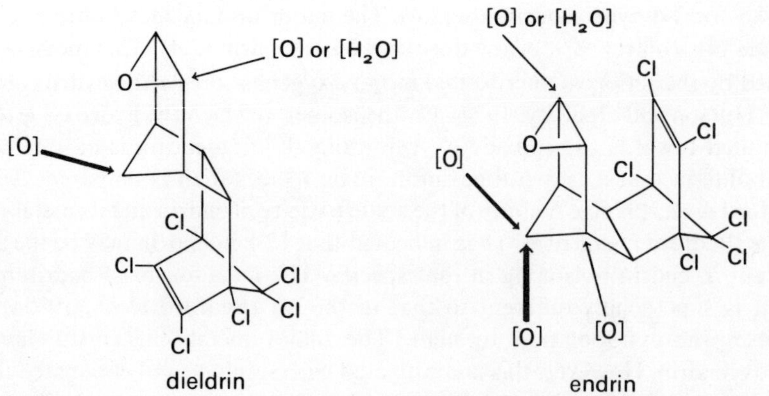

Figure 7.4 Comparative oxidative biotransformations of dieldrin and endrin

analysed with high sensitivity, could not be detected. 3-Hydroxyendrin and the diol (2.6), neither of which can be analysed with the high sensitivity associated with endrin or 12-ketoendrin, were not detected in faeces or urine. Thus, although the data are not quantitative in relation to exposure, the available evidence suggests that endrin is metabolized by man as it is in experimental animals.

DDT

Strictly speaking, DDT is an approximately 3:1 mixture of 1,1-*bis*-(4-chlorophenyl)-2,2,2-trichloroethane (Figure 7.5, 5.1) and 1-(2-chlorophenyl)-1-(4-chlorophenyl)-2,2,2-trichloroethane. These are commonly referred to as *pp'*DDT and *op'*DDT, respectively. DDT is the now classic example of a persistent organochlorine insecticide which has saved the lives of millions of people through disease vector control and crop protection. In the process it has become widely distributed throughout the environment. The concentration of DDT or rather its metabolite DDE (5.2) in human fat, for example, is many fold higher than the concentration of dieldrin. The difference was found to be 34-fold in New Orleans around 1960 (Hayes *et al.*, 1965). Undoubtedly DDT has been responsible for some large-scale kills of birds, fish, and other non-target species and other effects (e.g. eggshell thinning) have been produced in species near the top of the food chain. However, adverse effects of DDT in man, even when deliberately ingesting the insecticide, have been very difficult to demonstrate. The ensuing argument, which has fuelled a gargantuan scientific effort, is outside the scope of this chapter. It has been reviewed in a long, but very entertaining article by Dr. D. L. Gunn (1975) entitled 'Uses and abuse of DDT and dieldrin'.

A number of studies of the fate of ingested DDT in human volunteers were carried out by W. J. Hayes and co-workers in the late 1950s and 1960s. The results have been summarized conveniently in one paper (Hayes et al., 1971). The full metabolic fate of the compound was not known at the time but dose dependent storage of DDT and DDE in adipose tissue was demonstrated. The excretion of the urinary metabolite DDA (5.3) was found to be proportional to the dose of DDT. At 35 mg of pure pp′DDT per day, urinary DDA reached a plateau (c. 0.2 mg h^{-1}) after approximately 20 weeks. Storage in adipose tissue reached 200–300 µg g^{-1} at 52–94 weeks. Elimination of DDT from the fat was very slow, with a half-life of about 2 years. There were no compound-related effects on the health of the volunteers and a high degree of safety of DDT for the general population (at that time ingesting 500–1000-fold less) was inferred. A more recent study was reported in 1974 by Morgan and Roan. The metabolic pathway proposed by Peterson and Robison (1964) was used as background for the study. This scheme (Figure 7.5) is probably essentially correct, if incomplete (phenols have recently been demonstrated as metabolites in rats and sheep). op′DDT,

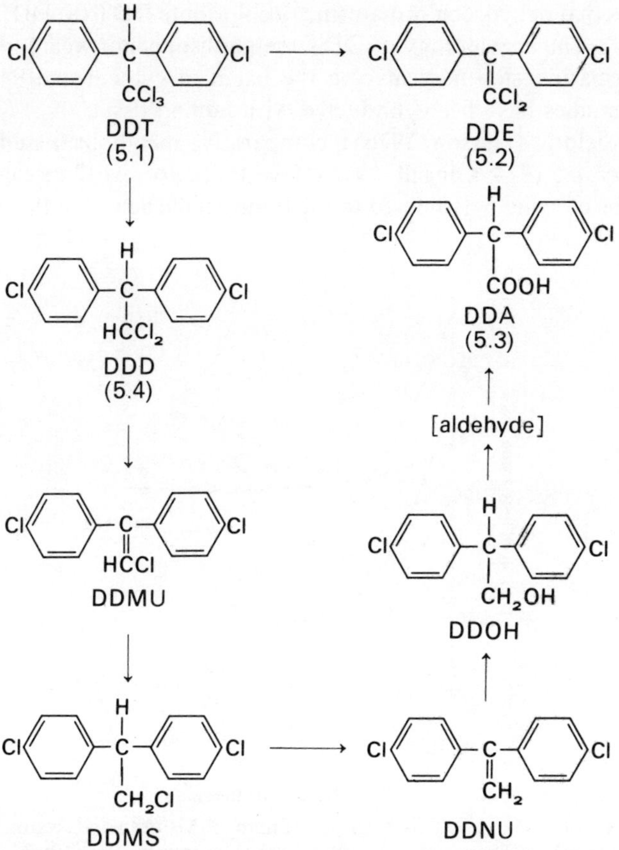

Figure 7.5 Biotransformations of DDT in mammals (Peterson and Robison, 1964)

pp'DDE (5.2), and pp'DDD (5.4) in blood and adipose tissue were measured. Urinary pp'DDA (5.3) was also measured but it is difficult to judge what proportion of the conjugates would have been hydrolysed during the isolation of the DDA. The results of the study indicated that DDT was readily absorbed from the human intestine. Some is then converted slowly into DDE and some into DDD, thence into DDA which was eliminated in the urine. The rates of depletion of DDT and its metabolites from the body lay in the order: p,p'DDA $\gg p,p'$DDD $> o,p'$DDT $> p,p'$DDT $> p,p'$DDE. It was concluded that the body burden of DDT (25 mg) in the average American adult resident carrying 17 kg of fat would be lost within 1–2 decades. The elimination of DDE (75 mg), however, will require most of man's natural life-span.

There are marked species differences in the rate of loss of DDT from adipose tissue. These are compared in Figure 7.6 for rat (Datta and Nelson, 1968), dog (Deichman et al., 1969), monkey (Durham et al., 1963), and man (Hayes et al., 1971; Morgan and Roan, 1974). These rates must reflect the activity/efficiency of the enzymes that dehydrochlorinate and dechlorinate DDT (to DDE and DDD, respectively). The enzymology of DDT metabolism is not well understood but the dechlorination step may involve the haem of cytochrome P450 (Stotter, 1977). No studies have been conducted with human tissue.

As with dieldrin (Hutson, 1976a), comparative metabolism studies of DDT (Wallcave et al., 1974; Gingell, 1976) have failed to reveal species differences which can be convincingly related to the tumorigenic action of the insecticide in

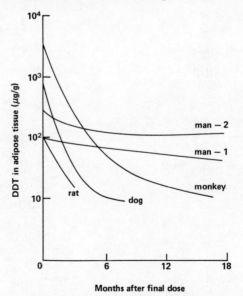

Figure 7.6 Depletion of DDT from the fat of man (1, Morgan and Roan, 1974; 2, Hayes et al., 1971), monkey (Durham et al., 1963), dog (Deichman et al., 1969), and rat (Datta and Nelson, 1968)

mice. Therefore, metabolism studies in man with this end in view are unlikely to be helpful at this time. The relatively recently discovered aryl hydroxylation of DDT at the 2, 3, and 4 (with NIH shift) positions (Sundström et al., 1977) found in rats and pigs may be of more interest in this respect. These hydroxylations probably proceed via arene oxide formation and could conceivably relate to the tumorigenic action of DDT.

The conjugates of DDA deserve some further study, particularly in man. Carboxylic acids are conjugated with glucuronic acid and/or a variety of amino acids. These processes are notoriously species dependent. The development of a monitoring method based on urine analysis would necessitate at least a hydrolysability study of the various conjugates.

This survey of the metabolism of three organochlorine insecticides in man indicates that metabolism proceeds via the same major pathways as found in experimental animals. The striking feature is the slow rate of metabolism in man. Dieldrin and endrin are metabolized by cytochrome P450-dependent monooxygenase (which is reasonably effective in the case of the isomer containing the unhindered C-12-H bond); DDT may also be metabolized by a reaction involving cytochrome P450. It appears that human cytochrome P450 is less effective (towards these particular xenobiotics) than that of rat, mouse, etc.

ORGANOPHOSPHORUS INSECTICIDES

The organophosphorus insecticides are neutral esters of phosphoric acid or its thio analogues. Their nomenclature is illustrated with typical examples in Figure 7.7. The phosphonates possess a carbon–phosphorus bond which appears to be

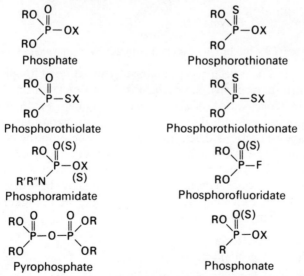

Figure 7.7 Structures of some organophosphorus compounds

biologically very stable (Menn and McBain, 1974). In other respects, however, they are very similar to the other organophosphorus insecticides. The organophosphates are toxic to mammals and insects by interference with the nervous system. The target enzyme is acetylcholinesterase, the normal function of which is to remove the neurotransmitter, acetylcholine, after it has functioned in the synaptic transmission of a nerve impulse. The enzyme is inhibited by dialkylphosphorylation of the serine hydroxyl group at its active site. The resulting phosphorylated enzyme may be stable, it may undergo regeneration by spontaneous hydrolysis, or it may 'age' (by dealkylation) to a very stable, permanently inhibited, species. Other esterases, e.g. chymotrypsin, lipases, carboxylesterases, and the 'cholinesterases' in plasma and red blood cells, are also inhibited, but without acute effects.

The inhibitory power of an organophosphorus ester derives from the affinity between its electrophilic phosphorus atom and the nucleophilic serine hydroxyl group of the enzyme (Figure 7.8). The electrophilicity of phosphorus may be adjusted by the use of suitable electron-withdrawing groups in the OX moiety (Figure 7.7). The nitro group of paraoxon (Figure 7.8, 8.2) is an example. Figure 7.8 also illustrates the principle of the oxidative bioactivation of the phosphorothionates.

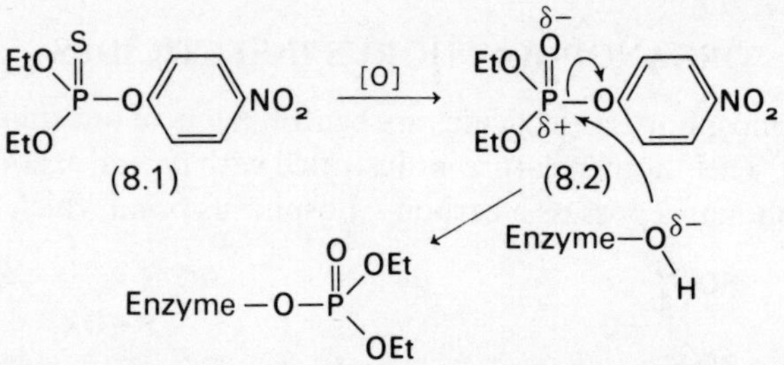

Figure 7.8 Inhibition of a cholinesterase by a phosphate triester

Insect/mammalian selectivity is obtained by variation of lipophilic character, stability, intrinsic reactivity, stereochemistry, and metabolism. Metabolism is important in bioactivation of the thionates (Figure 7.8). In the case of the phosphates, however, with the exception of certain variations in the OX side chain, metabolism affords detoxification. This is because the conversion of a triester (Figure 7.9, 9.1) into a diester (9.2) destroys the electrophilic character of the phosphorus atom and hence the inhibitory potency of the insecticide.

Various enzymatic mechanisms of detoxification have been discovered for the organophosphates. Some important ones are illustrated in Figure 7.10 for ethyl methyl p-nitrophenyl phosphate:

(1) Hydrolytic dearylation
(2) Glutathione-dependent dearylation
(3) Glutathione-dependent demethylation
(4) Oxidative de-ethylation

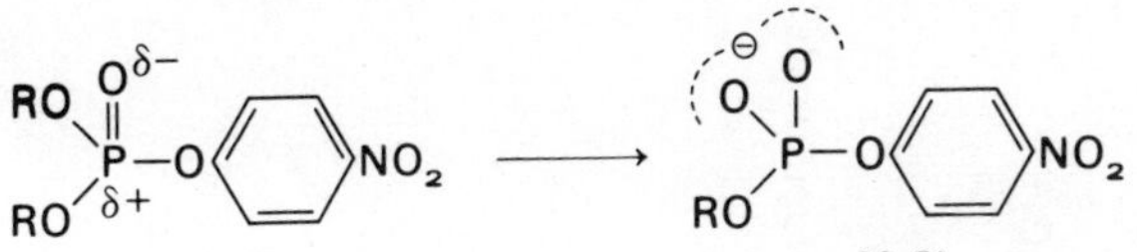

Figure 7.9 Detoxification of a phosphate triester to a diester

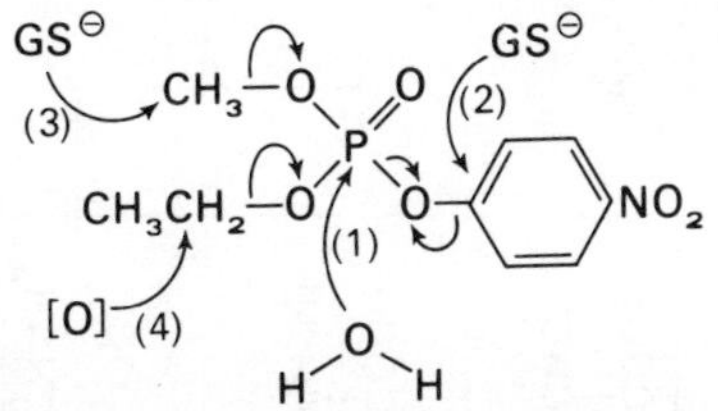

Figure 7.10 Detoxification mechanisms operating on a phosphoric acid triester

All or some of these general reactions may occur simultaneously, together with various compound-specific reactions on the periphery of the esterifying groups (e.g. nitroreduction in the example cited in Figure 7.10; side chain de-esterification of malathion). Both metabolism and its effects are difficult to predict in many cases. A safe prediction, however, is that with the variety of mechanisms available for biotransformation, the organophosphorus insecticides should be readily degraded in animals and man. Many studies in experimental animals have confirmed rapid metabolism. The situation in man is, of course, less well studied but a certain amount of work has been done. The study of the elimination of radioactivity derived from intravenously administered pesticides cited above (Feldman and Maibach, 1974) included monocrotophos, ethion, azinphosmethyl, malathion, and parathion. The urinary elimination of radioactivity was 40–90% in 5 days. These results are typical of those obtained in experimental animals and demonstrate that man has a high capacity for the metabolism of the organophosphorus pesticides.

One reason for the limited amount of information on the metabolism of the organophosphorus pesticides in man stems from the convenience and sensitivity of measurements of blood cholinesterase depression. Indeed, the automatic titrator is to organophosphorus compounds what the electron-capture detector is to the organochlorines. Unfortunately the relationship between cholinesterase depression and the concentration of the insecticide *in vivo* does not have the precision of g.l.c. analysis and no information on metabolites can be obtained using the procedure. Metabolism studies using ^{14}C-labelled materials have been carried out in human volunteers; some of these are described below.

Dichlorvos

This insecticide and anthelmintic possesses a very simple structure (Figure 7.11, 11.1) and is a typical member of the phosphoric acid triester class. It is an

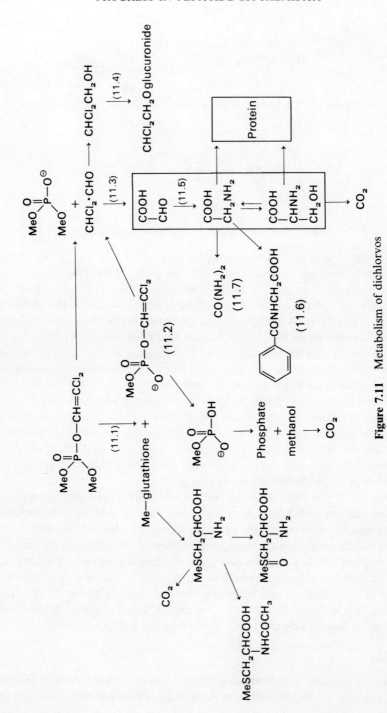

Figure 7.11 Metabolism of dichlorvos

inhibitor of insect and mammalian cholinesterases and it is a very active insecticide. Its fate in mammals is characterized by very rapid degradation (Hutson et al., 1971a; Hutson and Hoadley, 1972a; Blair et al., 1975) by hydrolysis and by glutathione-dependent demethylation. Its biochemical reactivity, particularly in relation to its weak methylating reactivity, has been reviewed recently (Wright et al., 1979). The radioactive metabolites derived from ^{14}C-methyl and ^{14}C-vinyl labelled dichlorvos are shown in Figure 7.11. They include CO_2, hippuric acid (11.6), and urea (11.7), all derived from the efficient incorporation of the vinyl carbon atoms into the 2-carbon pathway, and desmethyldichlorvos (11.2). The yields of these metabolites and of urinary and faecal radioactivity were measured in rats, mice, hamsters, and in a human volunteer (Hutson and Hoadley, 1972b). The results (Table 7.1) show that, in spite of the various dosages used, the yields of metabolites from the human were similar to those from other species. In a separate experiment dichloroethanol (11.4) (0.4 $\mu g\, ml^{-1}$) was detected in the urine of a subject who had inhaled dichlorvos (38 $\mu g\, l^{-1}$) for 105 min. Thus the major biotransformation pathways of dichlorvos in animals (Figure 7.11) were shown to be operative in man. Of the species studied, the mouse clearly emerged as atypical.

Table 7.1 The yields of radioactivity and metabolites 24 h after the oral ingestion of [^{14}C-vinyl]dichlorvos

Metabolite	Elimination route	Per cent dose in following species:			
		Rat (3.6)[a]	Mouse (7.3)[a]	Hamster (1.5–3.7)[a]	Man (0.07)[a]
^{14}C	Urine	9.8	27.4	14.7	7.6
^{14}C	Faeces	1.5	3.2	2.9	—
$^{14}CO_2$	Respiration	28.8	23.1	33.5	27.0[b]
Hippuric acid	Urine	1.7	0.6	1.0	0.4
Demethyl dichlorvos	Urine	2.2	18.5	—	0.15
Urea	Urine	0.6	0.6	—	0.1

[a] Dose in $mg\, kg^{-1}$.
[b] Eight hours only.

Chlorfenvinphos

This insecticide (Figure 7.12, 12.1) is, like dichlorvos, a member of the vinyl phosphate series. Metabolism studies in rats and dogs (Hutson et al., 1967) and *in vitro* studies with liver enzymes (Donninger et al., 1972; Hutson et al., 1976) revealed the biotransformation pathways shown in Figure 7.12. The primary detoxification reaction was oxidative de-ethylation to afford de-ethyl-chlorfenvinphos (12.2) part of which was excreted and part of which was further

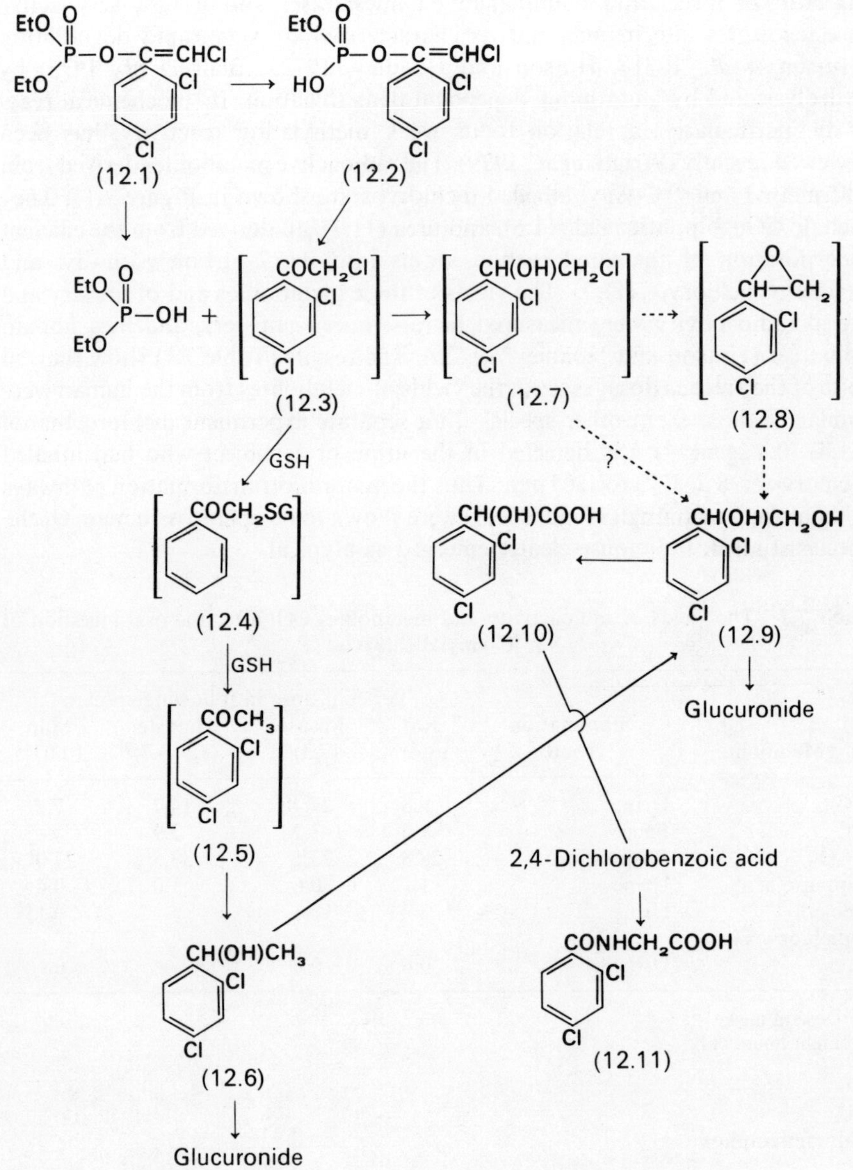

Figure 7.12 Metabolism of chlorfenvinphos in mammals

metabolized via 2,4-dichlorophenacyl chloride (12.3) and 2,4-dichloroacetophenone (12.5) to 2,4-dichlorophenylethanol (12.6), the ethandiol (12.9), and 2,4-dichloromandelic acid (12.10). The biotransformation of [^{14}C]chlorfenvinphos was then studied in a human volunteer. An oral dose of 0.18 mg kg^{-1}

was almost totally eliminated (72% in 4.5 h; 94% in 26.5 h). The yields of the urinary metabolites are compared with those from rat, dog, and rabbit in Table 7.2. Each species had its distinctive metabolite profile, part of which may have been a feature of dose dependency. It was clear that de-ethylchlorfenvinphos, 2,4-dichloromandelic acid, or 2,4-dichlorophenylethanol glucuronide could be used to monitor occupational exposure. A further volunteer study was carried out using non-radioactive chlorfenvinphos in order to relate the yield of a specific urinary metabolite (de-ethylchlorfenvinphos) to amount ingested (Hunter et al., 1972). Fourteen male volunteers were each given 3 mg of chlorfenvinphos daily for 53 days. During the exposure, the average daily excretion of the metabolite was 120 ±5 µg. This was equivalent to only 4.7 ±0.2% of the dose. It could not be detected in the post-exposure period. The relatively low yield could have been a feature of the lower dose given in the second study. It is always possible that compounds subject to metabolism by several routes or steps exhibit dose dependent metabolic profiles. Given a similar subcellular location, the enzyme with the lowest K_m value will be the most effective at low doses. The results with chlorfenvinphos in humans indicate that, as is the case with dichlorvos in animals (Wright et al., 1979), hydrolytic cleavage of the P–O–vinyl bond predominates at low dose levels.

Table 7.2 Comparative metabolism of [^{14}C]chlorfenvinphos

Metabolite (Figure 7.12)	Per cent dose in following species:			
	Rat (2.0)[a]	Dog (0.3)[a]	Rabbit (4.0)[a]	Man (0.18)[a]
Chlorfenvinphos (12.1)	0	0	0	0
De-ethylchlorfenvinphos (12.2)	27[c]	65[c]	47	22[c]
2,4-dichloromandelic acid (12.10)	5	12	3	22[c]
2,4-Dichlorophenylethanol (12.6)[b]	30	3	28	21
2,4-Dichlorophenylethandiol (12.9)[b]	2	2	3	14
2,4-Dichlorobenzoylglycine (12.11)	3	<1	1	4

[a] Dose in mg kg^{-1}.
[b] As glucuronide conjugates.
[c] Isotope dilution analysis on whole urine.

Ethyl Parathion and Methyl Parathion

Ethyl parathion (8.1) (1 and 2 mg) was given to two male volunteers daily for 5 days; similarly methyl parathion (1) (2 and 4 mg) was given to two other volunteers (Morgan et al., 1977). Thus the dose range was 0.014–0.06 mg kg^{-1}. There were no symptoms of intoxication but a depression of blood cholinesterase was noted. In each case the elimination of p-nitrophenol was essentially complete within 24 h of stopping the ingestion (60% in 4 h; 86% in 8 h). Urinary elimination of dimethyl phosphate (DMP) and diethyl phosphate (DEP) was more prolonged, peaking at 4–8 h after ingestion; DEP could be detected up to 72 h. Diethyl thiophosphate (2) (a minor metabolite) was eliminated within 4 h. In terms of absolute yield, *p*-nitrophenol (free and conjugated) accounted for 37% of the dose of ethyl parathion, and the total alkyl phosphates for more than 50% of the dose. These values were maintained over the 1–4 mg dose range. Thus most of a dose of ethyl parathion was oxidized to the oxon (8.2) and hydrolysed before elimination.

$$\begin{array}{cc} \underset{MeO}{\overset{MeO}{\diagdown}}\!\!\overset{\overset{S}{\|}}{P}\!-\!O\!-\!\!\!\left\langle\!\!\bigcirc\!\!\right\rangle\!\!-\!NO_2 & \underset{EtO}{\overset{EtO}{\diagdown}}\!\!\overset{\overset{S}{\|}}{P^{\ominus}}\!\!\underset{O}{} \quad H^{\oplus} \\ (1) & (2) \end{array}$$

The yield of dimethyl phosphate from methyl parathion was only about 12% in 24 h. This may be due to the operation of a competing route, glutathione-dependent demethylation to methyl *p*-nitrophenyl phosphate which, of course, cannot afford DMP on further hydrolysis.

General

General methods involving the measurement of urinary dialkyl phosphates and thionates (Shafik and Bradway, 1976) are still under development involving improved extraction (Lores and Bradway, 1977) and derivatization procedures (Blair and Roderick, 1976). Specific methods involving measurement of the OX leaving group (Figure 7.7) (e.g. *p*-nitrophenol, etc.) are also used. These methods are useful for the experimental testing of the relationship between metabolite yield and dose. However, as they rely on the study of predicted metabolites, their development and use very rarely reveal novel pathways.

A novel metabolite of chlorpyrifos (3) has been found by a thorough analysis of the liver from a human who had received a lethal dose of the compound (Lores *et al.*, 1978b). The metabolite (4) was formed by apparently substituting a methylmercapto group on to the electrophilic aromatic carbon atom. The mechanism of formation of this type of metabolite is not yet clear but such metabolites are being found with increasing frequency in animal excreta as methods improve.

$$\underset{(3)}{\underset{EtO}{\overset{EtO}{\diagdown}}\!\!\overset{\overset{S}{\|}}{P}\!-\!O\!\!\left\langle\!\!\underset{N}{\overset{Cl}{}}\!\!\right\rangle\!\!\underset{Cl}{\overset{Cl}{}}} \longrightarrow \underset{(4)}{\underset{EtO}{\overset{EtO}{\diagdown}}\!\!\overset{\overset{S}{\|}}{P}\!-\!O\!\!\left\langle\!\!\underset{N}{\overset{Cl}{}}\!\!\right\rangle\!\!\underset{Cl}{\overset{SMe}{}}}$$

The few studies of the metabolism of organophosphorus insecticides in humans tend to confirm the rapid metabolism found in animals. However, while the major portion of ingested insecticide may be rapidly eliminated, it is possible, especially with the more stable and lipophilic phosphorothionates, that the retained dose (e.g. that in adipose tissue) may be eliminated only very slowly. An unusual clinical picture encountered following the ingestion of dichlorofenthion (**5**) prompted some further study (Davies *et al.*, 1975). This insecticide is in widespread horticultural use in Florida.

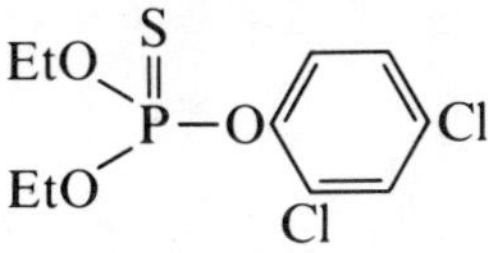

$$\text{(EtO)}_2\text{P(S)}-\text{O}-\text{C}_6\text{H}_3\text{Cl}_2$$

(**5**)

A study of five suicidal ingestions of dichlorofenthion, in which three patients survived, revealed that the parent compound could still be detected in the adipose tissue of one patient after 54 days. It could be detected in the blood of another patient 75 days later. Even here though, the half-life of the elimination of the insecticide from the blood was of the order of 2–10 days. 2,4-Dichlorophenol was detected in the urine of these patients.

A thorough approach to cases of this type is very valuable. Further examples are described by Lores *et al.* (1978a) involving poisoning cases with dicrotophos, chlorpyrifos, malathion, and parathion. Not only do such studies identify the pesticide involved in the poisoning, they also provide data on toxicity and metabolism which cannot be obtained by experimentation.

CARBAMATE INSECTICIDES

Some of the earliest human toxicity trials must have been those conducted on the banks of the Cross River estuary in S.E. Calibar Province, Nigeria. Calibar bean seeds were macerated with water and the extract was swallowed by prisoners accused of witchcraft (and by their accusers). If the prisoner vomited and lived he was judged innocent. If he died he was judged guilty. The active principle was later shown to be the *N*-methylcarbamate, physostigmine (**6**). The

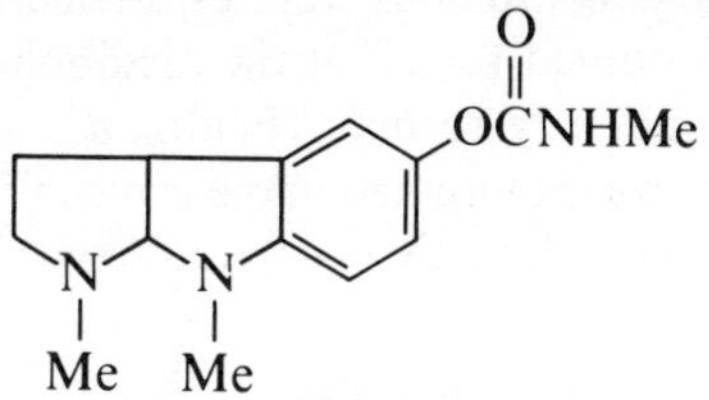

(**6**)

development of the *N*-alkylcarbamate insecticides from physostigmine has been described concisely by Kuhr and Dorough (1976) in an excellent monograph dealing with the chemistry, biochemistry, and toxicology of this class. The carbamates, like the organophosphates, act via the inhibition of acetylcholinesterase. The active site of the enzyme is *N*-alkylcarbamoylated. As with the organophosphorus insecticides, toxicity of the carbamates is controlled by a number of factors and their oral LD_{50} values in the rat vary widely (e.g. aldicarb (**7**), 1 mg kg^{-1}; butacarb (**8**), >4000 mg kg^{-1}). The metabolism of the carbamates has been reviewed by Ryan (1972), Fukuto (1973), and by Kuhr and Dorough (1976).

$$\text{MeS}-\underset{\underset{\text{Me}}{|}}{\overset{\overset{\text{Me}}{|}}{\text{C}}}-\text{CH}=\text{NO}\overset{\overset{\text{O}}{\|}}{\text{C}}\text{NHMe}$$

(**7**)

(**8**) [3,5-bis(2-methyl-2-propyl)phenyl N-methylcarbamate structure with OCNHMe group]

Carbaryl (Sevin)

Carbaryl (Figure 7.13, 13.1) has a very simple structure and it makes an ideal model for the study of carbamates. In addition, since its commercial introduction in 1958 (Back, 1965), it has been highly successful and it is used on more than 100 crops. Major uses are on cotton, soya beans, and corn and it is estimated that 40 million pounds are used annually. The metabolism of carbaryl is complex (Dorough, 1970) and has not yet been fully elucidated. The known pathways are shown in Figure 7.13. Decarbamoylation to 1-naphthol is a major reaction; its mechanism is not necesssarily hydrolytic: NADPH-dependent microsomal oxidation is certainly involved in the ester cleavage of some members of this class (Douch et al., 1971). Oxidation also occurs at various points around the naphthol ring, affording products which retain the N-methylcarbamoyl group (and, therefore, anticholinesterase action). Most animals excrete between 65 and 75% of the radioactivity derived from [1-naphthyl-^{14}C]carbaryl in the urine within 24 h of dosing. A study in man showed that 28–38% of the insecticide was eliminated in the urine (Knaak et al., 1968). However, non-radioactive carbaryl was used and the urinary metabolites were estimated by colorimetic methods which may not have accounted for all of the metabolites. When [^{14}C]carbaryl was administered intravenously to man (Feldman and Maibach, 1974), only 7.4% of the radioactivity was eliminated in the urine in 5 days. In the same series of experiments, 84% of the radioactivity derived from [^{14}C]propoxur (**9**) was so eliminated.

(**9**) [2-isopropoxyphenyl N-methylcarbamate structure with iPrO and OCNHMe groups]

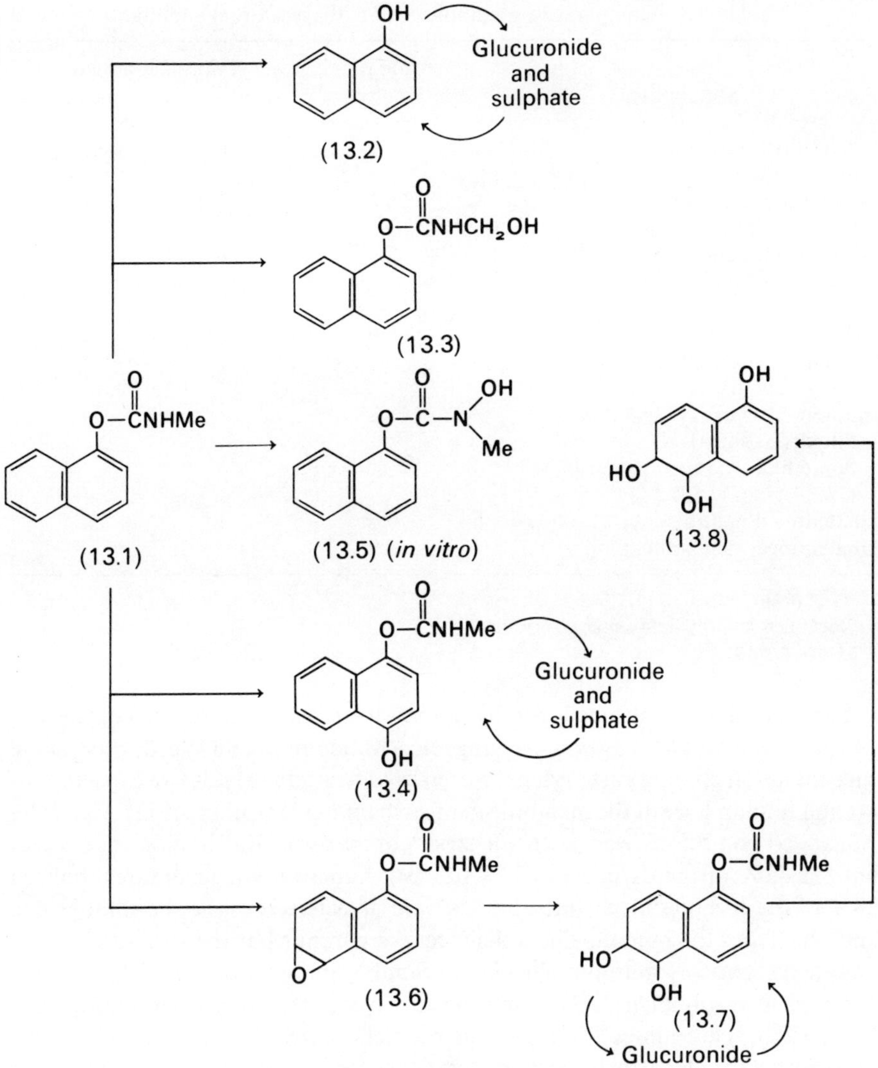

METABOLISM OF INSECTICIDES IN MAN

Figure 7.13 Metabolism of carbaryl in mammals

The main quantitative difference between the various species studied by Sullivan and co-workers (Knaak *et al.*, 1965, 1968) (rat, guinea pig, man, monkey, pig, and sheep) was the extent to which carbaryl was cleaved to 1-naphthol. The reaction was unimportant in monkey and pig but occurred extensively in rat, sheep, and man. Here is an example of the invalidation of the 'classical' animal models for man. The comparative data for rat and man are shown in Table 7.3.

Table 7.3 Comparative metabolism of carbaryl in rats and man

Metabolite	Male rat (8.5–26 mg kg^{-1})[a]	Two male humans (2 mg kg^{-1})[a]	(2 mg kg^{-1})[a]
1-Naphthyl glucuronide (13.2—gluc)	11.2	10.4	15.5
1-Naphthyl sulphate (13.2—sulph)	16.2	11.0	6.1
4-(Methylcarbamoyloxy)-1-naphthyl glucuronide (13.4—gluc)	7.2	4.3	5.8
4-(Methylcarbamoyloxy)-1-naphthyl sulphate (13.4—sulph)	5.0	0	0
trans-5,6-Dihydroxy-5,6-dihydronaphth-1-yl N-methylcarbamate glucuronide[b] (13.7—gluc)	18.2	+[c]	+[c]
Unidentified neutrals	9.2	+[c]	+[c]
Unidentified metabolites	2.9	+[c]	+[c]

[a] Oral dosage.
[b] Identified later by Sullivan et al. (1972b).
[c] Trace present.

Carbaryl is metabolized slowly by human liver homogenate, the enzymes of which were sensitive to ageing, freezing, etc. (Matsumura and Ward, 1966). Five carbamates, including carbaryl, have been used to study the relative capacities of rat and human liver in the metabolism of carbamates (Strother, 1972). Carefully controlled conditions were used for strict comparison. Rat homogenate metabolized 42% of the carbaryl (0.33×10^{-6} M); human homogenate metabolized 28% of the compound. Six metabolites were detected from the rat homogenate and nine from the human. The differences were minor but the structures of the unique metabolites were not elucidated. Similar amounts of 4- and 5-hydroxycarbaryl were obtained from both species; 1-naphthyl-N-hydroxymethyl carbamate (13.3) accounted for 4% of the rat metabolites, but less than 1% of the human metabolites. Carbaryl is metabolized by human embryonic lung cells (Baron and Locke, 1970; Lin et al., 1975). Under the conditions used, total metabolism occurred in 3 days, affording 1-naphthol, 4- and 5-hydroxycarbaryl, the 5,6-dihydrodiol (13.6) and acid hydrolysable conjugates of these metabolites.

An organ-maintenance technique for tissue explants, developed by Sullivan and co-workers (Sullivan et al., 1972a), has been applied to the study of carbaryl metabolism in man (Chin et al., 1974). The technique used 100 µg of carbaryl and 500 mg of tissue (a 'dose' equivalent to 7 mg kg^{-1} in vivo) incubated for 18 h at 37 °C. The anionic metabolites were analysed by chromatography on DEAE–cellulose. The profiles so obtained qualitatively and semiquantitatively reflected the *in vivo* metabolic processes in animals including man. The analogy proved to be much closer than that obtained with homogenates. Activity for carbaryl metabolism by rat tissues lay in the order: liver > kidney > lung; the activity of the human tissues lay in the order: liver > lung > kidney > placenta > vaginal mucosa > uterus. Hepatic tissue effected de-esterification, hydroxylation, and conjugation. The kidney afforded 1-naphthyl glucuronide, whereas the uterus, lung, and placenta afforded the sulphate. A comparison of the metabolism of carbaryl from liver explants from rat (Chin *et al.*, 1979) and human (Chin *et al.*, 1974) revealed that the total metabolism in the rat tissue (75% in 18 h) was higher than that in the human tissue (40–60%). The major metabolites in rat liver, 1-naphthyl sulphate and the 5,6-dihydrodiol glucuronide, were minor in the human liver which afforded mainly 1-naphthyl glucuronide. The difference in the conjugation of 1-naphthol is unimportant; the significant finding is that the arene oxide/diol pathway is about ten times more effective in rat than in man.

Propoxur (Baygon)

This carbamate, 2-isopropoxyphenyl-*N*-methylcarabamate (**9**), has been the subject of a human volunteer study carried out to validate a monitoring procedure (Dawson *et al.*, 1964). The study involved nine males who ingested 50–92.2 mg (*c.* 1 mg kg^{-1}) of non-radiolabelled compound. About 30% of the dose was eliminated as 2-isopropoxyphenol (**10**) (most of this within the first 8–10 h). Once again, however, this study confirmed expectations derived from animal studies and was not designed to discover new information specific to the human.

(**10**)

Other Carbamates

Strother (1972), in addition to studying carbaryl, included aminocarb (**11**, Matacil, 4-dimethylamino-3-methylphenyl *N*-methylcarbamate), mexacarbate (**12**, Zectran, 4-dimethylamino-3,5-dimethylphenyl *N*-methylcarbamate),

(**11**)   (**12**)

methiocarb (**13**, Mesurol, 4-methylthio-3,5-dimethylphenyl *N*-methylcarbamate), and carbanolate (**14**, Banol, 2-choro-4,5-dimethylphenyl *N*-methylcarbamate) in his study using rat liver homogenate. An estimate of the rate of metabolism of the four insecticides is given in Table 7.4. Both species afforded the same major metabolites from the carbamates but quantitative differences were apparent. Ring hydroxylation and *N*-dealkylation were the main reactions observed. Side chain oxidation to the *N*-hydroxymethylcarbamates was slower in man than in rat. Ester cleavage to the phenol was minor in both species. In view of the findings *in vivo*, the low amounts of ester cleavage suggest that the homogenate is lacking in some respects.

Table 7.4 Comparative biodegradation of some carbamate insecticides in rat liver homogenate

Insecticide	Percentage degraded	
	Man	Rat
Matacil	22	67
Zectran	67	75
Mesurol	88	98
Banol	50	63

PYRETHROID INSECTICIDES

The synthetic pyrethroid insecticides have been developed from the natural pyrethrins by structural modifications aimed at retaining bioactivity and increasing photostability (Elliott, 1977). This process has been successful, and, together with some attention to stereochemistry, has also led to increased activity. For example, deltamethrin (**15**), in which a 1 *R cis* acid is esterified with an *S*-cyanohydrin (Elliott *et al.*, 1974), is 1000 times more active than pyrethrin I and has an LD_{50} to house flies of 0.0003 μg. Its acute toxicity to rats (25–63 mg kg^{-1}, Barnes and Verschoyle, 1974) suggests a very favourable mammal–insect toxicity ratio. This compound together with fenvalerate (16), permethrin (17), and cypermethrin (18) form a new generation of pyrethroid insecticides which are now used commercially on cotton, vegetables, and top fruit and as veterinary products (see also Chapter 3 in this volume). Their metabolic fate has been studied in rat, mouse, dog, goat, and cows (Casida et al., 1979; Hutson, 1979; Casida and Ruzo, 1980) and a clear, if complex, picture is emerging. The pyrethroids are carboxylic esters and it is no surprise to find that they are metabolized by ester cleavage to the constituent acid and alcohol. In the case of the cyanopyrethroids (e.g. (15), (16), and (18)) the alcohol is unstable and is rapidly converted into an aldehyde which is largely oxidized *in vivo* to 3-phenoxybenzoic acid. Hydroxylation of the pyrethroids, before and after ester cleavage, occurs but this is apparently a major reaction only in rats (and to a lesser extent in mice).

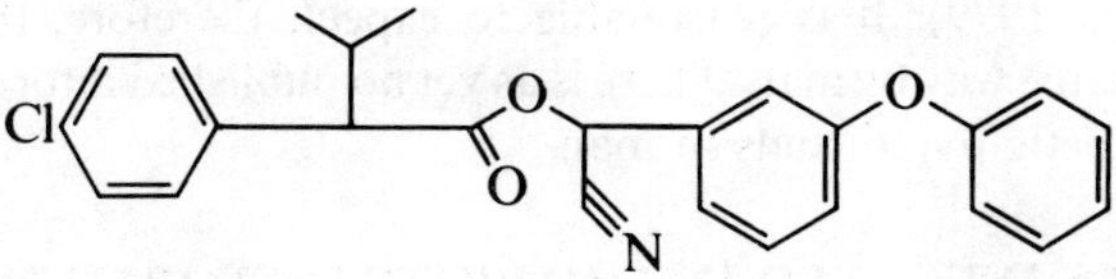

(16)

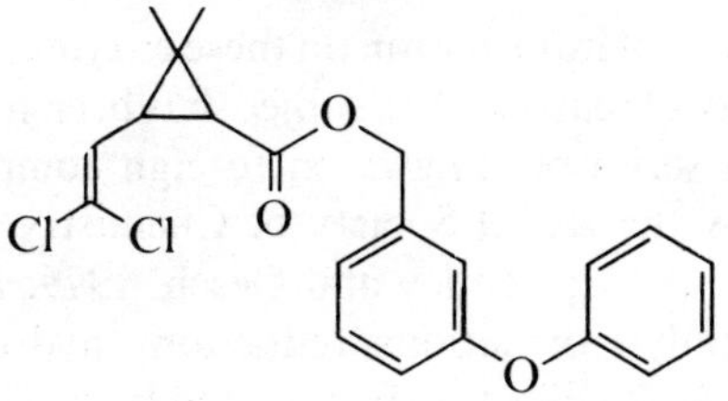

(17)

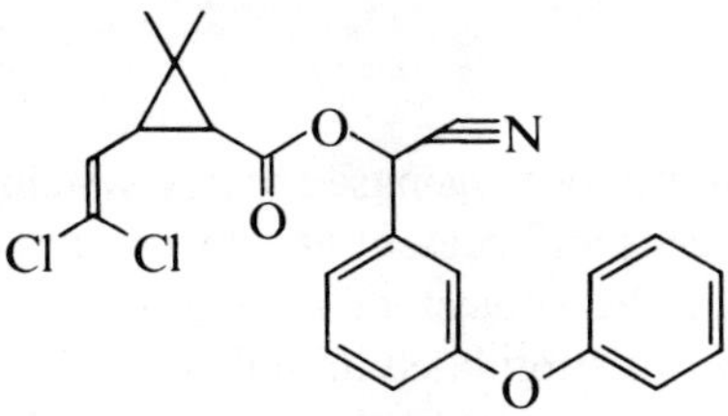

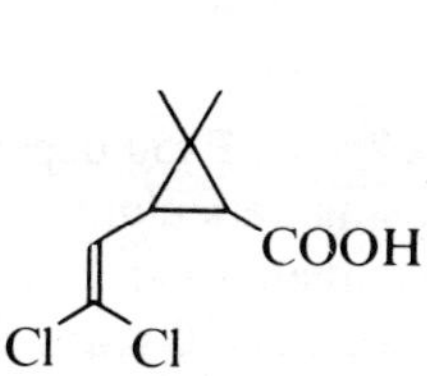

(18)   (19)

Studies of the metabolism of the synthetic pyrethroids in man have not been reported but the animal data should be of considerable help in the selection of metabolites for use in human monitoring. In the four species studied in detail to date, the acid moieties are excreted (as glucuronides) in the urine in 50–70% yields and these are, therefore, the clear choice as the basis for a monitoring method. Moreover, use of the 3-phenoxybenzyl moiety is clearly unsuitable without further study. 3-Phenoxybenzoic acid exhibits a remarkable species specificity in its conjugation. The major metabolites from each of the four species are:

(1) from rat, 3-(4-hydroxyphenoxy)benzoic acid sulphate (Gaughan *et al.*, 1977);
(2) from mouse, 3-phenoxybenzoyltaurine (Hutson and Casida, 1978);
(3) from dog, 3-phenoxybenzoylglycine (Hutson, 1979);
(4) from cow, 3-phenoxybenzoylglutamic acid (Gaughan *et al.*, 1978).

This array of metabolites and the very different behaviour of benzoic acid and 3-phenoxybenzoic acid in this respect render the conjugation of 3-phenoxybenzoic acid in man totally unpredictable.

The fate of the cyclopropanecarboxylic acid (**19**) is not so complicated in that it is excreted mainly as its ester glucuronide in the urine of all species so far studied (Hutson, 1979). It is reasonable to expect, therefore, that it will be excreted in the same way by man. There is as yet no published information on the fate of the synthetic pyrethroids in man.

ENZYMES INVOLVED IN THE BIOTRANSFORMATION OF FOREIGN COMPOUNDS

A considerable amount of information on these enzymes has been generated in recent years for a variety of reasons. The subject has been included regularly over the past 10 years in a series of reviews on foreign compound metabolism in mammals published by the Royal Society of Chemistry (Specialist Periodical Report series, Hutson, 1977a; Bentley and Oesch, 1979; Hirom and Millburn, 1979). The enzymes catalyse oxidations, reductions, hydrolyses, and a series of conjugation reactions. A classification by reaction type affords a suitable system for a brief review with particular reference to information about human enzymes.

Oxidation

The cytochrome P450-dependent mono-oxygenase enzyme system present in the endoplasmic reticulum of liver cells and present at a lower activity in most other mammalian tissues has been the subject of an enormous amount of research in the past 20 years. Study was stimulated initially out of curiosity and then maintained because of the central role of the enzyme in drug metabolism. It has been stimulated further by the realization that oxidative metabolism is often a determinant in chemical mutagenesis, carcinogenesis, and other forms of toxicity.

Studies mostly on the rat and rabbit liver mono-oxygenase have shown it to be a complex of cytochromes P450 and b_5 with their corresponding reductases embedded in phospholipid which is essential to the mechanism of action. This action results in the insertion of an atom of oxygen at a C–H or C=C bond and also to oxidation at nitrogen and sulphur. The mechanism involves:

(1) binding of the substrate to oxidized cytochrome P450 (ferric P450);
(2) electron transfer from NADPH via the reductase, forming a substrate–ferrous P450 complex;
(3) formation of an oxygen–substrate–ferrous P450 complex;
(4) transfer of a second electron;
(5) elimination of water, leaving a species of singlet oxygen bound to iron;
(6) transfer of this oxygen to substrate;
(7) dissociation of product from the (now) oxidized ferric-P450.

The cytochrome, the reductase, and the lipid are fundamental to the process; cytochrome b_5 also interacts in some way, possibly as an effector of cytochrome P450. Multiple forms of cytochrome P450 have been isolated by biochemical methods and different forms have been induced in response to the challenge of different chemicals *in vivo*. Various types of substrate interactions have been identified by difference spectroscopy (types I, II, etc.) (Kulkarni *et al.*, 1975), some of which are related to oxygenation and others to ligand interactions with the cytochrome which may, in some cases, inhibit its action. The rate of oxidation of a particular substrate in an animal depends on several factors including:

(1) partitioning of the substrate into the endoplasmic reticulum from the cytosol;
(2) the amount of cytochrome P450 in the liver cell;
(3) the nature of its interaction with substrate; and
(4) the activity of the NADPH-cytochrome P450 reductase.

Clearly the system is a very complex one and the measurement of no single parameter is a safe guide to the capacity of the enzyme in a particular tissue to effect a reaction. Even the measurement of the oxidation of a substrate of immediate interest can be the subject of dramatic species differences. For example, the oxidative de-ethylation of chlorfenvinphos (12.1) to the diester (12.2) in liver slices occurs at the following relative rates: rat, 1; mouse, 8; rabbit, 24; dog, 88 (Donninger *et al.*, 1972).

The presence of cytochrome P450 in human liver microsomes was first reported by Alvarez *et al.* (1969). In its main features the mono-oxygenase system appears to be similar in man and experimental animals with the exceptions that the human enzyme activity is very variable and the cytochrome P450 content of human liver microsomes is lower than that in the livers of the common laboratory animals. Using samples obtained following violent deaths (post-mortems 1–7 h later), Nelson *et al.* (1971) found a 20-fold variation in aminopyrine demethylase activity. This enzyme activity correlated with the cytochrome P450 reductase concentrations rather than with the cytochrome concentrations. Davies et al. (1973) have found a four-fold difference in the rate of demethylation of ethylmorphine. Ullrich and Kremers (1977) have reported similar variations in the rate of O-dealkylation of 7-ethoxycoumarin. The cytochrome P450 content of human liver microsomes is usually between 0.1 and 0.6 nmol mg^{-1} of protein (Nelson et al., 1971), though higher values, e.g. 0.54 (Beaune et al., 1979) and 0.67–1.02 (Davies et al., 1973) have been reported. These values compare with values of 0.8–1.2 for rats, mice, rabbits, and guinea pigs. The human liver enzyme has been solubilized and successfully reconstituted by methods similar to those used with animal livers (Kaschnitz and Coon, 1975; Björkhem et al., 1976; Beaune et al., 1979), though it was noted (Kaschnitz and Coon, 1975) that human cytochrome P450 was more labile than that of rat and rabbit. It was also more difficult to purify because of the presence of large amounts of lipid.

Jakobsson et al. (1978) have studied the capacity of human liver, lung, and kidney microsomes (from the same donors) for aminopyrine demethylation, benzphetamine demethylation, laurate hydroxylation, and benzo(a)pyrene hydroxylation. The use of this range of substrates was a commendable attempt to generate some generally applicable data. A 10-fold variation in enzyme activity (aminopyrine demethylation) was again noted and cytochrome P450 fell in the range 0.13–0.51 nmol mg^{-1} protein. The relationship between the cytochrome P450 content of liver biopsy samples and the antipyrine half-lives *in vivo* has been studied in 143 patients (Sotaniemi et al., 1978a). There was a reasonable inverse correlation, particularly between a group of 13 'control' patients and 6 patients known to be receiving inducing drugs. The antipyrine *clearance* (C) where

$$C = \text{dose} \div \text{integrated plasma concentration/time curve}$$

rather than the half-life, may be a better guide to the state of the mono-oxygenase *in vivo* (Sotaniemi et al., 1978b).

Walker (1978) has reviewed species differences in microsomal mono-oxygenase activity and its relation to biological half-life in 36 species/strains ranging from fish, birds, and amphibia to mammals. Only two substrates, aminopyrine and tolbutamide, were tested *in vivo* and *in vitro* in man. They were both metabolized more slowly by man than by other animals (see p. 279).

Ziegler and co-workers (Poulsen et al., 1974) have isolated an enzyme from liver microsomes which catalyses N- and S-oxygenation. It is a flavoprotein and the oxidations, while requiring NADPH, do not involve cytochrome P450. This enzyme is present in human liver microsomes at about one-fifth of the activity found in rat, guinea pig, rabbit, and pig (Gold and Ziegler, 1973). the S-oxygenation of the herbicide cyanatryn (**20**), however, is catalysed by cytochrome P450 and not by the flavoprotein (Crawford and Hutson, 1980). The enzyme of human liver microsomes (one sample) was more active than that from rat liver microsomes and a linear rate of oxidation could be maintained for 30 min at 37 °C (cf. only 8 min for rat microsomes).

$$\text{(20)}$$

Structure (20): 2-methylthio-4-(ethylamino)-6-[(2-cyanopropan-2-yl)amino]-1,3,5-triazine

In view of the variability of human mono-oxygenase, it is useful in certain circumstances to have assessments of the enzyme activity in individuals, preferably gained via a non-invasive technique. Aminopyrine half-lives (in blood) have been used and a recent modification has involved the use of the ^{14}C-methyl labelled form which affords $^{14}CO_2$ in the expired air when metabolized (Platzer *et al.*, 1978). A stable isotope (^{13}C) modification has also been reported (Schneider *et al.*, 1978).

Reduction

A variety of reductions of xenobiotics have been recognized via studies with experimental animals *in vivo* and *in vitro*. These include reductions of ketones, aldehydes, epoxides, nitro compounds, azo groups, *N*-hydroxylamines, *N*-oxides, and oximes. Some of these reactions have been confirmed by the study of drug metabolism in humans. For example, the cytosol of lung and liver of human and rat all contain an enzyme which reduces warfarin (21) to warfarin alcohol (22) (Moreland and Hewick, 1975). Aldehyde reductase has been purified from

(21) warfarin   (22) warfarin alcohol

human liver (Wermuth *et al.*, 1977). It is more active to aromatic than to aliphatic aldehydes. Reductive dechlorinations, e.g. DDT to DDE (see above), apparently occur in man *in vivo* but there are no reports of the reaction in human tissues *in vitro*. Similarly, the study of the metabolism of the organophosphate chlorfenvinphos (12.1) demonstrates that the reductive dechlorination of phenacyl halides to acetophenones (and their subsequent further reduction) occur in man. This reaction has been shown by studies with rat liver enzymes to involve the reaction (Figure 7.14) of the substrate with one molecule of glutathione (GSH) followed by the enzymatic reduction of the resultant phenacylglutathoione (14.2) with another molecule of GSH or other thiol (Hutson *et al.*, 1976; Akhtar, 1979). This reaction is an intermediate step in the metabolism of chlorofenvinphos and

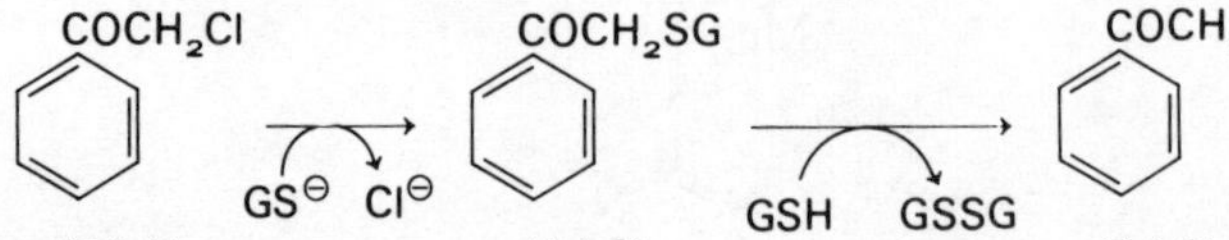

COCH₂Cl →(GS⁻, Cl⁻) COCH₂SG →(GSH, GSSG) COCH₃
(14.1)   (14.2)   (14.3)

Figure 7.14 Reductive dechlorination of phenacyl halides to acetophenones

tetrachlorvinphos. Like most reductions in pesticide biochemistry, it is not a primary detoxification reaction though it is certainly effective in destroying the electrophilic reactivity of the phenacyl chlorides.

Hydrolysis

Hydrolases of various types are widely distributed in mammals. They include:

(1) the carboxylesterases (EC 3.1.1.1) which preferentially hydrolyse aliphatic esters and are inhibited by organophosphates;
(2) the arylesterases (EC 3.1.1.2, A-esterases, paraoxonase) which preferentially hydrolyse aromatic esters and are not inhibited by organophosphates (in fact they catalyse their hydrolysis) but are inhibited by sulphydryl reagents;
(3) the cholinester hydrolases (EC 3.1.1.7 and 3.1.1.8) which act on choline esters and are inhibited by organophosphate and carbamate insecticides;
(4) lipases (e.g. pancreatic lipase, EC 3.1.1.3); and
(5) the acetylesterases (EC 3.1.1.6) which are not inhibited effectively by the insecticides or by sulphydryl reagents.

The esterases thus fall into approximately two classes (Aldridge and Reiner, 1972) the A esterases, which hydrolyse organophosphates, and the B esterases which are inhibited by them. They are fundamentally different in their mechanisms in that A esterases have an SH group at their active site and the B esterases possess an OH (serine) group. Most of the esterases referred to above are B esterases.

Carboxylesters

In the absence of factors such as steric hinderance, the metabolism of esters of carboxylic acids is dominated by ester bond cleavage. The hepatic enzyme is located mostly in the endoplasmic reticulum from whence the pig and rat enzyme have been purified to homogeneity by Krisch and co-workers (Heymann *et al.*, 1974). Much of the early interest of insecticide biochemists on carboxylesterase was generated by work on malathion (**23**) which happens to be detoxified via its action. As the enzyme is inhibited by organophosphates, the potentiation of

$$\begin{array}{c} \text{MeO} \diagdown \overset{S}{\underset{\|}{P}} - \text{SCHCOOEt} \\ \text{MeO} \diagup | \\ \text{CH}_2\text{COOEt} \end{array}$$

(23)

malathion toxicity can occur with insecticide mixtures and also by the self-generation of malaoxon.

Potentially, the most important class of ester insecticides is the pyrethroid group; Casida and co-workers (1976) discovered that these insecticides are hydrolysed by rat and mouse liver microsomal carboxylesterase. The *trans*-isomers are hydrolysed 20–50 times faster than the *cis*-isomers. Recently, Suzuki and Miyamoto (1978) have purified rat liver 'pyrethroid hydrolase' to homogeneity. They judge that malathion esterase and *p*-nitrophenyl acetate hydrolase are properties of the same protein. Ecobichon (1972) has studied the hepatic enzyme in a wide range of species using 1-naphthyl acetate as substrate. Enzyme activities (μmol min^{-1} mg^{-1} protein) lay in the following order: rabbit, 1.48; horse, 0.36; cat, 0.35; guinea pig, 0.32; hamster, 0.30; man, 0.29; pig, 0.24; mouse, 0.19; sheep, 0.19; dog, 0.16; rat, 0.15. That is, rabbit was outstanding, but man was well within the normal species variation. There is, therefore, no reason to suggest that man differs from the common animals (except rabbit) in this respect. The situation with serum carboxylesterase, however, is different. The enzyme is commonly studied in rat, however it is virtually immeasurable in human sera.

Organophosphorus compounds

Serum paraoxonase played a part in the pioneering work of Aldridge (1953) on the mechanism of action of the A esterases. The enzyme is particularly active in rabbit serum but this seems to be a unique source in that other rabbit organs contain unexceptional activities. Other species generally possess highest activities in the liver. Activity in human serum lies between those of the rat and the mouse. Serum arylesterase activity measured against thiophenyl acetate lies in the order sheep > guinea pig > human > rat > dog > monkey > cat (Mendoza *et al.*, 1976). Human serum paraoxonase is surprisingly specific. It hydrolyses methyl paraoxon and 2-chloromethyl paraoxon; however, derivatives containing *ortho*- or *meta*-nitro groups are not hydrolysed (Von Mallinckrodt *et al.*, 1973). The phosphorothionate analogues are very poor substrates, presumably because of the much reduced electrophilicity of the phosphorus. Human serum contains an efficient dichlorvos hydrolase (K_m value 3.2 μM) (Blair *et al.*, 1975). Its relationship with paraoxonase is unknown.

The *in vivo* studies of organophosphates in man described above indicate that the hydrolysis of these compounds is efficient; however, it must be stressed that oxidation and glutathione transferase action may often be more important than hydrolysis in specific cases.

The hydration of epoxides

Epoxide hydrolase has received an enormous amount of attention owing to its role in the destruction of mutagenic/carcinogenic arene oxides (Oesch, 1979). The enzyme which has received the most study in animal and human tissue (Oesch *et al.*, 1974; Kapitulnik *et al.*, 1977) is the hepatic microsomal form which is active in hydrating such substrates as benzo[a]pyrene-4,5-oxide and styrene oxide. Aliphatic epoxides containing tetra-, tri-, or 1,2-*cis*-disubstituents are poor substrates for the human enzyme (Oesch, 1974). This makes it unlikely that dieldrin and endrin are substrates for the enzyme and indeed, as has been discussed above, there is no need to invoke epoxide hydrolase in their metabolism (Bedford, 1975). Less hindered analogues, however, are hydrated by this enzyme (El Zorgani *et al.*, 1970) and it also must be assumed to participate in the metabolism of any insecticide which suffers aryl hydroxylation. Epoxide hydrolase is important in the metabolism of certain insect juvenile hormones and their synthetic analogues. A recent highlight has been the discovery that mouse liver contains significant amounts of non-microsomal epoxide hydrolase (Mumby and Hammock, 1979). However, no information is available on the human equivalent. The metabolism of the juvenile hormone analogues and other growth regulants is discussed in Chapter 1 of this volume.

Conjugation Reactions

These reactions occur as frequently in man as they do in experimental animals, though in many cases the occurrence of species differences makes prediction of the exact biotransformation pathways difficult. As a generalization, a conjugation reaction can be described as one in which a functional group of a xenobiotic reacts with an endogenous molecule forming a less toxic, water-soluble, excretable product. However, as is often the case in xenobiochemistry, there are instances in which a conjugation reaction can lead to precisely the opposite type of product. The generalization nevertheless holds. Another generalization emerges from work with a wide variety of mammals (Caldwell, 1978): glucuronidation has a high capacity, amino acid and glutathione conjugations have a medium capacity and methylation and sulphation have a low capacity. Acetylation has a variable capacity depending on the genetic status of the individual. Conjugation reactions of insecticides have recently been reviewed by Dorough (1979).

Glutathione conjugation

Glutathione (GSH) conjugation differs from other conjugation reactions in that the energy for the process is not derived from the biosynthesis of a reactive endogenous cofactor. Rather, it is derived from a powerful electrophile–nucleophile interaction. The enzyme (a GSH transferase) binds the substrates, enhances the nucleophilicity of GSH, and provides a dipolar aprotic environment for the reaction. Once thought to involve only electrophilic *carbon* atoms (a typical substrate would be benzyl chloride) the reaction is now known to proceed at any suitable electrophilic centre. Nitrogen and sulphur atoms, for example, are attacked in suitable substrates. The nature of the transferases has been reviewed by Jakoby (1978) and Chasseaud (1974) and their role in pesticide metabolism has been summarized (Hutson, 1976b). The enzymes are present in the cytosol of liver and other tissues. They have been isolated by Jacoby and co-workers (1976) and separated into several fractions (AA, A, B, C, D, E, and M) with differing but overlapping substrate specificities. Five distinct transferases were purified from human liver (Kamisaka *et al.*, 1975). These are possibly charge isomers of one protein arising as a result of deamidation *in vivo* (Habig *et al.*, 1976). An immunochemically distinct enzyme has been found in human erythrocytes (Marcus *et al.*, 1978) but it has a relatively low activity and is unlikely to be important in pesticide metabolism.

The major role of glutathione in insecticide metabolism is undoubtedly in the detoxification of organophosphorus esters. Dealkylation (path a) and dearylation (path b) are illustrated in Figure 7.15. Dealkylation is effective (in

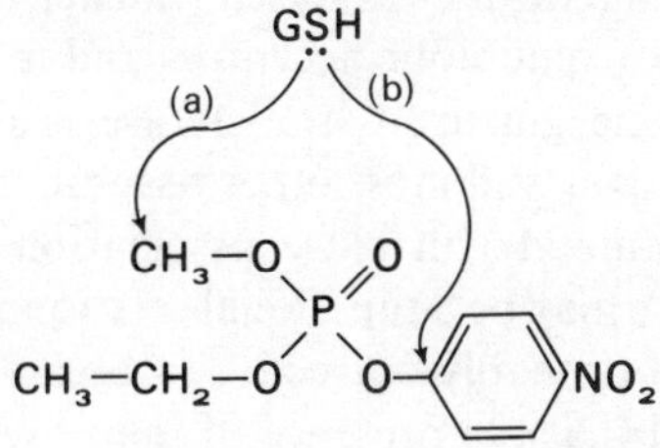

Figure 7.15 Alternative points for attack of glutathione on a phosphate triester

experimental animals) only with methyl esters (Hutson, 1977b); some of these, e.g. tetrachlorvinphos, are demethylated rapidly (Hutson *et al.*, 1972) and with stereochemical selectivity (Hutson, 1977b). The presence of demethylated insecticides in the urine of man following ingestion of the parent insecticide (e.g. dichlorvos) (Hutson and Hoadley, 1972b) indicates that the mechanism is operative. Further evidence was obtained using the model substrate dimethyl 1-naphthyl phosphate. The cytosol of one sample of human liver, fortified with GSH, possessed an activity of 0.1 nmol min^{-1} mg^{-1} protein, cf. rat liver, 0.5 (Hutson, 1980). There is little further information on human GSH transferases involved in the metabolism of organophosphorus pesticides.

The importance of dealkylation may be overestimated because at the low doses to which man is exposed, both occupationally and experimentally, demethylation may not be an important pathway of metabolism. On the other hand, neither of the products of demethylation, methylglutathione, and demethyl insecticide is a true terminal metabolite and, therefore, analysis of urinary metabolites may underestimate the importance of the pathway.

GSH-dependent dearylation is a well established pathway in experimental animals (Shishido *et al.*, 1972; Hollingworth *et al.*, 1973) but it has not yet been confirmed in man. The terminal metabolites of the glutathione conjugates are their N-acetylcysteinyl derivatives (mercapturic acids) and it is to these urinary metabolites that we must look for confirmation of the process in man.

Glutathione conjugation, and metabolism via the arylmercapturic acid pathway, must be expected in competition with epoxide hydrolase action in any biotransformation sequence involving aryl hydroxylation via an arene oxide. Carbaryl, for example, may undergo a minor amount of glutathione conjugation at the 3, 4, 5, or 6 carbon atom. Glutathione is not involved in the metabolism of the stable epoxides dieldrin and endrin.

Glucuronide formation

Conjugation of xenobiotic alcohols, phenols, thiols, amines, and carboxylic acids with glucuronic acid donated from UDPGA is probably a major reaction in man. Bilirubin is so conjugated (Heirwegh, 1978) as are many drugs and their metabolites (e.g. salicylic acid from aspirin). This conjugation reaction is a classic in terms of its mechanism (enzyme-catalysed transfer of endogenous conjugand from an active donor to a xenobiotic acceptor) and in its effects (detoxification and excretion). The enzyme, glucuronyltransferase, is located in the endoplasmic reticulum of the cells of liver and most other tissues, including the intestine and skin (of experimental animals) in close association with the hydroxylating enzymes with which there may be a functional relationship. Though glucuronide conjugation is assumed, probably correctly, to be a very common reaction in man, it must be stressed that the presence of many xenobiotic glucuronides in human urine has been inferred from acid hydrolysis experiments, etc., and from chromatographic comparison with rat metabolites. Many of these, in turn, have been characterized on the basis of lability towards impure mollusc, bacterial, or mammalian β-glucuronidase. Therefore the number of human glucuronic acid conjugates which have been isolated, derivatized *without* hydrolysis, and characterized by physical methods is very small indeed. None of the glucuronides reported in the sections above was so characterized. The improvements in separation technology offered by h.p.l.c., the power of mass spectrometry, and the increasing regulatory pressure for unambiguous identification will combine to alter this situation. Some methods for unambiguous identification have been collected and reviewed by Bakke (1976). In the absence of such methods, the demonstration of the UDPGA-dependent biosynthesis of a polar product from radioactive substrate and washed liver microsomes provides a reasonable substitute for an authentic standard compound, e.g. 1-naphthyl glucuronide (Dorough *et al.*, 1974) and 12-hydroxyendrin glucuronide (Bedford *et al.*,

1975a). These standards should then be compared with the metabolite by sophisticated, non-destructive techniques such as reverse phase radio-h.p.l.c. The use of β-glucuronidase in conjunction with appropriate controls and saccharonolactone inhibition provides an indication of the presence of a glucuronide. Ester glucuronides may prove difficult in this respect in some instances. Recent structural studies on the bilirubin glucuronides (Compernolle *et al.*, 1978) have shown that under unsympathetic storage conditions and also *in vivo* in man suffering cholestasis, acyl migration occurs from the 1 position of the sugar affording a mixture of 2-, 3-, and 4-*O*-acyl glucuronic acid derivatives. These are not glucuronides in the normal sense. That is, they do not possess a glycosidic linkage and would be stable to β-glucuronidase. For a totally different reason (steric hindrance by epoxide oxygen) the glucuronide of *syn*-12-hydroxydieldrin is not hydrolysed by β-glucuronidase (Matthews *et al.*, 1971). These factors must be considered when the cleavage of a glucuronide is part of a biomedical monitoring sequence and, if necessary, alternative methods of glucuronide cleavage (Baldwin and Hutson, 1980) should be investigated.

The enzyme which catalyses glucuronide biosynthesis in animals has received much study (Dutton and Burchell, 1977) but the human enzyme has received scant attention (Dutton, 1978).

Sulphation

Xenobiotic alcohols, phenols, and amines are sulphated (Paulson, 1976) as well as conjugated with glucuronic acid. An unusual substrate, an oxime (Figure 7.16, 16.2), derived from the metabolism of the carbamate (16.1) is both sulphated and conjugated with glucuronic acid in dogs and rats (Hutson *et al.*,

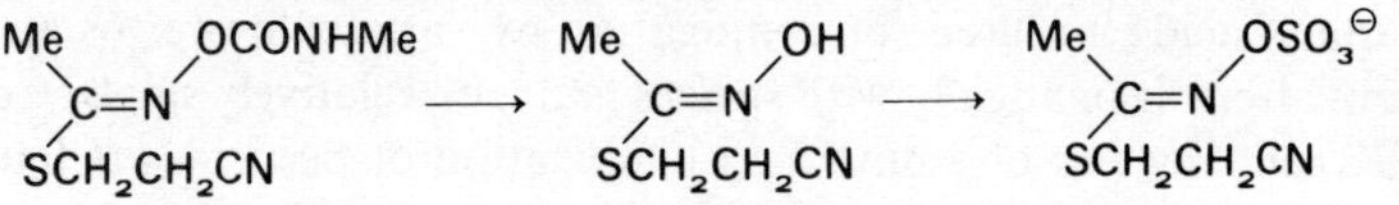

$$\underset{(16.1)}{\underset{SCH_2CH_2CN}{\overset{Me}{\diagdown}}C=N\overset{OCONHMe}{\diagup}} \longrightarrow \underset{(16.2)}{\underset{SCH_2CH_2CN}{\overset{Me}{\diagdown}}C=N\overset{OH}{\diagup}} \longrightarrow \underset{(16.3)}{\underset{SCH_2CH_2CN}{\overset{Me}{\diagdown}}C=N\overset{OSO_3^{\ominus}}{\diagup}}$$

Figure 7.16 Sulphation of an oxime

1971b) but the reaction has not been demonstrated in man. A recent study of the hindered phenol, 2,6-di-isopropylphenol in rat, mini pig, and man revealed that 4-hydroxylation occurred in all three species the product was mostly conjugated with glucuronic acid in man and pig but with sulphate in the rat (Rhodes *et al.*, 1978). It is not possible to generalize from this result but it supports an impression that very few sulphate conjugates have been unambiguously identified in man. The relative extent to which sulphation and glucuronidation occur is substrate, species, and dose dependent (Mehta *et al.*, 1978). The sulphotransferases, which catalyse the transfer of sulphate from the active donor 3'-phosphoadenosine-5'-phosphosulphate (PAPS), have apparently not been isolated from human tissues.

Acetylation

Acetylation is relatively unimportant in insecticide metabolism. However, the reaction must always be anticipated when an aromatic amine is liberated. Man and rabbit exhibit a genetic polymorphism in acetylation (for example, of 4-aminobenzoic acid and isoniazid) and individuals can be classed as 'rapid' or 'slow' acetylators (Weber et al., 1976).

Methylation

O-, S-, and N-Methylation reactions do not feature widely in insecticide metabolism. The monomethylation of catechols (by catechol O-methyltransferase) is the most common reaction of this type in foreign compound metabolism. It is usually encountered at the end of the sequence: arene → arene oxide → dihydrodiol → catechol → monomethyl catechol. The reaction occurs in man, for example in the metabolism of adrenaline and the drug isoprenaline (Persson and Persson, 1972).

Conjugation with amino acids

Organic carboxylic acids are usually conjugated either with amino acids or with glucuronic acid. Relatively few pesticides fall into this class but carboxylic acids may be produced by the metabolic hydrolysis of nitriles, amides, and esters or by the oxidation of alcohols; therefore, their occurrence in insecticide metabolism is quite common. Glycine and taurine are the most commonly used amino acids in conjugation. This is possibly a reflection of xenobiotic molecules participating in bile acid conjugating pathways. However, an astonishing variety of other amino acids are used for conjugation by various life forms and their involvement alters dramatically with species and with relatively small structural changes. Recent reviews of amino acid conjugation of pesticides (Climie and Hutson, 1979) and xenobiotics generally (Hirom et al., 1976) mention glycine, taurine, serine, glutamine, glutamic acid, cysteine, alanine, ornithine, aspartic acid, valine, leucine, phenylalanine, and tryptophan. In rare cases dipeptide conjugates appear. Carnitine conjugation has also been discovered recently (Quistad et al., 1978). Most of the research on species differences in amino acid conjugation has been performed on model compounds. Man conjugates benzoic acid with glycine, but phenylacetic acid with glutamine (James et al., 1972). The related compound, 1-naphthylacetic acid, is excreted as the glucuronide and taurine conjugates, and no glutamine conjugate is formed (Dixon et al., 1977). Some generalizations on the relationship between species and amino acids used in conjugation have been implied in the literature. For example, glycine is used in mammals, glutamine in man and primates, ornithine in birds, taurine in carnivores, and aspartic acid in plants. These relationships have been developed with only a few model compounds and they cannot be used on a predictive basis.

The fate of 3-phenoxybenzoic acid, derived from the synthetic pyrethroids discussed above, exemplifies this.

The enzymology of amino acid conjugation has not been studied in any great detail. Caldwell et al. (1976) have found that [^{14}C]benzoic acid is conjugated with glycine in human liver. The enzymes were reasonably stable in cadavars stored at 4 °C for 72 h. It is known that the xenobiotic acid is first activated to its coenzyme A derivative from which the acyl group is transferred to an amino acid via the action of an acyl transferase. It is not known what controls the status of these amino acids as acceptors. Control may be exercised via amino acid availability, permeability factors, or enzyme specificity.

CONCLUSIONS

As far as it is possible to ascertain from available data, man apparently possesses most of the metabolizing enzymes which have been demonstrated to exist in laboratory animals. Capacity for cytochrome P450-dependent oxidation is lower in man than in laboratory animals, not because the specific activity is less but because there appears to be less cytochrome P450 in human liver. The liver is the major source of pesticide-metabolizing enzymes in man as in laboratory animals, but reactions occur in other tissues with reasonable facility.

The metabolic fate of a particular insecticide in man cannot be predicted in terms of either rate or route unless research has already been carried out with closely related compounds. The generalizations which appear in the literature are founded on experience with only a limited number of compounds in most cases or have been developed by unjustified extrapolation from other species. There are undoubtedly a number of studies which have not been published in the open literature. Publication should be encouraged because there is barely enough information available from which to draw general conclusions.

REFERENCES

Akhtar, M. H. (1979). 'Sequential participation of glutathione and sulph-hydryl(s) in reductive dechlorination of 2,4-di-, and 2,4,5-trichloro phenacyl chlorides by soluble fraction (105,000 × g) of chicken liver homogenate', *J. Environ. Sci. Health* **B14**, 53–71.

Aldridge, W. N. (1953). 'Serum esterases', *Biochem. J.* **53**, 117–124.

Aldridge, W. N., and Reiner, E. (1972). In *Enzyme Inhibitors as Substrates*, North Holland Pub., Amsterdam.

Alvarez, A. P., Schilling, G., Levin, W., Kuntzman, R., Brand, L., and Mark, R. C. (1969). 'Cytochromes P-450 and b_5 in human liver microsomes', *Clin. Pharmacol. Ther.* **10**, 655–659.

Back, R. C. (1965). 'Significant developments in eight years with sevin insecticide', *J. Agr. Food Chem.* **13**, 198–199.

Bakke, J. E. (1976). 'Recent advances in the isolation and identification of glucuronide conjugates', in *Bound and Conjugated Pesticide Residues*, ACS Symposia Series No. 29, pp. 55–67.

Baldwin, M. K. (1980). Unpublished work.

Baldwin, M. K., and Hutson, D. H. (1980). 'Analysis of human urine for a metabolite of endrin by chemical oxidation and gas–liquid chromatography as an indicator of exposure to endrin', *Analyst* **105**, 60–65.

Baldwin, M. K., Robinson, J., and Parke, D. V. (1972). 'A comparison of the metabolism of HEOD (Dieldrin) in the CF1 mouse with that in the CFE rat', *Food Cosmet. Toxicol.* **10**, 333–351.

Barnes, J. M., and Heath, D. F. (1964). 'Some toxic effects of dieldrin in rats', *Brit. J. Ind. Med.* **21**, 280–282.

Barnes, J. M., and Verschoyle, R. D. (1974). 'Toxicity of new pyrethroid insecticide', *Nature* **248**, 711.

Baron, R. L., and Locke, R. K. (1970). 'Utilization of cell culture techniques in carbaryl metabolism studies', *Bull. Environ. Contam. Toxicol.* **5**, 287–291.

Beaune, Ph., Dansette, P., Flinois, J. P., Columelli, S., Mansuy, D., and Leroux, J. P. (1979). 'Partial purification of human liver cytochrome P450', *Biochem. Biophys. Res. Commun.* **88**, 826–832.

Bedford, C. T. (1974). 'Von Baeyer/IUPAC names and abbreviated chemical names of metabolites and artifacts of aldrin (HHDN), dieldrin (HEOD) and endrin', *Pestic. Sci.* **5**, 473–489.

Bedford, C. T. (1975). In *Foreign Compound Metabolism in Mammals*, Specialist Periodical Reports, Vol. 3, p. 403, The Chemical Society, London.

Bedford, C. T., Harrod, R. K., Hoadley, E. C., and Hutson, D. H. (1975a). 'The metabolic fate of endrin in the rabbit', *Xenobiotica* **5**, 485–500.

Bedford, C. T., and Hutson, D. H. (1976). 'The comparative metabolism in rodents of the isomeric insecticides dieldrin and endrin', *Chem. and Ind.* pp. 440–447.

Bedford, C. T., Hutson, D. H., and Natoff, I. L. (1975b). 'The acute toxicity of endrin and its metabolites to rats', *Toxicol. Appl. Pharmacol.* **33**, 115–121.

Bentley, P., and Oesch, F. (1979). In *Foreign Compound Metabolism in Mammals*, Specialist Perdiodical Reports Vol. 5, pp. 89–131, The Chemical Society, London.

Björkhem, I., Kager, L., and Wikvall, K. (1976). 'Catalytic properties of cytochrome P-450 from the human liver', *Biochem. Med.* **15**, 87–94.

Blair, D., Hoadley, E. C., and Hutson, D. H. (1975). 'The distribution of dichlorvos in the tissues of mammals after its inhalation or intravenous administration', *Toxicol. Appl. Pharmacol.* **31**, 243–253.

Blair, D., and Roderick, H. R. (1976). 'An improved method for the determination of urinary dimethyl phosphate', *J. Agr. Food Chem.* **24**, 1221–1223.

Caldwell, J. (1978). 'The conjugation reactions: the poor relations of drug metabolism', in *Conjugation Reactions in Drug Biotransformation* (Ed. A. Aitio) pp. 477–485, Elsevier/North Holland, Amsterdam.

Caldwell, J., Moffatt, J. R., and Smith, R. L. (1976). 'Post-mortem survival of hippuric acid formation in rat and human cadaver tissue samples', *Xenobiotica* **6**, 275–280.

Casida, J. E., Gaughan, L. C., and Ruzo, L. O. (1979). 'Comparative metabolism of pyrethroids derived from 3-phenoxybenzyl and α-cyano-3-phenoxybenzyl alcohols', in *Advances in Pesticide Science* (4th Internat. Congress Pestic. Chem., Zurich, 1978), Vol. 2, pp. 182–189, Pergamon Press, Oxford.

Casida, J. E., and Ruzo, L. O. (1980). 'Metabolic chemistry of pyrethroid insecticides', *Pestic. Sci.* **11**, 257–269.

Casida, J. E., Ueda, K., Gaughan, L. C., Jao, L. T., and Soderlund, D. M. (1976). 'Structure–biodegradability relationships in pyrethroid insecticides', *Arch. Environ. Contam. Toxicol.* **3**, 391–500.

Chasseaud, L. F. (1974). 'The nature and distribution of enzymes catalyzing the conjugation of glutathione with foreign compounds', *Drug Metab. Rev.* **2**, 185–220.

Chin, B. H., Eldridge, J. M., Anderson, J. H., and Sullivan, L. J. (1979). 'Carbaryl metabolism in the rat. A comparison of *in vivo, in vitro* (tissue explant), and liver perfusion techniques', *J. Agr. Food Chem.* **27**, 716–720.

Chin, B. J., Eldridge, J. M., and Sullivan, L. J. (1974). 'Metabolism of carbaryl by selected human tissues using an organ-maintenance technique', *Clin. Toxicol.* **7**, 37–56.

Climie, I. J. G., and Hutson, D. H. (1979). 'Conjugation reactions with amino acids including glutathoine', in *Advances in Pesticide Science* (Ed. H. Geissbühler), Part 3, pp. 537–546, Pergamon Press, Oxford.

Cole, J. F., Klevay, L. M., and Zavon, M. R. (1970). 'Endrin and dieldrin: a comparison of hepatic excretion in the rat', *Toxicol. Appl. Pharmacol.* **16**, 547–555.

Compernolle, F., Van Hees, G. P., Blanckaert, N., and Heirwegh, K. P. M. (1978). 'Glucuronic acid conjugates of bilirubin-IXα in normal bile compared with post-obstructive bile. Transformation of the 1-O-acylglucuronide into 2-, 3-, and 4-O-acylglucuronides', *Biochem. J.* **171**, 185–201.

Crawford, M. J., and Hutson, D. H. (1980). 'S-Oxygenation of an alkylmercaptotriazine herbicide by rat, rabbit and human liver enzymes', *Xenobiotica* **10**, 187–192.

Datta, P. R., and Nelson, M. D. (1968). 'Enhanced metabolism of methyprylon, meprobamate, and chlordiazepoxide hydrochloride after chronic feeding of a low dietary level of DDT to male and female rats', *Toxicol. Appl. Pharmacol.* **13**, 346–352.

Davies, J. E., Barquet, A., Freed, V. H., Haque, R., Morgade, C., Sonneborn, R. E., and Vaclavek, C. (1975). 'Human pesticide poisonings by a fat-soluble organophosphate insecticide', *Arch. Environ. Health* **30**, 608–613.

Davies, D. S., Thorgeirsson, S. S., Breckenridge, A., and Orme, M. (1973). 'Interindividual differences in rates of drug oxidation in man', *Drug Metab. Disp.* **1**, 411–417.

Dawson, J. A., Heath, D. F., Rose, J. A., Thain, E. M., and Ward, J. B. (1964). 'The excretion by humans of the phenol derived *in vivo* from 2-isopropoxyphenyl N-methylcarbamate', *Bull. W.H.O.* **30**, 127–134.

Deichman, W. B., Keplinger, M., Dressler, I., and Sala, F. (1969). 'Retention of dieldrin and DDT in the tissues of dogs fed aldrin and DDT individually and as a mixture', *Toxicol. Appl. Pharmacol.* **14**, 205–213.

Dixon, P. A. F., Caldwell, J., and Smith, R. L. (1977). 'Metabolism of arylacetic acids. 1. The fate of 1-naphthylacetic acid and its variation with species and dose', *Xenobiotica* **7**, 695–706.

Donninger, C., Hutson, D. H., and Pickering, B. A. (1972). 'The oxidative dealkylation of insecticidal phosphoric acid triesters by mammalian liver enzymes', *Biochem. J.* **126**, 701–707.

Dorough, H. W. (1970). 'Metabolism of insecticidal methylcarbamates in animals', *J. Agr. Food Chem.* **18**, 1015–1022.

Dorough, H. W. (1979). 'Metabolism of insecticides by conjugation mechanisms', *Pharmacol. Ther.* **4**, 433–471.

Dorough, H. W., McManus, J. P., Kumer, S. S., and Cardona, R. A. (1974). 'Chemical and metabolic characteristics of 1-naphthyl β-D-glucoside', *J. Agr. Food Chem.* **22**, 642–645.

Douch, P. G. C., Smith, J. N., and Turner, J. C. (1971). 'NADPH-dependent cleavage of carbamates', *Life Sci.* **10II**, 1327–1333.

Durham, W. F., Ortega, P., and Hayes, W. J. (1963). 'The effect of various dietary levels of DDT on liver function, cell morphology, and DDT storage in the rhesus monkey', *Arch. Int. Pharmacodyn. Ther.* **141**, 111–129.

Dutton, G. J. (1978). In *Drug Metabolism in Man* (Eds. J. W. Gorrod and A. H. Beckett), pp. 81–96, Taylor and Francis Ltd, London.

Dutton, G. J., and Burchell, B. (1977). In *Progress in Drug Metabolism Vol. 2* (Eds. J. W. Bridges and L. F. Chasseaud), Vol. 2, pp. 1–70, Wiley, Chichester.

Ecobichon, D. J. (1972). 'Relative amounts of hepatic and renal carboxylesterases in mammalian species', *Res. Commun. Chem. Pathol.* **3**, 629–636.

Elliott, M. (1977). 'Synthetic pyrethroids', in *Synthetic Pyrethroids*, ACS Symposia Series No. 42, pp. 1–28.

Elliott, M., Farnham, A. W., Janes, N. F., Needham, P. H., and Pulman, D. A. (1974). 'Synthetic insecticide with a new order of activity', *Nature* **248**, 710–711.

El Zorgani, G. A., Walker, C. H., and Hassall, K. A. (1970). 'Species differences in the *in vitro* metabolism of HEOM a chlorinated cyclodiene epoxide', *Life Sci.* **9II**, 415–420.

Feil, V. J., Hedde, R. D., Zaylskie, R. G., and Zachrison, C. H. (1970). 'Identification of *trans*-6,7-dihydroxydihydroaldrin and 9-(*syn*-epoxy)hydroxy-1,2,3,4,10,10-hexachloro-6,7-epoxy-1,4,4a,5,6,7,8,8a-1,4-*endo*-5,8-*exo*-dimethanonaphthalene', *J. Agr. Food Chem.* **18**, 120–124.

Feldman, R. J., and Maibach, H. I. (1974). 'Percutaneous penetration of some pesticides and herbicides in man', *Toxicol. Appl. Pharmacol.* **28**, 126–132.

Fukuto, T. R. (1973). 'Metabolism of carbamate insecticides', *Drug Metab. Rev.* **1**, 117–151.

Gaughan, L. C., Ackerman, M. E., Unai, T., and Casida, J. E. (1978). 'Distribution and metabolism of *trans*- and *cis*-permethrin in lactating Jersey cows', *J. Agr. Food Chem.* **26**, 613–618.

Gaughan, L. C., Unai, T., and Casida, J. E. (1977). 'Permethrin metabolism in rats', *J. Agr. Food Chem.* **25**, 9–17.

Gingell, R. (1976). 'Metabolism of ^{14}C-DDT in the mouse and hamster', *Xenobiotica* **6**, 15–20.

Gold, M. S., and Ziegler, D. M. (1973). 'Dimethylamine *N*-oxidase and aminopyrine *N*-demethylase activities of human liver tissue', *Xenobiotica* **3**, 179–189.

Gunn, D. L. (1975). In *Foreign Compound Metabolism in Mammals*, Vol. 3, pp. 1–82, The Chemical Society, London.

Habig, W. H., Kamisaka, K., Ketley, J. N., Pabst, M. J., Arias, I. M., and Jakoby, W. B. (1976). 'The human hepatic glutathione S-transferases', in *Glutathione: Metabolism and Function* (Eds. I. M. Arias, and W. B. Jakoby), pp. 225–232, Raven Press, New York.

Hayes, W. J. (1976). 'Toxicity of pesticides to man: risks from present levels', *Proc. Roy. Soc. Lond. B* **167**, 101–127.

Hayes, W. J., Dale, W. E., and Birse, V. W. (1965). 'Chlorinated hydrocarbon pesticides in the fat of people in New Orleans', *Life Sci.* **4**, 1611–1615.

Hayes, W. J., Dale, W. E., and Pirkle, C. I. (1971). 'Evidence of safety of long-term, high, oral doses of DDT for man', *Arch. Environ. Health* **22**, 119–135.

Heirwegh, K. P. M. (1978). In *Conjugation Reactions in Drug biotransformation* (Ed A. Aitio), pp. 67–76, Elsevier/North Holland, Amsterdam.

Heymann, E., Junge, W., Krisch, K., and Marcussen-Wulf, G. (1974). 'Einfache und schnell Methoden zur Isolierung Hochgereinigter Carboxylesterasen (EC 3.1.1.1) aus Schweinleber, Rinderleber und Schweinenieren', *Z. Physiol. Chem.* **355**, 155–163.

Hirom, P. C., Idle, J. R., and Millburn, P. (1976). 'Aspects of biosynthesis and excretion of xenobiotic conjugates in mammals', in *Drug Metabolism from Microbes to Man* (Eds. D. V. Parke and R. L. Smith), pp. 299–329, Taylor and Francis Ltd., London.

Hirom, P. C., and Millburn, P. (1979). In *Foreign Compound Metabolism in Mammals*, Specialist Periodical Reports Vol. 5, pp. 132–158, The Chemical Society, London.

Hirom, P. C., Millburn, P., Smith, R. L., and Williams, R. T. (1972). 'Species variation in the threshold molecular-weight factor for the biliary excretion of organic anions', *Biochem. J.* **129**, 1071–1077.

Hollingworth, R. M., Alstott, R. L., and Litzenberg, R. D. (1973). 'Glutathione S-aryl transferase in the metabolism of parathion and its analogs', *Life Sci.* **13**, 191–199.

Hunter, C. G., and Robinson, J. (1967). 'Pharmacodynamics of dieldrin (HEOD). I. Ingestion by human subjects for 18 months', *Arch. Environ. Health* **15**, 614–626.

Hunter, C. G., Robinson, J., Bedford, C. T., and Lawson, J. M. (1972). 'Exposure to chlorfenvinphos by determination of a urinary metabolite', *J. Occup. Med.* **14**, 119–122.

Hunter, C. G., Robinson, J., and Roberts, M. (1969). 'Pharmacodynamics of dieldrin (HEOD). Ingestion by human subjects for 18 to 24 months, and postexposure for eight months', *Arch. Environ. Health* **18**, 12–21.

Hutson, D. H. (1976a). 'Comparative metabolism of dieldrin in the rat (CFE) and two strains of mouse (CF1 and LACG)', *Food Cosmet. Toxicol.* **14**, 577–591.

Hutson, D. H. (1976b). 'Glutathione conjugates', in *Bound and Conjugated Pesticide Residues*, ACS Symposia Series No. 29, pp. 103–131.

Hutson, D. H. (1977a). In *Foreign Compound Metabolism in Mammals* Specialist Periodical Reports Vol. 1 (1970), Vol. 2 (1972), Vol. 3 (1975), Vol. 4 (1977), The Chemical Society, London.

Hutson, D. H. (1977b). 'Some observations on the chemical and stereochemical specificity of the de-alkylation of organophosphorus esters by a hepatic glutathione transferase', *Chem-Biol. Interact.* **16**, 315–323.

Hutson, D. H. (1979). 'The metabolic fate of synthetic pyrethroid insecticides in mammals', in *Progress in Drug Metabolism* (Eds. J. W. Bridges and L. F. Chasseaud), Vol. 3, pp. 215–252, Wiley, Chichester.

Hutson, D. H. (1980). Unpublished work.

Hutson, D. H., Akintonwa, D. A. A., and Hathway, D. E. (1967). 'The metabolism of 2-chloro-1-(2′,4′-dichlorophenyl)vinyl diethyl phosphate (chlorfenvinphos) in the dog and rat', *Biochem. J.* **102**, 133–142.

Hutson, D. H., Baldwin, M. K., and Hoadley, E. C. (1975). 'Detoxication and bioactivation of endrin in the rat', *Xenobiotica* **5**, 697–714.

Hutson, D. H., and Casida, J. E. (1978). 'Taurine conjugation in metabolism of 3-phenoxybenzoic acid and the pyrethroid insecticide cypermethrin in the mouse', *Xenobiotica* **8**, 565–571.

Hutson, D. H., and Hoadley, E. C. (1972a). 'The metabolism of [^{14}C-methyl]dichlorvos in the rat and mouse', *Xenobiotica* **2**, 107–116.

Hutson, D. H., and Hoadley, E. C. (1972b). 'The comparative metabolism of [^{14}C-vinyl]dichlorvos in animals and man', *Arch. Toxicol.*, **30**, 9–18.

Hutson, D. H., and Hoadley, E. C. (1974). 'The oxidation of a cyclic alcohol (12-hydroxyendrin) to a ketone (12-keto-endrin) by microsomal mono-oxygenation', *Chemosphere* **3**, 205–210.

Hutson, D. H., Hoadley, E. C., and Pickering, B. A. (1971a). 'The metabolic fate of [vinyl-1-^{14}C]dichlorvos in the rat after oral and inhalation exposure', *Xenobiotica* **1**, 593–611.

Hutson, D. H., Hoadley, E. C., and Pickering, B. A. (1971b). 'The metabolism of S-2-cyanoethyl-N-(methylcarbamoyl)oxy]thioacetimidate, an insecticidal carbamate, in the rat', *Xenobiotica* **1**, 179–191.

Hutson, D. H., Holmes, D., and Crawford, M. J. (1976). 'The involvement of glutathione in the reductive dechlorination of a phenacyl halide', *Chemosphere* **5**, 79–84.

Hutson, D. H., Pickering, B. A., and Donninger, C. (1972). 'Phosphoric acid triester—glutathione alkyltransferase. A mechanism for the detoxification of dimethyl phosphate triesters', *Biochem. J.* **127**, 285–293.

Jager, K. W. (1970). *Aldrin, Dieldrin, Endrin, Telodrin. An epidemiological Study of Long-term Occupational Exposure*, Elsevier Publishing Company, Amsterdam.

Jakobsson, S. V., Okita, R. T., Prough, R. A., Mock, N. I., Buja, L. M., Graham, J. W., Petty, C. A., and Masters, B. S. S. (1978). 'Studies on the cytochrome P-450-dependent monooxygenase systems of human liver, lung, and kidney microsomes', in *Industrial and Environmental Xenobiotics* (Eds. J. R. Fouts and I. Gut), pp. 71–73, Excerpta Medica, Amsterdam.

Jakoby, W. B. (1978). 'The glutathione S-transferases: a group of multifunctional detoxification proteins', *Adv. Enzymol.* **46**, 383–414.

Jakoby, W. B., Habig, W. H., Keen, J. H., Ketley, J. N., and Pabst, M. J. (1976). 'Glutathione S-transferases: catalytic aspects', in *Glutathione: Metabolism and Function* (Eds. I. M. Arias and W. B. Jakoby), pp. 189–211, Raven Press, New York.

James, M. O., Smith, R. L., Williams, R. T., and Reidenberg, M. (1972). 'The conjugation of phenylacetic acid in man, sub-human primates and some non-primate species', *Proc. Roy. Soc. Lond. B* **182**, 25–35.

Kamisaka, K., Habig, W. H., Ketley, J. N., Arias, I. M., and Jakoby, W. B. (1975). 'Multiple forms of human glutathione S-transferase and their affinity for bilirubin', *Eur. J. Biochem.* **60**, 153–161.

Kapitulnik, J., Levin, W., Lu, A. Y. H., Morecki, R., Dansette, P. M., Jerina, D. M., and Conney, A. H. (1977). 'Hydration of arene and alkene oxides by epoxide hydrase in human liver microsomes', *Clin. Pharmacol. Ther.* **21**, 158–165.

Kaschnitz, R. M., and Coon, M. J. (1975). 'Drug and fatty acid hydroxylation by solubilized human liver microsomal cytochrome P-450—phospholipid requirement', *Biochem. Pharmacol.* **24**, 295–297.

Knaak, J. B., Tallant, M. J., Bartley, W. J., and Sullivan, L. J. (1965). 'The metabolism of carbaryl in the rat, guinea pig, and man', *J. Agr. Food Chem.* **13**, 537–543.

Knaak, J. B., Tallant, M. J., Kozbelt, S. J., and Sullivan, L. J. (1968). 'The metabolism of carbaryl in man, monkey, pig, and sheep', *J. Agr. Food Chem.* **16**, 465–470.

Kuhr, R. J., and Dorough, H. W. (1976). *Carbamate Insecticides: Chemistry, Biochemistry and Toxicology*, CRC, Cleveland, Ohio.

Kulkarni, A. P., Mailman, R. B., and Hodgson, E. (1975). 'Cytochrome P-450 optical difference spectra of insecticides. A comparative study', *J. Agr. Food Chem.* **23**, 177–183.

Lin, T. H., North, H. H., and Menzer, R. E. (1975). 'Metabolism of carbaryl (1-naphthyl N-methylcarbamate) in human embryonic lung cell cultures', *J. Agr. Food Chem.* **23**, 253–256.

Lores, E. M., and Bradway, D. E. (1977). 'Extraction and recovery of organophosphorus metabolites from urine using an anion exchange resin', *J. Agr. Food Chem.* **25**, 75–79.

Lores, E. M., Bradway, D. E., and Moseman, R. F. (1978a). 'Organophosphorus pecticide poisonings in humans: determination of residues and metabolites in tissues and urine', *Arch. Environ. Health* **33**, 270–276.

Lores, E. M., Sovocool, G. W., Harless, R. L., Wilson, N. K., and Moseman, R. F. (1978b). 'A new metabolite of chlorpyrifos: isolation and identification', *J. Agr. Food Chem.* **26**, 118–122.

Marcus, C. J., Habig, W. H., and Jakoby, W. B. (1978). 'Glutathione transferase from human erythrocytes', *Arch. Biochem. Biophys.* **188**, 287–293.

Matsumura, F., and Ward, C. T. (1966). 'Degradation of insecticides by the human and the rat liver', *Arch. Environ. Health* **13**, 257–261.

Matthews, H. B., McKinney, J. D., and Lucier, G. W. (1971). 'Dieldrin metabolism, excretion and storage in male and female rats', *J. Agr. Food Chem.* **19**, 1244–1248.

Mehta, R., Hirom, P. C., and Millburn, P. (1978). 'The influence of dose on the pattern of conjugation of phenol and 1-naphthol in non-human primates', *Xenobiotica* **8**, 445–452.

Mendoza, C. E., Shields, J. B., and Augustinsson, K.-B. (1976). 'Arylesterases from various mammalian sera in relation to cholinesterases, carboxylesterases and their activity towards some pesticides', *Comp. Biochem. Physiol.* **55C**, 23–26.

Menn, J. J., and McBain, J. B. (1974). 'New aspects of organophosphorus pesticides. IV. Newer aspects of the metabolism of phosphonate insecticides', *Res. Rev.* **53**, 35–51.

Moreland, T. A., and Hewick, D. S. (1975). 'Studies on a ketone reductase in human and rat liver and kidney soluble fraction using warfarin as a substrate', *Biochem. Pharmacol.* **24**, 1953–1957.

Morgan, D. P., Hetzler, H. L., Slach, E. F., and Lin, L. I. (1977). 'Urinary excretion of paranitrophenol and alkyl phosphates following ingestion of methyl or ethyl parathion by human subjects', *Arch. Environ. Contam. Toxicol.* **6**, 159–173.

Morgan, D. P., and Roan, C. C. (1974). 'The metabolism of DDT in man', in *Essays in Toxicology* (Ed. W. J. Hayes), Vol. 5, pp. 39–97, Academic Press, New York.

Müller, W., Nohynek, G., Woods, G., Korte, F., and Coulston, F. (1975). 'Comparative metabolism of dieldrin-^{14}C in mouse, rat, rabbit, rhesus monkey and chimpanzee', *Chemosphere* **4**, 89–92.

Mumby, S. M., and Hammock, B. D. (1979). 'Substrate selectivity and stereochemistry of enzymatic epoxide hydration in the soluble fraction of mouse liver', *Pestic. Biochem. Physiol.* **11**, 275–284.

Nelson, E. B., Raj, P. P., Belfi, K. J., and Masters, B. S. S. (1971). 'Oxidative drug metabolism in human liver microsomes', *J. Pharmacol. Exp. Ther.* **178**, 580–588.

Oesch, F. (1974). 'Purification and specificity of a human microsomal epoxide hydratase', *Biochem. J.* **139**, 77–88.

Oesch, F. (1979). 'Epoxide hydratase', in *Progress in Drug Metabolism* (Eds. J. W. Bridges and L. F. Chasseaud), Vol. 3, pp. 253–301, Wiley, Chichester.

Oesch, F., Thoenen, H., and Fahrlaender, H. (1974). 'Epoxide hydrase in human liver biopsy specimens: assay and properties', *Biochem. Pharmacol.* **23**, 1307–1317.

Paulson, G. D. (1976). 'Sulfate ester conjugates—their synthesis, purification, hydrolysis, and chemical spectral properties', in *Bound and Conjugated Pesticide Residues*, ACS Symposia Series No. 29, pp. 86–102.

Persson, K., and Persson, K. (1972). 'The metabolism of terbutaline *in vitro* by rat and human liver *O*-methyltransferases and monoamine oxidases', *Xenobiotica* **2**, 375–382.

Peterson, J. E., and Robison, W. H. (1964). 'Metabolic products of p,p'-DDT in the rat', *Toxicol. Appl. Pharmacol.* **6**, 321–327.

Platzer, R., Galeazzi, R. L., Karlaganis, G., and Bircher, J. (1978). 'Rate of drug metabolism in man measured by ^{14}CO$_2$-breath analysis', *Eur. J. Clin. Pharmacol.* **14**, 293–299.

Poulsen, L. L., Hyslop, R. M., and Ziegler, D. M. (1974). 'S-Oxidation of thioureylenes catalyzed by a microsomal flavoprotein mixed-function oxidase', *Biochem. Pharmacol.* **23**, 3431–3440.

Quistad, G. B., Staiger, L. E., and Schooley, D. A. (1978). 'Environmental degradation of the miticide cycloprate (hexadecyl cyclopropanecarboxylate)', *J. Agr. Food Chem.* **26**, 60–80.

Rhodes, C., Provan, W. M., and Roberts, H. (1978). 'Biotransformation of 2,6-diisopropyl phenol (ICI 35,868) by the rat, mini pig and man', in *Conjugation Reactions in Drug Biotransformation* (Ed. A. Aitio), p. 153, Elsevier/North Holland, Amsterdam.

Richardson, A., Baldwin, M. K., and Robinson, J. (1968). 'Identification of metabolites of dieldrin (HEOD) in the faeces and urine of rats', *J. Sci. Food Agr.* **19**, 524–529.

Richardson, A., and Robinson, J. (1971). 'The identification of a major metabolite of HEOD (dieldrin) in human faeces', *Xenobiotica* **1**, 213–219.

Richardson, L. A., Lane, J. R., Gardner, W. S., Peeler, J. T., and Campbell, J. E. (1967). 'Relationship of dietary intake to concentration of dieldrin and endrin in dogs', *Bull. Environ. Contam. Toxicol.* **2**, 207–219.

Ridgeway, R. L., Tinney, J. C., MacGregor, J. T., and Slater, N. J. (1978). 'Pesticide use in agriculture', *Environ. Health Perspect.* **27**, 103–112.

Robinson, J., Roberts, M., Baldwin, M. K., and Walker, A. I. T. (1969). 'The pharmacokinetics of HEOD (dieldrin) in the rat', *Food Cosmet. Toxicol.* **7**, 317–332.

Ryan, A. J. (1972). 'The metabolism of pesticidal carbamates', *Crit. Rev. Toxicol.* **1**, 33–54.

Schneider, J. F., Schoeller, D. A., Nemchausky, B., Boyer, J. L., and Klein, P. (1978). 'Validation of $^{13}CO_2$ breath analysis as a measurement of demethylation of stable isotope labelled aminopyrine in man', *Clin. Chim. Acta* **84**, 153–162.

Shafik, M. T., and Bradway, D. E. (1976). 'Worker re-entry safety. VIII. The determination of urinary metabolites—an index of human and animal exposure to nonpersistent pesticides', *Res. Rev.* **62**, 59–77.

Shishido, T., Usui, K., Soto, M., and Fukami, J. (1972). 'Enzymatic conjugation of diazinon with glutathione in rat and American cockroach', *Pestic. Biochem. Physiol.* **2**, 51–63.

Sotaniemi, E. A., Pelkonen, R. O., Ahokas, J., Ahlqvist, J., and Hokkanen, O. T. (1978a). 'Drug metabolism in man: *in vivo* and *in vitro* correlations', *Arch. Toxicol.* Suppl. **1**, 339–342.

Sotaniemi, E. A., Pelkonen, R. O., Ahokas, J. T., Pirttiaho, H. I., and Ahlqvist, J. (1978b). 'Relationship between *in vivo* and *in vitro* drug metabolism in man', *Eur. J. Drug Metab. Pharmacokinet.* **1**, 39–45.

Stooter, D. A. (1977). 'Metal centres and DDT', *J. Inorg. Nucl. Chem.* **39**, 721–727.

Strother, A. (1972). '*In vitro* metabolism of methylcarbamate insecticides by human and rat liver fractions', *Toxicol. Appl. Pharmacol.* **21**, 112–129.

Sullivan, L. J., Chin, B. H., and Carpenter, C. P. (1972a). '*In vitro* vs *in vivo* chromatographic profiles of carbaryl anionic metabolites in man and lower animals', *Toxicol. Appl. Pharmacol.* **22**, 161–174.

Sullivan, L. J., Eldridge, J. M., Knaak, J. B., and Tallant, M. J. (1972b). '5,6-Dihydro-5,6-dihydroxycarbaryl glucuronide as a significant metabolite of carbaryl in the rat', *J. Agr. Food Chem.* **20**, 980–985.

Sundström, G., Hutzinger, O., Safe, S., and Platonow, N. (1977). In *Fate of Pesticides in Large Animals* (Eds. G. W. Ivie and H. W. Dorough), pp. 175–182, Academic Press, New York.

Suzuki, T., and Miyamoto, J. (1978). 'Purification and properties of pyrethroid carboxylesterase in rat liver microsomes', *Pestic. Biochem. Physiol.* **8**, 186–198.

Ullrich, V., and Kremers, P. (1977). 'Multiple forms of cytochrome P450 in the microsomal monooxygenase system', *Arch. Toxicol.* **39**, 41–50.

Von Mallinckrodt, M. G., Pétényi, M., Flügel, M., Burgis, H., Dietzel, B., Metzner, H., Nirschl, H., and Renner, O. (1973). 'Zur spezifität der menschlichen Serumparaoxonase', *Z. Physiol. Chem.* **354**, 337–340.

Walker, C. H. (1978). 'Species differences in microsomal monooxygenase activity and their relationship to biological half-lives', *Drug Metab. Rev.* **7**, 295–323.

Wallcave, L., Bronlzyk, S., and Gingell, R. (1974). 'Excreted metabolites of 1,1,1-tricholor-2,2-bis(*p*-chlorophenol)ethane in the mouse and hamster', *J. Agr. Food Chem.* **22**, 904–908.

Weber, W. W., Miceli, J. N., Hearse, D. J., and Drummond, D. J. (1976). 'N-Acetylation of drugs. Pharmacogenetic studies in rabbits selected for their acetylator characteristics', *Drug Metab. Dis.* **4**, 94–101.

Wermuth, B., Münch, J. D. B., and von Wartburg, J. P. (1977). 'Purification and properties of NADPH-dependent aldehyde reductase from human liver', *J. Biol. Chem.* **252**, 3821–3828.

Wright, A. S., Hutson, D. H., and Wooder, M. F. (1979). 'The chemical and biochemical reactivity of dichlorvos', *Arch. Toxicol.* **42**, 1–18.

Note Added in Proof:

It has recently been reported that when 0.5 mg of the pyrethroid insecticide, cypermethrin, was ingested orally by man, the *trans*- and *cis*-dichlorovinyldimethylcyclopropane carboxylic acids (see p. 314) were excreted in the urine in 78% and 32% yields, respectively, in 24 hours. (Prinsen, G. H., and Van Sittert, N. J. (1980). 'Exposure and medical monitoring study of a new synthetic pyrethroid after one season of spraying on cotton in Ivory coast'. In *Field Worker Exposure During Pesticide Application* (Eds. W. F. Tordoir and E. A. H. van Heemstra), pp. 105–120, Elsevier, Amsterdam).

Subject Index

'A' esterases, 277
  in mammalian/avian selective toxicity, 277
  in selective toxicity of diazinon, 277
Acetylation, 324
  genetic polymorphism, 324
Acetylcholine, 236, 300
  increases, in type II avian teratogenesis, 236
Acetylcholinesterase, 237, 300, 307
  effect of organophosphates on chicken embryo enzyme, 237
Acetylesterases, 318
3-Acetylpyridine, 225
  avian teratogenesis, 225
Aerobic soil metabolism, 92
  apparatus for, 92
'Agroecosystem', 97
AI-3-36206(MV 678), 10
Alkylation, 46
  by juvenoids, 46
Allethrin, 115
Amino acid conjugates, 171
  conversion to sugar conjugates, 171
Amino acid conjugation, 324
  species differences, 324
  variety of amino acids utilized, 324
Aminocarb (Matacil), 311
6-Aminonicotin, 225
  avian teratogenesis, 225
Aminopyrine demethylase, 315, 316
  individual variation in man, 315, 316
Aminopyrine metabolism, 266
  effect of acetaminophen, 266
  effect of carbon tetrachloride, 266
Anaerobic soil conditions, 132, 134, 139, 142
Anti-juvenile hormones, 50–52
Antipyrine, 270
  in vivo/in vitro comparison, 270
Antipyrine half-life, 276
  correlation with hepatic cytochrome P450 content, 276
Antipyrine kinetics *in vivo*, 270
  correlation with hepatic cytochrome P450 content, 270
Apoplast, 150
  movement of herbicides in, 150
Arylesterases, 318
Autoradiography, 119, 178–179
  of plants, 119, 178–179
Auxin-binding proteins, 183
Auxin receptors, 183–184
Auxins, 175, 182, 185, 188, 189, 191, 192
  amino acid conjugation of, 175, 189
  and cell wall changes, 188
  biochemical effects of, 185
  mode of action, 182
  plant cell response to, 192
  primary reactions for, 191
  specific enzyme effects with, 189
  sugar conjugates, 175
Avian teratogenesis, 219, 226, 227, 236, 237, 240
  alleviation of type II with oximes, 237, 240
  effects of alleviating agents on embryo NAD levels, 227
  production of type II by cholinergic effectors, 240
  role of cholinergic system in type II, 236
  structure–activity considerations, 226
Azinphosmethyl, 301

Bartha–Pramer flask, 88, 91
  for soil studies, 88, 91
Benzo(a)pyrene hydroxylation, 316
Benzoylphenyl ureas, 52–57, 61
  degradation in soil, 61
  mode of action of, 52–57
  structure of, 53
Benzoylpropethyl, 198
  selective activity of, 198
Benzphetamine demethylation, 316

SUBJECT INDEX

Biliary excretion, 249, 250, 257
  of organic anions, 249
  of organic cations, 250
  molecular weight thresholds for, 249
  re-entrant cannulae, 257
Biliary secretion, 295
  of anions, 295
Biomedical monitoring, 289
Bound residues, 137, 145, 167–168
  in soil, 137
  mineralization of, 137
  of phenoxyacetic acids in plants, 167–168
  selective enzyme hydrolysis of, 168
Bound residues in plants, 168
  release of, 168
O-Bromophenoxymethyl imidazole, 26
  inhibition of JH biosynthesis, 26
Bromophos, 277

Carbamate insecticides, 307
Carbanolate (Banol), 312
Carbaryl, 221, 229, 258, 259, 308, 310
  avian teratogenesis, 221
  comparative metabolism in rats and man, 310
  effect on NAD levels in chicken embryos, 229
  in vivo/in vitro comparison, 259
  metabolism in human and rat liver homogenate, 310
  metabolism in tissue explants, 258, 310
  metabolism of, 260
Carboxylesterases, 318, 319
  in serum of rat and man, 319
  hydrolysis in endoplasmic reticulum, 318
Carnitine, 324
Cell cultures, 255
Cell suspension, 255
Cell suspension cultures, 181
Chitin, 54
  biosynthesis of, 54
Chitin biosynthesis, 53, 54–56
  assay of, 55
  inhibition of, 53, 54–56
Chitin synthesis inhibitors, 63
Chlorfenvinphos, 220, 276, 303, 304, 305, 317
  biotransformation in mammals, 304
  comparative metabolism, 305
  selective toxicity in mammals, 276

4-Chloroaniline, 59
4-Chloro-2-carboxyphenoxyacetic acid, 165
4-Chloro-2-methylphenoxyacetic acid (MCPA), 158, 163, 165, 167, 169, 170, 173, 187, 196
  binding to proteins, 169
  hydroxymethyl derivative, 165
  metabolism in plants, 163, 165, 167
  protein binding of, 173
  sugar ester conjugates, 170
  translocation of, 158
4-(4-Chloro-2-methylphenoxy)butyric acid (MCPB), 187
4-Chlorophenol, 59
2-(4-Chlorophenyl)-3-methyl butyric acid (Cl-Vacid), 130, 140, 142
4-Chlorophenylurea, 59, 61, 62
Chloroplast function, 186–187
  control of in plants, 186–187
Chlorpyriphose, 220, 277, 278, 306, 307
  avian teratogenesis, 220
Chlorpyriphos oxon, 278
Chlortoluron, 106, 107
Cholinester hydrolase, 318
Cholinesterase, 300
  inhibition by phosphate triesters, 300
Citrate synthetase, 190
  activation by IAA, 190
Clearance, 252, 271, 280, 316
  dose-dependency, 280
  in assessment of rate of metabolism, 252
  of antipyrine in man, 316
Coumaphose, 221, 277, 278
Coumaphos oxon, 278
Cuticle, 150
  penetration of herbicides by, 150
Cyanatryn, 316
  S-oxygenation, 316
Cyanogenic glycosides, 125
Cyclopropane dicarboxylic acid, 123, 139
  in plants, 123
  in soils, 139
Cypermethrin, 116–117, 118, 125–128, 129, 137–140
  amide analogue of, 127
  degradation in soil, 137–139
  degradation pathway in soil, 140
  half-life in soil, 137
  metabolism in man, 314, 332
  metabolism in plants, 125–128, 129

SUBJECT INDEX

Cytochrome P448, 255
Cytochrome P450, 45, 255, 270, 299, 314, 315, 316, 325
  amount in human liver, 325
  mechanism of action, 315
  stability *in vitro*, 316
Cytosolic fraction, 45

DDA, 297, 299
DDD, 297
DDE, 296, 297, 298, 317
  elimination rate from human fat, 298
DDMS, 297
DDMU, 297
DDNU, 297
DDOH, 297
DDT, 296, 297, 298, 299, 317
  biotransformation in mammals, 297
  dechlorination, 298
  elimination rate from human fat, 298
  half-life in human fat, 297
  species differences in depletion from fat, 298
Decamethrin, 116–117, 118, 123–125, 126
  half-life in plants, 123
  metabolism in plants, 123–125, 126
De-ethylchlorfenvinphos, 303, 305
Denitrification, 86
  in soil, 86
Desmethyldichlorvos, 303
DHHC, 260, 261, 262
Diallate, 108
  degradation in soil, 108
Diazinon, 220, 224, 229, 277, 278
  avian teratogenesis, 220, 224
  effect on NAD levels in chicken embryo, 229
  selective toxicity in vertebrates, 277
Diazoxon, 224, 278
  avian teratogenesis, 224
Dibromovinylcyclopropane carboxylic acid (Br_2CA), 124, 125
Dichlobenil, 53
  structure of, 53
2,4-Dichlorobenzoylglycine, 305
Dichlorofenthion, 307
2,5-Dichloro-4-hydroxyphenoxyacetic acid (4-OH-2,5-D), 163, 171, 173
  glucosylation of, 173
  glycoside of, 171
2,4-Dichloromandelic acid, 305

2,4-Dichlorophenoxyacetic acid (2,4-D), 153, 158, 160, 163–164, 167, 168, 170, 171, 177, 179, 180, 187, 194, 202
  accumulation of, 202
  binding to pectic polymers, 168
  bound residues from, 167
  distribution in roots, 180
  ethyl ester, 170
  hydroxylation of, 173
  metabolism in plants, 163–164
  metabolites in soya bean, 171
  selectivity of, 194
  subcellular distribution of, 177
  sugar ester conjugates, 170
  transport in plants, 153, 158
  uptake by roots, 160
2,4-D aspartate, 171
2,4-D glutamate, 171
2,4-D glycosides, 171
4-(2,4-Dichlorophenoxy) butyric acid (2,4-DB), 175, 187, 197
  β-oxidation of, 197
  selective toxicity of, 197
2,4-Dichlorophenylethandiol, 305
2,4-Dichlorophenylethanol, 305
Dichlorovinylcyclopropane carboxylic acid (Cl_2CA), 120, 121, 123, 127, 134, 137, 139
  conjugates of, 127
  tetra-acetyl glucose ester of, 127
Dichlorvos, 220, 301, 302, 303
  biotransformations in mammals, 302
  glutathione-dependent demethylation, 303
Dichlorvos hydrolase, 319
Dicrotophos, 220, 224, 225, 307
  avian teratogenesis, 220
  metabolism in avian teratogenesis, 224
  metabolism to monocrotophos in eggs, 224
  mixed types I and II avian teratogenesis, 225
Dieldrin, 265, 266, 268, 269, 289, 290, 291, 292, 294, 296
  biliary excretion of metabolites, 265, 266, 294
  biotransformations in mammals, 290
  elimination kinetics, 290
  half-life in man, 291
  half-life in rats, 292
  human volunteer study, 291

Dieldrin (cont.)
  hydroxylation, 291
  LD_{50} in rats, 292
  metabolism by rat liver microsomes, 269
  metabolism of, 268
  oxidative biotrandormation, 296
[^{14}C]Dieldrin, 292
  metabolism in man, 292
Dieldrin metabolism, 267
  effect of phenobarbital, 267
Dieldrin pentachloroketone, 291
Diethyl phosphate, 305
  from ethyl parathion in man, 305
Diethyl thiophosphate, 306
  from ethyl parathion in man, 306
Diflubenzuron, 53–54, 55, 56, 57–58, 59, 60, 61, 62
  and chitin synthetase inhibition, 54
  degradation by aquatic organisms, 61
  degradation by fish, 61
  effects on lipoprotein synthesis, 57
  environmental fate of, 60
  hydrolysis of, 58
  hydroxylation of, 58
  inhibition of chitin biosynthesis, 53–54
  insect metabolism of, 57–58, 59
  mammalian metabolism, 58, 60
  metabolism in plants, 62
  photodegradation of, 60
  resistance to, 57–58
  secondary effects, 56
  structure of, 53
Diflubenzuron analogues, 56
  structure activity correlations of, 56
2,6-Difluorobenzamide, 61
2,6-Difluorobenzoic acid, 59, 61, 62
4,5-trans-Dihydroaldrindiol, 289, 291
  excretion in bile as conjugate, 291
5,6-Dihydro-5,6-dihydroxycarbaryl, 260
5,6-Dihydro-5,6-dihydroxycarbaryl glucuronide, 258
4,5-trans-Dihydroisodrindiol, 295
trans-5,6-Dihydroxy-5,6-dihydronaphth-1-yl N-methylcarbamate glucuronide, 310
2,6-Di-isopropylphenol, 323
Dimethoate, 221, 277, 278
  selective toxicity, 277
Dimethoxon, 278
4-(4,8-Dimethyldecyloxy)-1,2-(methylenedioxy) benzene, 47

Dimethyl phosphate, 305, 306
  from methyl parathion in man, 305, 306
Dinitroaniline herbicides, 96
Dodecadienoates, 30
DU-19111, 53, 54, 56
  structure of, 53

Ectodesmata, 150, 155
  and herbicide movement, 150
Efflux analysis, 179–180
Embryo, 222, 223, 240
  protection by placenta in mammals, 240
  sensitivity to chemicals during development of avian, 222
  sensitivity to dicrotophos with time, 223
Embryo NAD, 228, 229. 234
  correlation of lowering with kynurenine formamidase inhibition, 234
  effect of organophosphorus compounds and methylcarbamates, 228
  time course of effect of diazinon, 229
Endoplasmic reticulum, 256, 314, 315
Endrin, 292, 293, 294, 296
  biliary secretion of metabolites, 294
  biotransformations in mammals, 293
  LD_{50} in rats, 292
  oxidative biotransformation, 296
Environmental degradation, 34
  of methoprene, 34
Environmental fate, 33, 37, 47–50, 60
  of diflubenzuron, 60
  of hydroprene, 37
  of juvenoids, 47–50
  of methoprene, 33
Enzyme preparations, 257
  solubilization of membrane-bound, 257
  substrate concentrations, 257
Epidermis, 6
  hydrolysis of JH in, 6
  site of action of JH, 6
EPN, 220
Epoxide hydration, 45, 49
  of R-20458, 45
Epoxide hydrolase, 14, 15, 16, 18, 320
  activity on R-20458, 16
  changes during insect development, 16
  induction in insects by pentamethylbenzene, 15

SUBJECT INDEX

induction in insects by phenobarbital, 15
inhibition by glycidyl ethers, 14
in insect haemolymph, 18
in juvenoid metabolism, 14
of insects, 14
of mammals, 14
pH optima of insect enzymes, 15
solubilization of insect enzyme, 16
subcellular location in insects, 15
(2E,6E)-10,11-Epoxyfarnesyl[2,3-^{3}H]propenyl ether, 10
Eserine, 221, 225
avian teratogenesis, 221
mixed types I and II avian teratogenesis, 225
Ethion, 301
Ethirimol, 107
7-Ethoxycoumarin dealkylation, 316
Ethyl 4-[2-(tert-butylcarbonyloxy)]butoxybenzoate (ETB), 50
Ethyl dichlorofornesoate, 10
radiosynthesis, 10
Ethyl parathion, 305, 307
Ethylene production, 188
auxin induced, 188
Ethylmorphine demethylation, 316
1-(4-Ethylphenoxy)-6,7-epoxy-3,7-dimethyl-2-octene (R-20458), 10, 11, 38, 39, 40, 41, 42–45, 46, 47
alkylation of, 46
2,3-epoxide, 38
6,7-epoxide, 38, 44
half-life in houseflies, 40
hydroxylation of, 38, 44
mammalian metabolism of, 42–45
metabolic pathways in insects, 38, 39
metabolism by hydrolases, 38
metabolism by oxidases, 38
persistence of, 41
radiosynthesis, 10, 11
reductive hydroxylation, 43
toxicity to algae, 47
1-(4-Ethylphenoxy)-6,7-epoxy-3,7-dimethyl-2-octene diene, 48
photodecomposition of, 48
1-(4-Ethylphenoxy)-6,7-epoxy-3,7-dimethyl-2-octene diol, 40, 48
2,3-olefin of, 40, 48
photodecomposition of, 48
1-(4-Ethylphenoxy)-6,7-epoxy-3,7-dimethyl-2-octene epoxide, 38, 45, 48

hydration of 38, 45, 48
photodecomposition of, 48
Etrimfos, 220
avian teratogenesis, 220
Excreted metabolites, 249
in assessment of metabolism, 249
Excretion of metabolites, 250, 251
in bile, 251
in eggs, 251
in faeces, 250
in milk, 251
in urine, 250
via lymph, 251
via lungs, 251

Fenchlorphos, 277
Fenpropathrin, 116–117, 118, 139, 140, 141
degradation in soil, 139–140
degradation pathway in soil, 141
half-life in soil, 139
Fenvalerate, 116–117, 118, 128, 130, 131, 140–142, 313
amide analogue of, 141
decarboxy derivative of, 130
degradation in soil, 140–142
degradation pathway in soil, 143
metabolism in plants, 128, 130, 131
photochemical conversion of, 130
soil leaching of, 142
Flamprop, 166, 170
sugar conjugates of, 166, 170
Flamprop glucose ester, 172
Flavoprotein, 316
N-oxygenation, 316
S-oxygenation, 316

Gas chromatography–mass spectrometry, 123, 139
of permethrin metabolites, 123
Gene expression, 185–186
control of in plants, 185–186
Glucan synthetase, 190
Glucuronic acid, 322
Glucuronic acid conjugates, 322, 323
acyl migration, 323
identification of, 322
of bilirubin, 323
β-Glucuronidase, 322
in identification of glucuronides, 322
Glucuronide formation, 282
coupling with mono-oxygenase, 282

Glutathione, 300, 317, 320, 321
  in chlorfenvinphos metabolism, 317
  in dearylation of organophosphates, 300, 321
  in demethylation of organophosphates, 300 321
  in organophosphate, dealkylation, 321
Glutathione conjugation, 320
Glutathione-*S*-epoxide transferase, 45
  in rat hepatocytes using JH, 45
Glutathione transferase, 321
  of human erythrocytes, 321
  of human liver, 321
Glycine *N*-acyltransferase, 325
  stability in cadavers, 325
Glycoside conjugation, 125
Glycosides, 166
  ester-linked, 166
  ether-linked, 166
Griseofulvin metabolism, 264
  in bile duct cannulated rats, 264
  is isolated perfused liver, 264
  in liver microsomes, 264

Half-life, 40, 106, 119, 123, 130, 134, 137, 139, 142, 252, 280
  dose-dependency, 280
  in assessment of rate of metabolism, 252
  of cypermethrin in soil, 137
  of decamethrin, 123
  of fenopopathrin in soil, 139
  of permethrin in soil, 134
  of permethrin isomers, 119
  of phenothrin in plants, 130
  of phenothrin in soil, 142
  of R-20458 in houseflies, 40
  of thiazafluron in soil, 106
Hanau Suntest apparatus, 103, 104, 110
HCE, 259, 260, 262, 265, 266, 267, 269
  biliary excretion of metabolites, 265, 266
  in vivo/in vitro comparative metabolism, 259
  metabolism by rat liver microsomes, 269
  metabolism of, 260
  metabolites in rat bile and liver microsomes, 262
HCE metabolism, 267
  effect of phenobarbital, 267
HCE-*trans*-diol, 260, 261

HEOD (dieldrin), 268
  metabolism of, 268
HEOM, 265, 266, 268, 269
  biliary excretion of metabolites, 265, 266
  metabolism by rat liver microsomes, 269
  metabolism of, 268
Herbicide conjugates, 169, 172
  extraction from plants, 169
  stability in plants, 172
Herbicide movement in plants, 157
  via plasmodesmata, 157
Herbicide–protein complexes, 170
Herbicides, 154
  movement to the vacuole, 154
  uptake to the symplast, 154
HHC, 259, 260, 261
endo-HHC, 260, 261, 262
Hill reaction, 187
  inhibition of, 187
Hippuric acid, 303
  in dichlorvos metabolism, 303
Hplc, 128
  of polar conjugates, 128
Hydroprene, 5, 10, 36, 45
  effect on JH esterase, 5
  effect on JH III synthesis, 5
  environmental fate, 36
  metabolism in insects, 36
  photosensitized oxidation, 36
  radiosynthesis, 10
4-Hydrocarbaryl, 260
5-Hydroxycarbaryl, 260
Hydroxycyclopropane carboxylic acid, 121, 134
4'-Hydroxycypermethrin(4'-HO-cyper), 128, 137
4'-Hydroxydecamethrin (4'-HO-dec), 124
trans-Hydroxydibromovinyl carboxylic acid, 124
12-Hydroxydieldrin, 289
syn-12-Hydroxydieldrin, 291, 292
  excretion in bile as glucuronide, 291
3-Hydroxyendrin, 294, 296
anti-12-Hydroxyendrin, 294, 295
  in urine and faeces of plant operatives, 295
syn-12-Hydroxyendrin, 295
4'-Hydroxyfenvalerate, 140
2'-Hydroxypermethrin (2'-HO-per), 120, 121

SUBJECT INDEX

4'-Hydroxypermethrin (4'-HO-per), 120, 121, 134, 136
4'-Hydroxyphenothrin, 143
4-Hydroxyphenoxyacetic acid (4-OH-POA), 168
4'-Hydroxy-3-phenoxybenzoic acid, 124, 128, 130, 137, 143
2'-Hydroxy-3-phenoxybenzyl alcohol (2'-HO-PBALc), 120
4'-Hydroxy-3-phenoxybenzyl alcohol (4'-HO-PBAlc), 120

Indole-3-acetic acid (IAA), 153, 158–159, 161, 173, 174, 179, 193
  conjugation with aspartic acid, 173
  glucosylation of, 174
  levels in plants, 193
  plant metabolites of, 161
  polar transport of, 158–159
  transport of, 153
IAA-aspartate, 161
Insect growth regulators, (IGRs), 2, 42, 63
  future of, 63
  resistance to, 42
In vitro metabolism, 258
  correlation with *in vivo* metabolism, 258
In vivo metabolism, 258
  correlation with *in vitro* metabolism, 258
2-Isopropoxyphenol, 311

Juvenile hormone (JH), 2, 3, 4, 8, 11–14, 27, 28, 29
  biosynthesis, 3
  hydrolase in insects, 12
  metabolism in insects, 11–14
  metabolism of JH I, 3
  metabolism of JH II, 3
  metabolism of JH III, 3
  mode of action, 4
  partition between water and fat *in vivo*, 28
  protein binding, 29
  protein binding, effect on hydrolysis, 28
  protein binding in haemolymph, 27
  radiosynthesis, 8
  stereospecificity, 28, 29
  structure of JH I, 2
  transport in haemolymph, 4
Juvenile hormone (JH) carrier protein, 27, 28
  of *M. sexta*, 27, 28
  role of, *in vivo*, 28
Juvenile hormone (JH) esterase, 13, 16–23
  assay methods, 13, 16–23
  changes during development, 17, 20
  distinction from serine esterases and proteases, 21
  effect of inhibition *in vivo*, 19
  identity with malathion esterase, 21
  identity with α-naphthyl acetate hydrolase, 17, 21
  in insect fat body, 27
  in insect haemolymph, 18
  inhibition, 20
  inhibition by O,O-di-isopropyl phosphofluoridate, 21
  inhibition by O-ethyl-S-phenyl phosphoramidothiolate, 19
  inhibition by paraoxon, 21
  inhibition by Triton X-100, 22
  inhibition of, in insect control, 23
  inhibition, species differences, 22
  inhibition as synergists, 22
  microsomal, 13
  regulation, 23
  soluble, 13
  species differences, 20
  stereochemical specificity, 16
  tissue distribution in insects, 20
Juvenoids (juvenile hormone analogues), 4, 5, 6, 7, 8, 17, 47–50
  assay by gas-liquid chromatography, 17
  assay by h.p.l.c., 17
  effects on JH degradation, 4
  effects on JH synthesis, 5
  effect on macromolecular biosynthesis, 7
  effects on site of action, 6
  environmental fate, 47–50
  mode of action, 4
  radiosynthesis, 8

12-Ketoendrin, 295, 296
Kynurenine formamidase, 228, 229, 230, 231, 232, 234, 235, 236

Kyurenine formamidase (cont.)
  correlation of inhibition with embryo NAD lowering, 234
  distribution in nature, 230
  effect of organophosphorus compounds and methylcarbamates on, 228
  kinetics of inhibition by organophosphates and carbamates, 235
  inhibition, 232
  inhibition in type I avian teratogenesis, 230
  inhibition of yolk sac membrane enzyme by diazoxon, 233
  inhibition of yolk sac membrane enzyme by dicrotophos amide, 233
  inhibition of yolk sac membrane enzyme by eserine, 233
  properties of, 230
  role in conversion of tryptophan to NAD, 231
  role of yolk sac membrane enzyme in avian teratogenesis, 230
  spontaneous reactivation of inhibited enzyme, 236
  time course of effect of diazinon on embryo enzyme, 229

Laurate hydroxylation, 316
LD_{50}, 253
  in assessment of rate of metabolism, 253
Leaf apoplast, 151
Leptophos, 220
Lineweaver–Burke plots, 256
Lipases, 318
Lysimeters, 99, 100

Malathion, 221, 301, 307, 318
  hydrolysis by carboxylesterases, 318
Mammalian teratogenesis, 238, 240
  induced by carbaryl 238
  induced by chlorfenvinphos, 238
  induced by diazinon, 238
  induced by dimethoate, 238
  induced by phosmet, 238
  induced by trichlorphon, 238
Methamidophos, 221
  avian teratogenesis, 221
Methiocarb (Mesurol), 312

Methoprene, 5, 7, 8, 10, 30, 31, 32, 33, 34, 36, 42, 45, 63
  5-^{14}C labelled, 10
  cholesteryl ester in hens, 36
  cross-resistance with dimethoate, 30
  effect on DNA and RNA synthesis, 8
  effect on JH esterase, 5
  environmental fate, 33, 34
  10-^{3}H labelled, 10
  in stored products, 33
  incorporation into triglycerides in hens, 36
  incorporation of 5-^{14}C in high molecular weight plant components, 33
  metabolism by O-demethylation, 30
  metabolism by microsomal oxidases, 30
  metabolism in fish, 33
  metabolism in insects, 30
  metabolism in mammals, 31
  metabolism in various species, 34
  metabolism to endogenesis products, 31, 32
  microbial degradation, 33
  photochemical degradation, 33
  resistance, 30
  vitellogenin synthesis and storage, 7
N-Methyl diflubenzuron, 53, 62
  metabolic fate of, 53, 62
  structure of, 53
Methyl epoxyfarnesoate, 10
  radiosynthesis, 10
Methyl epoxystearates, 45
Methyl p-nitrophenyl phosphate, 306
Methyl parathion, 220, 305, 306
Methylcarbamate, 225
  avian teratogenesis, 225
4-(Methylcarbamoyloxy)-1-naphthyl glucuronide, 310
4-(Methylcarbamoyloxy)-1-naphthyl sulphate, 310
Methylglutathione, 322
Mevinphos, 220
Mexacarbate (Zectran), 311
Mitochondrial function, 186–187
  control in plants, 186–187
Mode of action, 50–51
  of precocene, 50–51
Monocrotophos, 220, 224, 301
  avian teratogenesis, 220
  formation from dicrotophos in eggs, 224

SUBJECT INDEX

Mono-oxygenase, 30, 279, 282, 314, 317
  as a determinant of acute toxicity, 279
  coupling with glucuronidation, 282
  measurement in man by non-invasive techniques, 317
  metabolism in methoprene, 30
  phylogenetic distribution, 279
  species differences in vertebrates, 279

NAD, 225, 227
  analysis of level in embryo 225, 227
  lowered level in type I avian teratogenesis, 225
NADH-dehydrogenase, 190
  inhibition by 2,4-D and 2,4,5-T, 190
NADPH-cytochrome P450 reductase, 315
1-Naphthol, 260, 308
  metabolite of carbaryl, 308
1-Naphthyl glucuronide, 310
1-Naphthyl sulphate, 310
1-Naphthylacetic acid (NAA), 173
  aspartic acid conjugation, 173
Nitrification, 86
  in soil, 86
p-Nitrophenol, 305, 306
  from metabolism of parathion in man, 305, 306

Organ perfusion, 254
Organochlorine insecticides, 289, 299
Organophosphorus compounds, 224, 319
  avian teratogenesis, 224
  enzymic hydrolysis, 319
Oxidation, 314
  in pesticide metabolism in mammals, 314
β-Oxidation, 175, 197
  of 2,4-DB, 197
Oxidative metabolism, 43
  of R-20458, 43
Oxidative phosphorylation, 8
  uncoupling by JH I, JH II, and JH III, 8
Ozonolysis, 132

Paraoxon, 277, 278
Parathion, 108, 220, 301, 307
  avian teratogenesis, 220

Penfluron, 53, 62
  insect metabolism of, 62
  structure of, 53
Periodate cleavage, 295
  of hydroxyendrin glucuronide, 295
Permethrin, 116, 118, 119–123, 132–136, 313
  degradation in soil, 132–136
  glucoside metabolites of, 120
  half-life in soil, 134
  metabolic pathways in plants, 122
  metabolism in plants, 119–123
  soil leaching of, 134, 136
Permethrin isomers, 119
  half-life in plants, 119
Pesticide transformation, 109
  influence of other chemicals on, 109
Pesticides, 85, 86
  microbial degradation of, 85, 86
Pharmacokinetic models, 281
  limitations, 281
Pharmacokinetics, 268
Phenothrin, 115, 118, 130–133, 142–145
  degradation in soil, 142–145
  formyl derivatives, 132
  half-life in plants, 130
  half-life in sil, 142
  metabolism in plants, 130–133
Phenoxyacetate anion, 157
  movement in plants, 157
Phenoxyacetic acid, 173
  conjugation with aspartic acid, 173
Phenoxyacetic acid metabolites, 174–176
  biological activity of, 174–176
  turn over of, 201
Phenoxyacetic acids, 147–202
  adsorption to plant cell walls, 151
  bound residues in plants, 167–168
  conjugation of, 163
  'decarboxylation' of, 163
  diffusive uptake, 153
  entry to the symplast, 152
  hydroxylation of, 163, 175
  kinetics of metabolism in plants, 170–172
  location within plants, 176–181
  metabolic inactivation of, 175
  metabolism in plants, 161–162
  mode of action, 182–194
  movement in the symplast, 152
  movement of un-ionized acid in plants, 157

Phenoxyacetic acids (*cont.*)
  movement within plants, 156–163
  pathways of metabolism in plants, 163–169
  phloem transport of, 159–160
  phytotoxicity, 182
  protein binding, 168–169
  quantitative metabolism, 169–170
  root excretion of, 160
  selectivity of, 194–202
  side chain cleavage of, 163
  side chain degradation, 167
  toxic action of, 192–193
  uptake by leaves, 149–156
Phenoxy ethers, 38
  metabolism of, 38
3-Phenoxybenzaldehyde (PB ald), 124, 128
3-Phenoxybenzaldehyde cyanohydrin (PB cy), 124, 130
  conjugates, 130
3-Phenoxybenzoic acid (PB Acid), 121, 124, 128, 130, 134, 136, 137, 139, 143, 166, 313, 314
  conjugation in plants, 128
  glucosylarabinose ester of, 166
  species differences in conjugation with amino acids, 314
3-Phenoxybenzyl alcohol (PB Alc), 120, 121, 124, 128, 130, 134, 143
Phenyl saligenin cyclic phosphate, 221
Phenytoin kinetics 271
Phorate, 221
  avian teratogenesis, 221
Phosmet, 221
Phosphamidon, 220
  avian teratogenesis, 220
Phospholipid, 314
  role in mono-oxygenase action, 314
Photochemical conversion, 32, 48, 103–105, 117, 130, 132, 145
  in soil, 103–105
  of juvenoids, 48
  of methoprene, 32
  of pesticides on soil surfaces, 104
  of phenothrin, 132
  of pyrethroids, 130
  on soil surfaces, 105
Photolysis in/on soil, 103
Phylogenetic distribution of enzymes, 277
Physostigmine, 307

Pirimiphos-methyl, 220, 278
  avian teratogenesis, 220
Pirimiphos-methyl oxon, 277, 278
Pirimiphos-ethyl, 220, 278
  avian teratogenesis, 220
Plasma membrane, 155
  permeability of, 155
Polychlorobiphenyls, 273 *et seq.*
  effect of substitution on rate of metabolism, 275
  kinetics and metabolism, 273 *et seq*
Precocene, 50–51
  mode of action, 50–51
Precocene II, 50, 51, 52
  3,4-dihydrodiol of, 51
  glucosides of, 52
  insect metabolism of, 51, 52
  metabolism of, 51
  structure of, 52
Precocene II 3,4-epoxide, 51
  key role in mode of action, 51
Propoxur (Baygon) 308, 311
Protein binding, 27, 155, 201
  of herbicides, 155
  of JH and juvenoids, 27
Prothoracicotropic hormone (PTTH), 6
Pyrethrin I, 116
Pyrethroids, 119–145, 312, 313, 319
  degradation in soil, 132–145
  ester cleavage, 313
  hydrolysis by carboxylesterase, 319
  hydroxylation, 313
  metabolism in man, 314, 332
  metabolism in plants, 119–132
Pyrethroid hydrolase, 319
  identity with malathion esterase, 319
  identity with *p*-nitrophenyl acetate hydrolase, 319

Quinazolinedione, 61

Radiochemical enclosure, 99
  outdoor, 99
Resistance, 195, 198, 253
  of herbicides, 195, 198
Resmethrin, 115
RNA synthesis, 185
Ro-10-3108, 10, 11, 41, 47, 49
  acute toxicity of, 47
  degradation in water, 49

SUBJECT INDEX

epoxide hydration of, 49
ether cleavage of, 49
metabolism of, 41
oxidation of, 49
photostability of, 49
radiosynthesis, 11

Salicylic acid, 322
Selective toxicity, 253
Selectivity, 195
   mechanisms of, 195
SIR-8514, 53, 62
   structure of, 53
Soil, 85, 86–101, 102, 103, 106, 107, 109–110, 118
   autoclaving of, 102
   biodegradative capacity of, 107
   biotransformation studies, 106
   chemical sterilization of, 102
   chemical transformations in, 85, 101–103
   degradation studies, 118
   γ-irradiation of, 102
   in vitro experiments, 109–110
   microbial transformations in, 101–103
   microbiological activity of, 106–107
   reactivation of, 107
   sterilization of, 101, 110
   transformation studies in, 86–101
   volatilization of pesticides from, 100
Soil 'balance' studies, 98
Soil biometer flasks, 86, 87, 109
Soil degradation studies, 98–101
   field experiments, 98–101
Soil flow-through systems, 95
   advantages and disadvantages, 95
Soil metabolism, 97
   apparatus for, 97
Soil perfusion systems, 86–89, 90, 109
   advantages and disadvantages of, 90
Soil photolysis studies, 103–104
   design of, 103–104
Soil–plant incubation system, 97
Soil sterilization methods, 102
   shortcomings of, 102
Soil studies, 91, 94, 96, 109
   flow through systems, 91, 94, 96, 109
Soil thin-layer plates, 104
Stable isotopes, 101, 317
   in measurement of human monooxygenase activity *in vitro*, 317

Standard soils concept, 107
Sterile soil conditions, 132
Sterilization, 101
   of soils, 101
Storage, 252
   in adipose tissue, 252
Subcellular fractionation, 177–178
   of plant homogenates, 177–178
Subcellular fractions, 256
Sugar conjugates, 171
   formation from amino acid conjugates, 171
Sulphation, 323
Sunlight, 105
   spectral energy distribution, 105

Teratogenesis, 219–225, 240
   avian, 219
   in mammals, 240
   species of bird used for testing, 222
   structure–activity relationships in avian, 223–225
   test procedures for study of avian, 222
   type I, 220
   type II, 220
Teratogenicity, 219–222
   signs of avian, 219–222
Terrestrial–aquatic ecosystems, 98
Tetrachlorvinphos, 318, 321
Tetramethyl cyclopropanecarboxylic acid, 139
Thiazafluron, 106
   half-life in soil, 106
Tissue explants, 255, 310
   use in carbaryl metabolism, 310
Tissue slices, 255
Tocopherol oxidase, 175, 190, 196
   control by auxins, 175, 190, 196
   regulation by auxins, 190
TOCP, 221
Transesterification, 127, 170
   of sugar ester conjugates, 170
2,4,5-Trichlorophenoxyacetic acid (2,4,5-T), 163, 165, 175, 178, 187
   metabolism in plants, 163, 165
   subcellular distribution of, 178
2,4,5-T aspartate, 175
2,4,5-T glutamate, 175
2,4,6-Trichlorophenoxyacetic acid (2,4,6-T), 187
Trichlorphon 221

2,3,5-Tri-iodobezoic acid (TIBA), 154, 159

Urea, 303
  in dichlorvos metabolism, 303
Urine, 250
  elimination of metabolites in, 250

V_d (apparent volume of distribution), 271
Vitellogenin, 7
  effects of methoprene on synthesis and storage, 7

Warfarin, 317